Handbook of Technical Writing

most typical means of communication are discussed in selecting the medium.

Research

The only way to be sure that you can write about a complex subject is to thoroughly understand it. To do that, you must conduct adequate research, whether that means conducting an extensive investigation for a major proposal—through interviewing, library and Internet research, careful note-taking, and documenting sources—or simply checking a company Web site and jotting down points before you send an e-mail to a colleague.

Methods of Research. Researchers frequently distinguish between primary and secondary research, depending on the types of sources consulted and the method of gathering information. *Primary research* refers to the gathering of raw data compiled from interviews, direct observation, surveys, experiments, questionnaires, and audio and video recordings, for example. In fact, direct observation and hands-on experience are the only ways to obtain certain kinds of information, such as the behavior of people and animals, certain natural phenomena, mechanical processes, and the operation of systems and equipment. *Secondary research* refers to gathering information that has been analyzed, assessed, evaluated, compiled, or otherwise organized into accessible form. Such forms or sources include books, articles, reports, Web documents, e-mail discussions, correspondence, minutes of meetings, operating manuals, and brochures. Use the methods most appropriate to your needs, recognizing that some projects will require several types of research and that collaborative projects may require those research tasks to be distributed among team members.

Sources of Information. As you conduct research, numerous sources of information are available to you.

- Your own knowledge and that of your colleagues
- The knowledge of people outside your workplace, gathered through interviewing for information
- Internet sources, including Web sites, directories, archives, and discussion groups
- Library resources, including databases and indexes of articles as well as books and reference works
- Printed and electronic sources in the workplace, such as brochures, memos, e-mail, and Web documents

Consider all sources of information when you begin your research and use those that are appropriate and useful. The amount of research you will need to do depends on the scope of your project.

Organization

Without organization, the material gathered during your research will be incoherent to your readers. To organize information effectively, you need to determine the best way to structure your ideas; that is, you must choose a primary method of development.

Methods of Development. An appropriate method of development is the writer's tool for keeping information under control and the readers' means of following the writer's presentation. As you analyze the information you have gathered, choose the method that best suits your subject, your readers' needs, and your purpose. For example, if you were writing instructions for assembling office equipment, you would naturally present the steps of the process in the order readers should perform them: the sequential method of development. If you were writing about the history of an organization, your account would most naturally go from the beginning to the present: the chronological method of development. If your subject naturally lends itself to a certain method of development, use it—do not attempt to impose another method on it.

Often you will need to combine methods of development. For example, a persuasive brochure for a charitable organization might combine a specific-to-general method of development with a cause-and-effect method of development. That is, you could begin with persuasive case histories of individual people in need and then move to general information about the positive effects of donations on recipients.

Outlining. Once you have chosen a method of development, you are ready to prepare an outline. Outlining breaks large or complex subjects into manageable parts. It also enables you to emphasize key points by placing them in the positions of greatest importance. By structuring your thinking at an early stage, a well-developed outline ensures that your document will be complete and logically organized, allowing you to focus exclusively on writing when you begin the rough draft. An outline can be especially helpful for maintaining a collaborative-writing team's focus throughout a large project. However, even a short letter or memo needs the logic and structure that an outline provides, whether the outline exists in your mind or on-screen or on paper.

At this point, you must begin to consider layout and design elements that will be helpful to your readers and appropriate to your subject and purpose. For example, if visuals such as photographs or tables

EIGHTH EDITION

HANDBOOK OF TECHNICAL WRITING

Gerald J. Alred
University of Wisconsin–Milwaukee

Charles T. Brusaw
Consultant

Walter E. Oliu
Consultant

ST. MARTIN'S PRESS New York

FOR BEDFORD/ST. MARTIN'S

Developmental Editor: Caroline Thompson
Editorial Assistant: Kaitlin Hannon
Senior Production Supervisor: Nancy J. Myers
Project Management: Books By Design, Inc.
Cover Design: Donna L. Dennison, Kim Cevoli
Cover Art: *Computer Graphics over Pipe*. Wayne Calabrese, Photonica.
Composition: Pine Tree Composition, Inc.
Printing and Binding: R. R. Donnelley & Sons Company

President: Joan E. Feinberg
Editorial Director: Denise B. Wydra
Editor in Chief: Karen S. Henry
Director of Marketing: Karen Melton Soeltz
Director of Editing, Design, and Production: Marcia Cohen
Manager, Publishing Services: Emily Berleth

Library of Congress Control Number: 2005921350

Manufactured in the United States of America.

1 0 9 8 7 6

f e d c b a

For information, write

St. Martin's Press
175 Fifth Avenue
New York, NY 10010

ISBN: 0-312-43613-0 (paperback)
0-312-35267-0 (hardcover)
EAN: 978-0-312-43613-1

ACKNOWLEDGMENTS

Figure D-3: Entry for "regard" from *The American Heritage Dictionary of the English Language*, Fourth Edition. Copyright © 2000 by Houghton Mifflin Company. Reprinted with the permission of the publisher. All rights reserved.
Figure D-12: From Xerox Corporation, *WorkCentre XD Series User Guide*, page 6. Copyright © Xerox Corporation. Used with permission.

Acknowledgments and copyrights are continued at the back of the book on pages 586–87, which constitutes an extension of the copyright page.

Contents

Preface

The eighth edition of the *Handbook of Technical Writing*, like previous editions, is a comprehensive resource for both academic and professional audiences. The *Handbook*'s nearly 400 entries cover effective print, oral, and online technical communication, offering advice for writing and designing many different types of professional documents in addition to thorough explanations of grammar, style, and usage. This edition includes many new sample documents and visuals, as well as expanded advice for integrating visuals, making appropriate rhetorical choices, and considering ethics in technical writing.

How to Use This Book

The *Handbook of Technical Writing* is made up of alphabetically organized entries with color tabs. Within each entry, underlined cross-references such as "formal reports" link readers to related entries that contain further information. Many entries present advice and guidelines in the form of convenient Writer's Checklists.

The *Handbook*'s alphabetical organization enables readers to find specific topics quickly and easily; however, readers with general questions will discover several alternate ways to find information in the book.

- **Contents by Topic.** The complete Contents by Topic on pages xxiii–xxvi groups the alphabetical entries into topic categories. This topical key can help a writer focusing on a specific task or problem browse all related entries; it is also useful for instructors who want to correlate the *Handbook* with standard textbooks or their own course materials.
- **Commonly Misused Words and Phrases.** The list of Commonly Misused Words and Phrases on pages 635–36 extends the Contents by Topic by listing all of the usage entries, which appear in italics throughout the book.
- **Model Documents and Figures by Topic.** The topically organized list of model documents and figures on pages 637–39 makes it easier to browse the book's abundant sample documents

and visuals to find specific examples of technical communication genres.

- **Checklist of the Writing Process.** The checklist on pages xxi–xxii helps readers to reference key entries in a sequence useful for planning and carrying out a writing project.
- **Comprehensive Index.** The index lists all the topics covered in the book, including those topics that are not main entries in the alphabetical arrangement.

Helpful Features

ESL Tips boxes throughout the book offer special advice for writers of English as a second language. In addition, the Contents by Topic on the inside back cover includes a list of entries—ESL Trouble Spots—that may be of particular interest to ESL writers.

Digital Tips and Web Link boxes throughout the book direct readers to specific related resources on the companion Web site at <*bedfordstmartins.com/alredtech*>. Digital Tips in the book suggest ways to use workplace technology and word-processing software to simplify complex writing tasks such as designing a document or creating an index. Expanded Digital Tips on the Web site offer step-by-step instructions for completing each task. Web Links in the book point students to related resources on the companion site such as model documents, tutorials, and links to hundreds of useful related Web sites.

New to This Edition

Our focus in revising the *Handbook* for this edition has been to give greater attention to the rhetorical and ethical choices writers face—both in writing and in designing professional documents. In response to suggestions from many instructors and workplace professionals who use the *Handbook*, we have replaced, revised, and redesigned many sample documents—in fact, this edition contains more new and updated examples than any previous edition. We have added new entries and updated information throughout the book to better reflect the choices today's technical writers face. In addition to thoroughly updated coverage of grammar, usage, and style, readers will find the following improvements.

- **New and redesigned sample documents and visuals.** New sample documents reflect the prominence of e-mail in the workplace and draw from a greater range of writing situations in disciplines such as engineering and health sciences. Abundant illustrations—many of them new—include charts, graphs, drawings, tables, internationally recognized symbols, illustrated descriptions and instructions, brochure and newsletter pages, presentation slides, and more.

- **A new entry on the context of the writing situation** discusses questions writers must ask to understand and respond effectively to different writing situations. This entry also shows writers how to convey enough background information to help their readers understand the value and purpose of a document. To complement this discussion, most examples and model documents throughout the book are now introduced with a description of the rhetorical situation that prompted them.
- **Ethics Notes** throughout the book describe ethical issues in technical writing and help writers understand and address them.
- **Expanded treatment of visual communication.** A new chart in the entry visuals compares the functions of various types of graphics and can help writers make the best choices for their audience, purpose, and context. This entry also offers expanded advice for effectively integrating, labeling, and crediting visuals, and cautions readers about using misleading visuals.
- **New entries on writing white papers and requests for proposals** offer professional topics not found in most technical writing textbooks.
- **Updated coverage of résumé writing.** Job seekers will find new advice for using résumé summary statements, combining chronological and functional formats, and sending résumés as e-mail attachments. New sample résumés designed by Kim Isaacs, the résumé expert at Monster.com, and new sample application letters provide up-to-date models.
- **Expanded discussions of copyright and plagiarism.** Revised and expanded entries explain when and how to seek permission to use copyrighted material, how to properly acknowledge sources, and how to distinguish common knowledge from sources that must be cited.
- **Streamlined coverage of research.** We have combined and revised the previous edition's separate entries on library and Internet research in one research entry with an expanded *Writer's Checklist: Evaluating Print and Online Sources.*
- **New coverage of IEEE style.** The *Handbook* now provides guidelines for using IEEE (Institute of Electrical and Electronics Engineers) style for documenting sources, in addition to up-to-date advice and documentation models for APA (American Psychological Association) and MLA (Modern Language Association) styles.

- **An updated companion Web site at <*bedfordstmartins.com/alredtech*>.** The Web site helps instructors take advantage of the *Handbook*'s potential as a classroom text by offering lesson plans, handouts, teaching tips, and assignment ideas. For students, the Web site includes additional sample documents, useful tutorials, expanded Digital Tips, and links to hundreds of useful Web sites keyed to the *Handbook*'s main entries.

Acknowledgments

For their invaluable comments and suggestions for this edition of the *Handbook of Technical Writing*, we thank the following reviewers who responded to our questionnaire: Craig Allen, University of Washington; Jennie Blankert, Purdue University; Roger Bourret, South Seattle Community College; John Brocato, Mississippi State University; Teresa Fishman, Clemson University; Robert Goldberg, Prince George's Community College; Lila Harper, Central Washington University; Noah Ilinsky, University of Washington; Karen Kasonic, University of Washington; Amy Koerber, Texas Tech University; John Lee, University of San Antonio; Lynn Lewis, University of Oklahoma; Barry Maid, Arizona State University; Tamara Powell, Louisiana Tech University; Cynthia Raisor, Texas A&M University; Rochelle Rodrigo, Mesa Community College; Nancy Schneider, University of Maine–Augusta; Keith Stearns, University of Wisconsin–Eau Claire; Philip Tietjen, Virginia Tech University; Karen Welch, University of Wisconsin–Eau Claire; Patrick White, University of Delaware; and Steven Zwickel, University of Wisconsin–Madison.

For their helpful reviews of new entries, we thank Julie Dyke Ford, New Mexico Institute of Mining and Technology; Ruth Gerik, University of Texas at Arlington; Melody DeMeritt, California Polytechnic State University; and Rochelle Rodrigo, Mesa Community College. For their detailed reviews of the model documents, we thank Joyce D. Brotton, Northern Virginia Community College; Elizabeth Fife, University of Southern California; Annette Gooch, Santa Rosa Junior College; Jane Hughey, Texas A&M University; Karen Kasonic, University of Washington; and David L. Major, Austin Peay State University. For their helpful reviews of the companion Web site, we thank Anne Bliss, University of Colorado–Boulder; Lesley Baker, Tulane University; Mary Connerty, Pennsylvania State University–Erie; Annette Gooch, Santa Rosa Junior College; Christina Grignon, University of Wisconsin–Milwaukee; Nancy Hightower, University of Denver; Matthias Jonas, Niceware International, LLC; Elizabeth Robinson, Texas A&M University; Charlotte Rosen, Cornell University; and Philip Tietjen, Virginia Tech University. We gratefully acknowledge Eva Brumberger of Virginia Tech University for her thorough review of the visuals coverage, and

Kim Isaacs of Advanced Career Systems, Inc., for her review of the résumés entry and her work on new sample résumés.

For this edition, we owe special thanks to Matthias Jonas, Niceware International, LLC, for developing the white papers entry; Richard C. Hay, University of Wisconsin–Milwaukee, for revising the entry on documenting sources and for reviewing new entries; Randy Thompson, Computer Graphics Design, for developing new visuals; and Renee Tegge, whose keen eye, impressive knowledge of grammar, and insightful suggestions improved many entries. In addition, we are very much indebted to the many reviewers and contributors not named here who helped us shape the first seven editions.

We wish to thank Bedford/St. Martin's for supporting this book, especially Joan Feinberg, President, and Karen Henry, Editor in Chief. We are grateful to Emily Berleth, Manager of Publishing Services at Bedford/St. Martin's, and Herb Nolan of Books By Design for their patience and expert guidance. We are pleased to acknowledge the enthusiastic assistance of Kaitlin Hannon, Editorial Assistant. Finally, we wish to thank Caroline Thompson, our developmental editor at Bedford/St. Martin's, whose editing, professionalism, and collegiality helped produce an outstanding edition.

We offer heartfelt thanks to Barbara Brusaw for her patience and time spent preparing the manuscript for the first five editions. We also gratefully acknowledge the ongoing contributions of many students and instructors at the University of Wisconsin–Milwaukee. Finally, special thanks go to Janice Alred for her many hours of substantive assistance and for continuing to hold everything together.

G. J. A.
C. T. B.
W. E.O.

Five Steps to Successful Writing

Successful writing on the job is not the product of inspiration, nor is it merely the spoken word converted to print; it is the result of knowing how to structure information using both text and design to achieve an intended purpose for a clearly defined audience. The best way to ensure that your writing will succeed—whether it is in the form of a memo, a résumé, a proposal, or a Web page—is to approach writing using the following steps:

1. Preparation
2. Research
3. Organization
4. Writing
5. Revision

You will very likely need to follow those steps consciously—even self-consciously—at first. The same is true the first time you use new software, interview a candidate for a job, or chair a committee meeting. With practice, the steps become nearly automatic. That is not to suggest that writing becomes easy. It does not. However, the easiest and most efficient way to write effectively is to do it systematically.

As you master the five steps, keep in mind that they are interrelated and often overlap. For example, your readers' needs and your purpose, which you determine in step 1, will affect decisions you make in subsequent steps. You may also need to retrace steps. When you conduct research, for example, you may realize that you need to revise your initial impression of the document's purpose and audience. Similarly, when you begin to organize, you may discover the need to return to the research step to gather more information.

The time required for each step varies with different writing tasks. When writing an informal memo, for example, you might follow the first three steps (preparation, research, and organization) by simply listing the points in the order you want to cover them. In such situations, you gather and organize information mentally as you consider your purpose and audience. For a formal report, the first three steps require well-organized research, careful note-taking, and detailed outlining. For a routine e-mail message to a coworker, the first four steps merge as you type the information on the screen. In short, the five steps expand, contract, and at times must be repeated to fit the complexity or context of the writing task.

Dividing the writing process into steps is especially useful for collaborative writing, in which you typically divide work among team members, keep track of a project, and save time by not duplicating effort. When you collaborate, you can use e-mail to share text and other files, suggest improvements to each other's work, and generally keep everyone informed of your progress as you follow the steps in the writing process.

Preparation

Writing, like most professional tasks, requires solid preparation.* In fact, adequate preparation is as important as writing a draft. In preparation for writing, your goal is to accomplish the following four major tasks:

- Establish your primary purpose.
- Assess your audience (or readers) and the context.
- Determine the scope of your coverage.
- Select the appropriate medium.

Establishing Your Purpose. To establish your primary purpose simply ask yourself what you want your readers to know, to believe, or to be able to do after they have finished reading what you have written. Be precise. Often a writer states a purpose so broadly that it is almost useless. A purpose such as "to report on possible locations for a new research facility" is too general. However, "to compare the relative advantages of Paris, Singapore, and San Francisco as possible locations for a new research facility so top management can choose the best location" is a purpose statement that can guide you throughout the writing process. In addition to your primary purpose, consider possible secondary purposes for your document. For example, a secondary purpose of the research-facilities report might be to make corporate executive readers aware of the staffing needs of the new facility so that they can ensure its smooth operation in whatever location is selected.

Assessing Your Audience and Context. The next task is to assess your audience. Again, be precise and ask key questions. Who exactly is your reader? Do you have multiple readers? Who needs to see or to use

*In this discussion, as elsewhere throughout this book, words and phrases shown as links—underlined and set in an alternate typeface—refer to specific alphabetical entries.

will be useful, this is a good time to think about where they may be deployed and what kinds of visual elements will be effective, especially if they need to be prepared by someone else while you are writing and revising the draft. The outline can also suggest where headings, lists, and other special design features may be useful.

Writing

When you have established your purpose, your readers' needs, and your scope and have completed your research and your outline, you will be well prepared to write a first draft. Expand your outline into paragraphs, without worrying about grammar, refinements of language usage, or punctuation. Writing and revising are different activities; refinements come with revision.

Write the rough draft, concentrating entirely on converting your outline into sentences and paragraphs. You might try writing as though you were explaining your subject to a reader sitting across from you. Do not worry about a good opening. Just start. Do not be concerned in the rough draft about exact word choice unless it comes quickly and easily—concentrate instead on ideas.

Even with good preparation, writing the draft remains a chore for many writers. The most effective way to get started and keep going is to use your outline as a map for your first draft. Do not wait for inspiration—you need to treat writing a draft as you would any on-the-job task. The entry writing a draft describes tactics used by experienced writers—discover which ones are best suited to you and your task.

Consider writing an introduction last because then you will know more precisely what is in the body of the draft. Your opening should announce the subject and give readers essential background information, such as the document's primary purpose. For longer documents, an introduction should serve as a frame into which readers can fit the detailed information that follows.

Finally, you will need to write a conclusion that ties the main ideas together and emphatically makes a final significant point. The final point may be to recommend a course of action, make a prediction or a judgment, or merely summarize your main points—the way you conclude depends on the purpose of your writing and your readers' needs.

Revision

The clearer a finished piece of writing seems to the reader, the more effort the writer has likely put into its revision. If you have followed the steps of the writing process to this point, you will have a rough draft that needs to be revised. Revising, however, requires a different frame of mind than does writing the draft. During revision, be eager to find and correct faults and be honest. Be hard on yourself for the benefit of your readers. Read and evaluate the draft as if you were a reader seeing it for the first time.

Check your draft for accuracy, completeness, and effectiveness in achieving your purpose and meeting your readers' needs and expectations. Trim extraneous information: Your writing should give readers exactly what they need, but it should not burden them with unnecessary information or sidetrack them into loosely related subjects.

Do not try to revise for everything at once. Read your rough draft several times, each time looking for and correcting a different set of problems or errors. Concentrate first on larger issues, such as unity and coherence; save mechanical corrections, like spelling and punctuation, for later proofreading. See also ethics in writing.

Finally, for important documents, consider having others review your writing and make suggestions for improvement. For collaborative writing, of course, team members must review each other's work on segments of the document as well as the final master draft. Use the Checklist of the Writing Process on pages xxi–xxii to guide you not only as you revise but also throughout the writing process.

Checklist of the Writing Process

This checklist arranges key entries of the *Handbook of Technical Writing* according to the sequence presented in Five Steps to Successful Writing, which begins on page xiii. This checklist is useful both for following the steps and for diagnosing writing problems.

CONTENTS BY TOPIC

Use this list as a quick reference for finding entries by topic. To search this book in more detail, use the index. For a list of figures and model documents, see pages 637–39.

TECHNICAL WRITING DOCUMENTS AND ELEMENTS

DESIGN AND VISUALS

CORRESPONDENCE

JOB SEARCH AND APPLICATION

PRESENTATIONS AND MEETINGS

RESEARCH AND DOCUMENTATION

ORGANIZATION, WRITING, AND REVISION

DIGITAL TIPS

ESL TROUBLE SPOTS

ESL TIPS

STYLE AND LANGUAGE

USAGE

See pages 635–36.

SENTENCES AND PARAGRAPHS

PARTS OF SPEECH AND GRAMMAR

PUNCTUATION AND MECHANICS

a / an

A and *an* are indefinite articles because the noun designated by the article is not a specific person, place, or thing but is one of a group.

- She installed *a* program.
  [not a specific program but an unnamed program]

Use *a* before words or abbreviations beginning with a consonant or consonant sound, including *y* or *w*.

- *A* manual has been written on that subject.
- It was *a* historic event for the laboratory.
- We received *a* DNA sample.
- The year's activities are summarized in *a* one-page report.
  [*One* begins with the consonant sound "wuh."]

Use *an* before words or abbreviations beginning with a vowel or a consonant with a vowel sound.

- The report is *an* overview of the year's activities.
- He bought *an* SLR digital camera.
  [*SLR* begins with a vowel sound "es."]
- The applicant arrived *an* hour early.
  [*Hour* begins with a silent *h*.]

Do not use unnecessary indefinite articles in a sentence.

- Fill with *a* half *a* pint of fluid.
  [Choose one article and eliminate the other.]

See also adjectives.

a lot

A lot is often incorrectly written as one word (*alot*).The phrase *a lot* is informal and normally should not be used in technical writing. Use *many* or *numerous* for estimates or give a specific number or amount.

- The staff raised ~~a lot of~~ *many* objections to the policy.

abbreviations

DIRECTORY

Abbreviations are shortened versions of words or combinations of the first letters of words (Corp./Corporation, URL/*U*niform *R*esource *L*ocator). Abbreviations, if used appropriately, can be convenient for both the reader and the writer. Like symbols, they can be important space savers in technical writing.

Abbreviations that are formed by combining the initial letter of each word in a multiword term are called initialisms. *Initialisms* are pronounced as separate letters (AC or ac/*a*lternating *c*urrent). Abbreviations that combine the first letter or letters of several words—and can be pronounced—are called *acronyms*. Acronyms are always written without periods (LAN/*l*ocal *a*rea *n*etwork, laser/*l*ight *a*mplification by *s*timulated *e*mission of *r*adiation).

Using Abbreviations

The most important consideration in the use of abbreviations is whether they will be understood by your readers. Even the same abbreviation, for example, can have two different meanings (NEA stands for both the National Education Association and the National Endowment for the Arts). In business, industry, and government, specialists and people working together on particular projects often use abbreviations. Like jargon, shortened forms will be easily understood within a group of specialists;

outside of the group, however, they might be incomprehensible. In fact, abbreviations can be easily overused, either as an affectation or in a misguided attempt to make writing concise, especially in e-mail. Remember that memos, e-mail, or reports addressed to specific people may be read by other people—you must consider those secondary readers as well. A good rule to follow: when in doubt, spell it out.

Writer's Checklist: Using Abbreviations

☑ Except for commonly used abbreviations (U.S., a.m.), spell out a term to be abbreviated the first time it is used, followed by the abbreviation in parentheses. Thereafter, the abbreviation may be used alone.

☑ In long documents, repeat the full term in parentheses after the abbreviation at regular intervals to remind readers of the abbreviation's meaning, as in "Remember to submit the CAR (Capital Appropriations Request) by. . . ."

☑ Do not add an additional period at the end of a sentence that ends with an abbreviation.

- The official name of the company is DataBase, Inc.

☑ For abbreviations specific to your profession or discipline, use a style guide recommended by your professional organization or company. (A list of style guides appears at the end of documenting sources.)

☑ Write acronyms in capital letters without periods. The only exceptions are acronyms that have become accepted as common nouns, which are written in lowercase letters, such as *scuba* (*s*elf-*c*ontained *u*nderwater *b*reathing *a*pparatus).

☑ Generally, use periods for lowercase initialisms (a.k.a., e.d.p., p.m.) but not for uppercase ones (GDP, IRA, UFO). Exceptions include geographic names (U.S., U.K., E.U.) and formal expressions of academic degrees (B.A., M.B.A., Ph.D.).

☑ Form the plural of an acronym or initialism by adding a lowercase *s*. Do not use an apostrophe (CARs, DVDs).

☑ Do not follow an abbreviation with a word that repeats the final term in the abbreviation (HIV transmission *not* HIV virus transmission).

☑ Do not make up your own abbreviations; they will confuse readers.

Forming Abbreviations

Names of Organizations. A company may include in its name a term such as *Brothers, Incorporated, Corporation,* or *Company*. If the term is abbreviated in the official company name that appears on letterhead stationery or on its Web site, use the abbreviated form: *Bros., Inc.,*

Corp., or *Co.* If the term is not abbreviated in the official name, spell it out in writing, except with addresses, footnotes, bibliographies, and lists where abbreviations may be used. Likewise, use an ampersand (&) only if it appears in the official company name. For names of divisions within organizations, terms such as *Department* and *Division* should be abbreviated only when space is limited (*Dept.* and *Div.*).

Measurements. Except for abbreviations that may be confused with words (*in.* for *inch* and *gal.* for *gallon*), abbreviations of measurement do not require periods (*yd* for *yard* and *qt* for *quart*). Abbreviations of units of measure are identical in the singular and plural: 1 *cm* and 15 *cm* (*not* 15 *cms*). Some abbreviations can be used in combination with other symbols (°*F* for *degrees Fahrenheit* and *Å* for *angstrom units*).

The following list includes abbreviations for the basic units of the International System of Units (SI), the metric system. This system not only is standard in science but also is used in international commerce and trade.

MEASUREMENT	UNIT	ABBREVIATION
length	meter	m
mass	kilogram	kg
time	second	s
electric current	ampere	A
thermodynamic temperature	kelvin	K
amount of substance	mole	mol
luminous intensity	candela	cd

For additional definitions and background, see the National Institute of Standards and Technology Web site at <http://physics.nist.gov/cuu/Units/units.html>. For information on abbreviating dates and time, see numbers.

Personal Names and Titles. Personal names generally should not be abbreviated: Thomas (*not* Thos.) and William (*not* Wm.). An academic, civil, religious, or military title should be spelled out and in lowercase when it does not precede a name. (The *captain* wanted to check the orders.) When they precede names, some titles are customarily abbreviated (Dr. Smith, Mr. Mills, Ms. Katz). See also Ms./Miss/Mrs.

An abbreviation of a title may follow the name; however, be certain that it does not duplicate a title before the name (Angeline Martinez, Ph.D. *or* Dr. Angeline Martinez). When addressing correspondence and including names in other documents, you normally should spell out titles (The Honorable Mary J. Holt; Professor Charles Matlin). Traditionally, periods are used with academic degrees, although they are sometimes omitted (M.A./MA, M.S./MS, Ph.D./PhD).

Common Scholarly Abbreviations. The following is a partial list of abbreviations commonly used in reference books and for documenting sources in research papers and reports. Other than in formal scholarly work, generally avoid such abbreviations.

anon.	anonymous
bibliog.	bibliography, bibliographer, bibliographic
ca., c.	*circa*, "about" (used with approximate dates: ca. 1756)
cf.	*confer*, "compare"
chap.	chapter
diss.	dissertation
ed., eds.	edited by, editor(s), edition(s)
e.g.	*exempli gratia*, "for example" (see e.g./i.e.)
esp.	especially
et al.	*et alii*, "and others"
etc.	*et cetera*, "and so forth" (see etc.)
ff.	and the following page(s) or line(s)
GPO	Government Printing Office, Washington, D.C.
i.e.	*id est*, "that is" (see e.g./i.e.)
l., ll.	line, lines
MS, MSS	manuscript, manuscripts
n., nn.	note, notes (used immediately after page number: 56n., 56n.3, 56nn.3–5)
N.B., n.b.	*nota bene*, "take notice, mark well"
n.d.	no date (of publication)
n.p.	no place (of publication); no publisher; no page
p., pp.	page, pages
proc.	proceedings
pseud.	pseudonym
pub.	published by, publisher, publication
rev.	revised by, revised, revision; review, reviewed by (Spell out "review" where "rev." might be ambiguous.)
rpt.	reprinted by, reprint
sec., secs.	section, sections
sic	so, thus; inserted in brackets ([*sic*]) after a misspelled or wrongly used word in quotations
supp., suppl.	supplement
trans.	translated by, translator, translation
UP	University Press (used in MLA style, as in Oxford UP)
viz.	*videlicet*, "namely"
vol., vols.	volume, volumes
vs., v.	*versus*, "against" (*v.* preferred in titles of legal cases)

WEB LINK USING ABBREVIATIONS

For links to Web sites specifying standard abbreviations and acronyms, including U.S. Postal Service abbreviations, see <*bedfordstmartins.com/alredtech*> and select *Links for Handbook Entries.*

above

Avoid using *above* to refer to a preceding passage or visual because its reference is often vague. The same is true of *aforesaid, aforementioned, the former,* and *the latter*. To refer to something previously mentioned, repeat the noun or pronoun, or construct your paragraph so that your reference is obvious.

- Please fill out and submit ~~the above~~ *your travel voucher* by March 1.

Using such references can also be an affectation. See also former/latter.

absolute words (*see* adjectives)

absolutely

Absolutely means "definitely," "entirely," "completely," or "unquestionably." Avoid it as a redundant intensifier to mean "very" or "much."

- We are ~~absolutely~~ sure we can meet the deadline.

abstract / concrete words

Abstract words refer to general ideas, qualities, conditions, acts, or relationships—intangible things that cannot be detected by the five senses (sight, hearing, touch, taste, and smell), such as *learning, leadership,* and *technology. Concrete* words identify things that can be perceived by the five senses, such as *diploma, manager,* and *keyboard.*

Abstract words must frequently be further defined or described.

- The investigative team needs freedom *to interview the maintenance technicians*.

Abstract words are best used with concrete words to help make intangible concepts specific and vivid.

- *Transportation* [abstract] was limited to *buses* [concrete] and *commuter trains* [concrete].

Just how concrete a particular context might require you to be is shown in Figure A–1, which goes from the most abstract, on the left, to the most concrete, on the right. The example in Figure A–1 represents seven levels of abstraction; the appropriate level depends on your purpose in writing and on the context in which you are using the word. See also word choice.

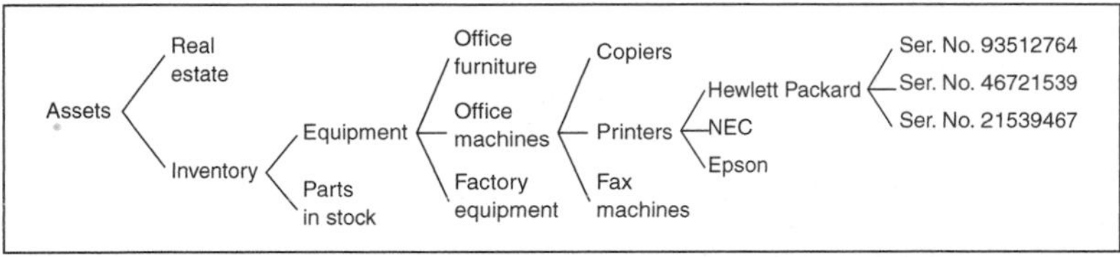

FIGURE A–1. Abstract-to-Concrete Words

abstracts

An abstract summarizes and highlights the major points of a formal report, trade journal article, dissertation, or other work. Its primary purpose is to enable readers to decide whether to read the work in full. For a discussion of how summaries differ from abstracts, see executive summaries.

Although abstracts, typically 200 to 250 words long, are published with the longer works they condense, they can also be published separately in periodical indexes and by abstracting services (see research). For this reason, an abstract must be readable apart from the original document.

Depending on the kind of information they contain, abstracts are often classified as descriptive or informative. A *descriptive abstract* summarizes the purpose, scope, and methods used to arrive at the reported findings. It is a slightly expanded table of contents in sentence and paragraph form. A descriptive abstract need not be longer than several sentences. An *informative abstract* is an expanded version of the descriptive abstract. In addition to information about the purpose, scope, and research methods used, the informative abstract summarizes the results, conclusions, and any recommendations. The informative abstract retains the tone and essential scope of the report, omitting its details. The first four sentences of the abstract shown in Figure A–2 alone would be descriptive; with the addition of the sentences that detail the conclusions of the report, the abstract becomes informative. Notice that this example includes "keywords" that are sometimes included to assist searches of abstract databases.

The type of abstract you should write depends on your readers and the organization or publication for which you are writing. Informative abstracts work best for wide audiences that need to know conclusions

ABSTRACT

"The Effects of Long-Distance Running on Male and Female Runners Aged 50 to 72 Years"

by Sandra Young

Purpose | Methods and scope | Findings | Conclusions | Recommendations | Descriptive section

The long-term effects of long-distance running on the bones, joints, and general health of runners aged 50 to 72 can help determine whether physicians should recommend long-distance running for their older patients. Recent studies conducted at Stanford University and the University of Florida tested and compared male and female long-distance runners aged 50 to 72 with a control group of runners and nonrunners. The groups were matched by sex, race, education, and occupation. The Florida study used only male runners who had run at least 20 miles a week for five years and compared them with a group of runners and nonrunners. Both studies based findings on medical histories and on physical and X-ray examinations. Both studies conclude that long-distance running is not associated with increased degenerative joint disease. Control groups were more prone to spur formation, sclerosis, and joint-space narrowing and showed more joint degeneration than runners. Female long-distance runners exhibited somewhat more sclerosis in knee joints and the lumbar spine area than matched control subjects. Both studies support the role of exercise in retarding bone loss with aging. The investigation concludes that the health risk factors are fewer for long-distance runners than for those less active aged 50 to 72. The investigation recommends that physicians recognize that an exercise program that includes long-distance running can be beneficial to their aging patients' health.

Keywords: bone loss, exercise programs, geriatric patients, health risk, joint and lumbar degeneration, long-distance running, runners, sclerosis.

iii

FIGURE A–2. Informative Abstract (for an Article)

and recommendations; descriptive abstracts work best for cc tions, such as proceedings and progress reports, that do not c conclusions or recommendations.

Writing Style

Write the abstract *after* finishing the report or document. Otherwise, the abstract may not accurately reflect the longer work. Begin with a topic sentence that announces the subject and scope of your report or document. Then, using the major and minor headings of your outline or table of contents to distinguish primary ideas from secondary ones, decide what material is relevant to your abstract. (See outlining.) Write with clarity and conciseness, eliminating unnecessary words and ideas. Do not, however, become so terse that you omit articles (*a*, *an*, *the*) and important transitional words and phrases (*however*, *therefore*, *but*, *in summary*). Write complete sentences, but avoid stringing together a group of short sentences end to end; instead, combine ideas by using subordination and parallel structure. Spell out all but the most common abbreviations. Typically, in a formal report, an abstract follows the title page and is numbered page iii.

accept / except

Accept is a verb meaning "consent to," "agree to take," or "admit willingly." (I *accept* the responsibility.) *Except* is normally used as a preposition meaning "other than" or "excluding." (We agreed on everything *except* the schedule.)

acceptance / refusal letters (for employment)

When you decide to accept a job offer, you can notify your new employer by telephone or in a meeting—but to make your decision official, you need to send an acceptance in writing to your new employer. See also correspondence and e-mail.

Figure A–3 shows an example of an acceptance letter written by a college student. Of course, the details you include in your own letter will vary, depending on your previous conversations with your new employer. Note that in the first paragraph of Figure A–3, the student identifies the job he is accepting and the salary he has been offered—doing so can avoid any misunderstandings about the job or the salary. In the second paragraph, the student details his plans for moving and reporting for work. Even if the student discussed these arrangements during

From: Chad Johnston <chjohnston@hotmail.com>
To: Helen Castro <castro@cdc.com>
Sent: Monday, April 17, 2006 8:39 AM
Subject: Acceptance of Software Designer Position

Dear Ms. Castro:

I am pleased to accept your offer of $35,000 per year as a software designer in your Software Development Department.

After graduation, I plan to leave Boston on Tuesday, June 6. I should be able to find suitable living accommodations within a few days and be ready to report for work on the following Monday, June 12. Please let me know if this date is satisfactory to you.

I look forward to working with the design team at CDC.

Sincerely,

Chad Johnston

FIGURE A–3. Acceptance Letter (Sent as E-mail)

earlier conversations, he needs to confirm them, officially, in this letter. The student then concludes with a brief but enthusiastic statement that he looks forward to working for the new employer.

When you decide to reject a job offer, send a job-refusal letter to make that decision official, even if you have already notified the employer during a meeting or on the phone. Writing a letter is a gesture that the employer will appreciate. Be especially tactful and courteous—the employer you are refusing has spent time and effort interviewing you and may have counted on your accepting the job. Remember, you may apply for another job at that company in the future. In Figure A–4, an example of a job-refusal letter, the applicant mentions something positive about her contact with the employer and refers to the specific job offered. She indicates her serious consideration of the offer, provides a logical reason for the refusal, and concludes on a pleasant note. For further strategy on handling refusals and negative messages generally, see refusal letters.

From: Emily Brazy <ebrazy@yahoo.com>
To: Roger Vallone <vallone@adservices.com>
Sent: Friday, January 27, 2006 3:47 PM
Subject: Copywriter Position

Dear Mr. Vallone:

I enjoyed talking with you about your opening for a copywriter, and I was gratified to receive your offer.

After giving the offer serious thought, however, I have decided to accept a position as a technical writer at a documentation company. I feel that the job I have chosen is better suited to my skills and long-term goals.

I appreciate your consideration and the time you spent with me. I wish you the best of luck in filling the position.

Sincerely,

Emily Brazy

FIGURE A–4. Refusal Letter (Sent as E-mail)

accuracy / precision

Accuracy means error-free and correct. *Precision*, mathematically, refers to the "degree of refinement with which a measurement is made." Thus, a measurement carried to five decimal places (5.28371) is more precise than one carried to two decimal places (5.28), but it is not necessarily more accurate.

acknowledgment letters

One way to build goodwill with colleagues and clients is to send an acknowledgment letter, letting them know that something they sent arrived and expressing thanks. It is usually a short, polite note. If you

From: Roger Vonblatz <vonblatz@omni-electric.com>
To: David Evans <evans239@evans-assoc.com>
Sent: Friday, February 24, 2006 5:52 PM
Subject: Mark II Report Received

Dear Mr. Evans:

I received your comprehensive report today. When I finish studying it in detail, I'll send you our cost estimate for the installation of the Mark II Energy Saving System.

Thank you for preparing such a thorough analysis.

Regards,

Roger Vonblatz

FIGURE A–5. Acknowledgment Letter (Sent as E-mail)

have established a working relationship with someone, an e-mail message is appropriate. (See correspondence.) The example shown in Figure A–5 is typical and could be sent as a letter or an e-mail.

acronyms and initialisms

Acronyms are formed by combining the first letter or letters of several words; they are pronounced as words and written without periods (ANSI/*A*merican *N*ational *S*tandards *I*nstitute). Initialisms are formed by combining the initial letter of each word in a multiword term; they are pronounced as separate letters (STC/*S*ociety for *T*echnical *C*ommunication). See abbreviations for the use of acronyms and initialisms.

activate / actuate

Both *activate* and *actuate* mean "make active," although *actuate* is usually applied only to mechanical processes.

- The relay *actuates* the trip hammer. [mechanical process]

- The electrolyte *activates* the battery. [chemical process]
- The governor *activated* the National Guard. [legal process]

active voice (*see* voice)

ad hoc

Ad hoc is Latin for "for this" or "for this particular occasion." An ad hoc committee is one set up to consider a particular issue, as opposed to a permanent committee. The term has been fully assimilated into English and thus does not have to be italicized. See also affectation and foreign words in English.

adapt / adept / adopt

Adapt is a verb meaning "adjust to a new situation." *Adept* is an adjective meaning "highly skilled." *Adopt* is a verb meaning "take or use as one's own."

- The company will *adopt* a policy of finding engineers who are *adept* managers and who can *adapt* to new situations.

adjectives

DIRECTORY

An adjective is any word that modifies a noun or pronoun. *Descriptive adjectives* identify a quality of a noun or pronoun. *Limiting adjectives* impose boundaries on the noun or pronoun.

- *hot* surface [descriptive]
- *three* phone lines [limiting]

Limiting Adjectives

Limiting adjectives include these categories:

- Articles (*a*, *an*, *the*)
- Demonstrative adjectives (*this*, *that*, *these*, *those*)
- Possessive adjectives (*my*, *your*, *his*, *her*, *its*, *our*, *their*)
- Numeral adjectives (*two*, *first*)
- Indefinite adjectives (*all*, *none*, *some*, *any*)

Articles. Articles (*a*, *an*, *the*) are traditionally classified as adjectives because they modify nouns by either limiting them or making them more specific. See also a/an, articles, and English as a second language.

Demonstrative Adjectives. A demonstrative adjective points to the thing it modifies, specifying the object's position in space or time. *This* and *these* specify a closer position; *that* and *those* specify a more remote position.

- *This* report is more current than *that* report, which Human Resources distributed last month.
- *These* sales figures are more recent than *those* reported last week.

Demonstrative adjectives often cause problems when they modify the nouns *kind*, *type*, and *sort*. Demonstrative adjectives used with those nouns should agree with them in number.

- *this* kind, *these* kinds; *that* type, *those* types

Confusion often develops when the preposition *of* is added (*this kind of*, *these kinds of*) and the object of the preposition does not conform in number to the demonstrative adjective and its noun. See also agreement and prepositions.

- *This kind of* human resources ~~policies are~~ *policy is* standard.
- *These kinds of* human resources ~~policy is~~ *policies are* standard.

Avoid using demonstrative adjectives with words like *kind*, *type*, and *sort* because doing so can easily lead to vagueness. Instead, be more specific. See also kind of/sort of.

Possessive Adjectives. Because possessive adjectives (*my*, *your*, *his*, *her*, *its*, *our*, *their*) directly modify nouns, they function as adjectives, even though they are pronoun forms (*my* idea, *her* plans, *their* projects). See also functional shift.

Numeral Adjectives. Numeral adjectives identify quantity, degree, or place in a sequence. They always modify count nouns. Numeral adjectives are divided into two subclasses: cardinal and ordinal. A *cardinal adjective* expresses an exact quantity (*one* pencil, *two* computers); an *ordinal adjective* expresses degree or sequence (*first* quarter, *second* edition).

In most writing, an ordinal adjective should be spelled out if it is a single word (*tenth*) and written in figures if it is more than one word (*312th*). Ordinal numbers can also function as adverbs. (John arrived *first*.) See also first/firstly and numbers.

Indefinite Adjectives. Indefinite adjectives do not designate anything specific about the nouns they modify (*some* CD-ROMs, *all* designers). The articles *a* and *an* are included among the indefinite adjectives (*a* chair, *an* application).

Comparison of Adjectives

Most adjectives in the positive form show the comparative form with the suffix *-er* for two items and the superlative form with the suffix *-est* for three or more items.

- The first report is *long*. [positive form]
- The second report is *longer*. [comparative form]
- The third report is *longest*. [superlative form]

Many two-syllable adjectives and most three-syllable adjectives are preceded by the word *more* or *most* to form the comparative or the superlative.

- The new media center is *more* impressive than the old one. It is the *most* impressive in the county.

A few adjectives have irregular forms of comparison (*much*, *more*, *most*; *little*, *less*, *least*).

Some adjectives (*round*, *unique*, *exact*, *accurate*), often called *absolute words*, are not logically subject to comparison.

- We modified our manual to follow more ~~exactly~~ *closely* the recommended changes.
- We modified our manual to follow ~~more~~ exactly the recommended changes.

TIPS FOR USING ADJECTIVES

Do not add *-s* or *-es* to an adjective to make it plural.

- the *long* trip
- the *long* trips

Capitalize adjectives of origin (city, state, nation, continent).

- the *Venetian* canals
- the *Texas* hat
- the *French* government
- the *African* deserts

In English, verbs of feeling (for example, *bore*, *interest*, *surprise*) have two adjectival forms: the present participle (*-ing*) and the past participle (*-ed*). Use the present participle to describe what causes the feeling. Use the past participle to describe the person who experiences the feeling.

- We heard the *surprising* election results.
  [The *election results* cause the feeling.]
- Only the losing candidate was *surprised* by the election results.
  [The *candidate* experienced the feeling of surprise.]

Adjectives follow nouns in English in only two cases: when the adjective functions as a subjective complement

- That project is not *finished*.

and when an adjective phrase or clause modifies the noun.

- The project *that was suspended temporarily* . . .

In all other cases, adjectives are placed before the noun.

When there are multiple adjectives, it is often difficult to know the right order. The guidelines illustrated in the following example would apply in most circumstances, but there are exceptions. (Normally do not use a phrase with so many stacked modifiers.) See also articles.

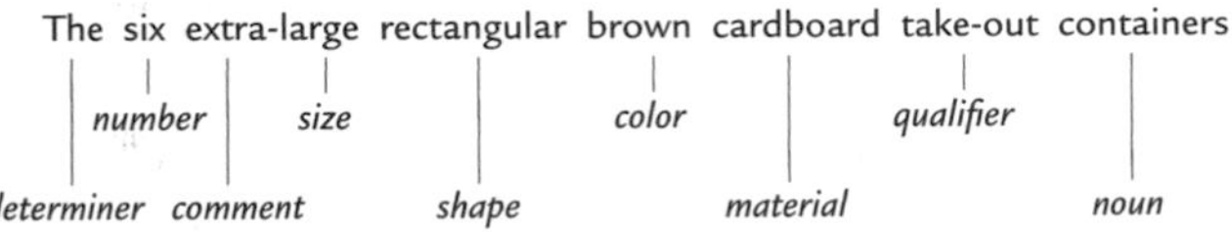

Placement of Adjectives

When limiting and descriptive adjectives appear together, the limiting adjectives precede the descriptive adjectives, with the articles usually in the first position.

- *The ten red* cars were parked in a row.
  [article (*The*), limiting adjective (*ten*), descriptive adjective (*red*)]

Within a sentence, adjectives may appear before the nouns they modify (the attributive position) or after the nouns they modify (the predicative position).

- *The small* jobs are given priority. [attributive position]
- The exposure is *brief*. [predicative position]

Use of Adjectives

Nouns often function as adjectives to make other nouns more precise.

- The *accident* report prompted a *product* redesign.

When adjectives modifying the same noun can be reversed and still make sense or when they can be separated by *and* or *or*, they should be separated by commas.

- The company seeks a *bright, energetic, creative* management team.

Notice that there is no comma after *creative*. Never use a comma between a final adjective and the noun it modifies. When an adjective modifies a phrase, no comma is required.

- We need an *updated Web page design*.
  [*Updated* modifies the phrase *Web page design*.]

Writers sometimes string together a series of nouns used as adjectives to form a unit modifier, thereby creating stacked (jammed) modifiers, which can confuse readers. See also word choice.

adjustment letters

An adjustment letter is written in response to a complaint letter and tells the customer what your company intends to do about the complaint. Although sent in response to a problem, an adjustment letter actually provides an excellent opportunity to build goodwill for your company. An effective adjustment letter, such as those shown in

Figures A–6 and A–7, can both repair any damage done and restore the customer's confidence in your company.

No matter how unreasonable the complaint, your response and tone should be positive and respectful. Avoid emphasizing the problem, but do take responsibility for it. Focus on what you are doing to correct the problem. You should settle such matters quickly and courteously, and always try to satisfy the customer at a reasonable cost to your company. See also refusal letters.

Full Adjustments

Before granting an adjustment to a claim for which your company is at fault, first determine what happened and what you can do to satisfy the customer. Be certain that you are familiar with your company's adjustment policy. In addition, be careful about your wording; for example, "We have just received your letter of May 7 about our *defective product*" could be ruled in a court of law as an admission that the product is in fact defective. Treat every claim individually, and lean toward giving the customer the benefit of the doubt.

Grant adjustments graciously; a settlement made grudgingly will do more harm than good. Not only must you be gracious, but you must also acknowledge the error in such a way that the customer will not lose confidence in your company. (See also tone and correspondence.) Emphasize early what the reader will consider good news.

- Yes, you were incorrectly billed for the delivery.
- Please accept our apologies for the error in your account.
- Enclosed is a replacement for the damaged part.

If an explanation will help restore your reader's confidence, explain what caused the problem. You might point out any steps you may be taking to prevent a recurrence of the problem. Explain that customer feedback helps your firm keep the quality of its product or service high. Close pleasantly, looking forward, not back. Avoid recalling the problem in your closing (Again, we apologize . . .).

The adjustment letter in Figure A–6, for example, begins by accepting responsibility and offers an apology for the customer's inconvenience (note the use of the pronouns *we* and *our*). The second paragraph expresses a desire to restore goodwill and describes specifically how the writer intends to make the adjustment. The third paragraph expresses appreciation to the customer for calling attention to the problem and assures him that his complaint has been taken seriously.

INTERNET SERVICES CORPORATION
10876 Crispen Way
Chicago, Illinois 60601

May 10, 2006

Mr. Jason Brandon
4319 Anglewood Street
Tacoma, WA 98402

Dear Mr. Brandon:

We are sorry that your experience with our customer support help line did not go smoothly. We are eager to restore your confidence in our ability to provide dependable, high-quality service. Your next three months of Internet access will be complimentary as our sincere apology for your unpleasant experience.

Providing dependable service is what is expected of us, and when our staff doesn't provide quality service, it is easy to understand our customers' disappointment. I truly wish we had performed better in our guidance for setup and log-on procedures and that your experience had been a positive one. To prevent similar problems in the future, we plan to use your letter in training sessions with customer support personnel.

We appreciate your taking the time to write us. It helps to receive comments such as yours, and we conscientiously follow through to be sure that proper procedures are being met.

Yours truly,

Inez Carlson

Inez Carlson, Vice President
Customer Support Services

www.isc.com

FIGURE A–6. Adjustment Letter (When Company Is at Fault)

ADDISON COMPUTERS
6409 East Seventh Street
Hartford, Connecticut 06156

September 21, 2006

Mr. Carlos Sanchez
570 Masterfield Avenue
Baton Rouge, LA 70805

Dear Mr. Sanchez:

Enclosed is your Addison Laptop Computer, which you shipped to us on August 29.

Our technical staff reports that the laptop was damaged by exposure to high levels of humidity. You stated in your letter that you often use your laptop on a covered patio. Doing so in a high-humidity environment, as is typical in Louisiana, can result in damage to the internal circuitry of your computer—as described on page 32 of your Addison Owner's Manual.

We have replaced the damaged circuitry and thoroughly tested your laptop. To ensure that a repetition of your recent experience does not occur, we recommend you avoid leaving your laptop exposed to high humidity for extended periods.

If you should find that the problem recurs, please call me at 800-555-0990. I will be glad to work with you to find a solution.

Sincerely,

Stephen Winton

Stephen Winton
Customer Service Representative

www.addison.com

FIGURE A–7. Partial Adjustment Letter (Accompanying a Product)

Partial Adjustments

You may sometimes need to grant a partial adjustment—even if a claim is not really justified—to regain the lost goodwill of a customer or client. If, for example, a customer incorrectly uses a product or service, you may need to help the reader better understand the correct use of that product or service. In such a circumstance, remember that your customer or client believes that his or her claim is justified. Therefore, you should give the explanation before granting the claim—otherwise, your reader may never get to the explanation. If your explanation establishes customer responsibility, do so tactfully. Figure A–7 is an example of a partial adjustment letter.

adverbs

DIRECTORY

An adverb modifies the action or condition expressed by a verb.

- The wrecking ball hit the side of the building *hard.*
  [The adverb tells *how* the wrecking ball hit the building.]

An adverb also can modify an adjective, another adverb, or a clause.

- The brochure design used *remarkably* bright colors.
  [*Remarkably* modifies the adjective *bright.*]
- The redesigned brake pad lasted *much* longer.
  [*Much* modifies the adverb *longer.*]
- *Surprisingly,* the engine failed.
  [*Surprisingly* modifies the clause *the engine failed.*]

Types of Adverbs

A simple adverb can answer one of the following questions:

Where? (adverb of place)

- Move the display *forward* slightly.

When? or *How often?* (adverb of time)

- Replace the thermostat *immediately*.
- I worked overtime *twice* this week.

How? (adverb of manner)

- Add the solvent *cautiously*.

How much? (adverb of degree)

- The *nearly* completed report was deleted from his disk.

An interrogative adverb can ask a question (*Where? When? Why? How?*):

- *How* many hours did you work last week?
- *Why* was the disk reformatted?

A conjunctive adverb can modify the clause that it introduces as well as join two independent clauses with a semicolon. The most common conjunctive adverbs are *however*, *nevertheless*, *moreover*, *therefore*, *further*, *then*, *consequently*, *besides*, *accordingly*, *also*, and *thus*.

- *Therefore*, I should finish the project by the deadline.
- I rarely work on weekends; *however*, this weekend will be an exception.

In the second example, note that a semicolon precedes and a comma follows *however*. The conjunctive adverb (*however*) introduces the independent clause (*this weekend will be an exception*) and indicates its relationship to the preceding independent clause (*I rarely work on weekends*). See also transition.

Comparison of Adverbs

Most one-syllable adverbs show comparison with the suffixes *-er* and *-est*.

- This copier is *fast*.
  [positive form]
- This copier is *faster* than the old one.
  [comparative form]
- This copier is the *fastest* of the three tested.
  [superlative form]

Most adverbs with two or more syllables end in *-ly*, and most adverbs ending in *-ly* are compared by inserting the comparative *more* or *less* or the superlative *most* or *least* in front of them.

- She moved *more quickly* than any other company's sales representative.
- *Most surprisingly*, the engine failed during the final test phase.

A few irregular adverbs require a change in form to indicate comparison (*well, better, best; badly, worse, worst; far, farther, farthest*).

- The training program functions *well.*
- Our training program functions *better* than most others in the industry.
- Many consider our training program the *best* in the industry.

Placement of Adverbs

An adverb usually should be placed in front of the verb it modifies.

- The pilot *methodically* performed the preflight check.

An adverb may, however, follow the verb (or the verb and its object) that it modifies.

- The system failed *unexpectedly*.
- They replaced the hard drive *quickly*.

An adverb may be placed between a helping verb and a main verb.

- In this temperature range, the pressure will *quickly* drop.

Adverbs such as *only*, *nearly*, *almost*, *just*, and *hardly* should be placed immediately before the words they limit. See also only and modifiers.

affect / effect

Affect is a verb that means "influence."

- The utility commission's decisions *affect* all state utilities.

Effect can function as a noun that means "result."

- The utility commission's decision had a positive *effect.*

Effect can also function as a verb that means "bring about" or "cause." However, avoid using *effect* as a verb. A less formal word, such as *make* or *produce*, is usually preferable.

- The technician will ~~effect~~ *make* changes to improve network security.

affectation

Affectation is the use of language that is more formal, technical, or showy than necessary to communicate information to the reader. Affectation is a widespread writing problem in the workplace because many people feel that affectation lends a degree of authority to their writing. In fact, affectation can alienate customers, clients, and colleagues.

Affected writing forces readers to work harder to understand the writer's meaning. It typically contains abstract, highly technical, or foreign words and is often liberally sprinkled with trendy buzzwords.

ETHICS NOTE Jargon and euphemisms can become affectation, especially if their purpose is to hide relevant facts or give a false impression of competence. See ethics in writing. ✦

Writers are easily lured into affectation through the use of long variants—words created by adding prefixes and suffixes to simpler words (*orientate* for *orient*; *utilization* for *use*). Unnecessarily formal words (such as *penultimate* for *next to last*), created words using *-ese* (such as *budgetese*), and outdated words (such as *aforesaid*) can produce affectation. (See also above.) Elegant variation—attempting to avoid repeating a word within a paragraph by substituting a pretentious synonym—is also a form of affectation.

- The use of robotics in the assembly process has increased
  , and it
  production~~. Robotic utilization~~ has cut costs.

Another type of affectation is gobbledygook, which is wordy, roundabout writing with many legal- and scientific-sounding terms (such as *herewith* and *quantum*).

Figure A–8 shows the original maintenance regulations for those leasing space in a university research park. The revision of those regulations, which eliminates various forms of affectation, is much clearer. (See Figure A–9.)

Understanding the possible reasons for affectation is the first step toward avoiding it. The following are some causes of affectation.

- *Impression.* Some writers use pretentious language in an attempt to impress the reader with fancy words instead of evidence and logic.
- *Insecurity.* Writers who are insecure about their facts, conclusions, or arguments may try to hide behind a smoke screen of pretentious words.
- *Imitation.* Perhaps unconsciously, some writers imitate poor writing they see around them.

In addition to performing interior housekeeping services, the lessee shall perform custodial maintenance on the exterior of the facility and grounds. Where a lessee shares a facility with one or more other lessees, exterior custodial maintenance responsibilities will be assigned by research park management on a fair and equitable basis. In those instances where the lessee's activity is located in a research park complex wherein predominant tenancy is by research park–operated activities, then research park management shall be responsible for exterior custodial maintenance except for those described in 1, 2, 3, and 4 below. The necessary equipment and labor to perform exterior custodial maintenance, when such a responsibility has been assigned to the lessee, shall be furnished by the lessee. Exterior custodial maintenance shall include the following tasks:

1. Clean entrance door and exterior of facility windows daily.
2. Sweep and clean the entrance and facility walks daily.
3. Empty and clean waste and manage hazardous-material disposal daily.
4. Check exterior lighting and report failures to members of the research park staff daily.

FIGURE A–8. Affected Writing

The lessee will, with his or her own equipment, perform the following duties daily:

1. Maintain a clean and neat appearance inside the facility.
2. Clean the entrance door and the outsides of the office windows.
3. Sweep the entrance and the sidewalk.
4. Empty and clean wastebaskets and dispose of hazardous materials.
5. Check the exterior lighting and report failures to the staff of the research park.

The lessee will also maintain the grounds surrounding the building, as specified in the lessee's contract. Where two or more lessees share the same building, the research park management will assign responsibility for the grounds. Where the lessees' facilities are in a building that is occupied predominantly by research park operations, research park management will be responsible for the grounds.

FIGURE A–9. Affected Writing Revised for Clarity

- *Intimidation.* A few writers, consciously or unconsciously, try to intimidate or overwhelm their readers with words, often to protect themselves from criticism.
- *Initiation.* Writers who are new to a field often feel that one way to prove their professional expertise is to use as much technical terminology and jargon as possible.
- *Imprecision.* Writers who are having trouble being precise sometimes find that an easy solution is to use a vague, trendy, or pretentious word.

See also clichés, conciseness, and nominalizations.

affinity

Affinity refers to the attraction of two persons or things to each other.

- The *affinity* between these two elements can be explained in terms of their valance electrons.

Affinity should not be used to mean "ability" or "aptitude."

- She has an ~~affinity~~ *aptitude* for problem solving.

agreement

DIRECTORY

Grammatical agreement is the correspondence in form between different elements of a sentence to indicate number, person, gender, and case.

A subject and its verb must agree in number.

- The *design is* acceptable.
  [The singular subject, *design*, requires the singular verb, *is*.]

- The new *products are* going into production soon.
  [The plural subject, *products*, requires the plural verb, *are*.]

A subject and its verb must agree in person.

- *I am* the designer.
  [The first-person singular subject, *I*, requires the first-person singular verb, *am*.]
- *They are* the designers.
  [The third-person plural subject, *they*, requires the third-person plural verb, *are*.]

A pronoun and its antecedent must agree in person, number, gender, and case.

- The *employees* report that *they* are more efficient in the new facility.
  [The third-person plural subject, *employees*, requires the third-person plural pronoun, *they*.]
- *Kaye McGuire* will meet with the staff on Friday, when *she* will assign duties.
  [The third-person singular subject, *Kaye McGuire*, requires *she*, the third-person feminine pronoun, in the subjective case.]

See also sentence construction.

Subject-Verb Agreement

Subject-verb agreement is not affected by intervening phrases and clauses.

- *One* in twenty hard drives we receive from our suppliers *is* faulty.
  [The verb, *is*, must agree in number with the subject, *one*, not *hard drives* or *suppliers*.]

The same is true when nouns fall between a subject and its verb.

- Only *one* of the emergency lights *was* functioning.
  [The subject of the verb is *one*, not *lights*.]
- *Each* of the switches *controls* a separate circuit.
  [The subject of the verb is *each*, not *switches*.]

Note that *one* and *each* are normally singular.

Indefinite pronouns such as *some*, *none*, *all*, *more*, and *most* may be singular or plural, depending on whether they are used with a mass noun (*Most* of the oil *has* been used) or with a count noun (*Most* of the drivers *know* why they are here). Mass nouns are singular, and count nouns are plural. Other words, such as *type*, *part*, *series*, and *portion*, take singular verbs even when they precede a phrase containing a plural noun.

- A *series* of meetings *was* held about the best way to market the new product.
- A large *portion* of most annual reports *is* devoted to promoting the corporate image.

Modifying phrases can obscure a simple subject.

- The *advice* of two engineers, one lawyer, and three executives *was* obtained prior to making a commitment.
  [The subject of the verb *was* is *advice*.]

Inverted word order can cause problems with agreement.

- From this work *have come* several important *improvements*.
  [The subject of the verb is *improvements*, not *work*.]

The number of a subjective complement does not affect the number of the verb—the verb must always agree with the subject.

- The *topic* of his report *is* employee benefits.
  [The subject of the sentence is *topic*, not *benefits*.]

A subject that expresses measurement, weight, mass, or total often takes a singular verb even when the subject word is plural in form. Such subjects are treated as a unit.

- *Four weeks is* the normal duration of the training program.

A verb following the relative pronoun *who* or *that* agrees in number with the noun to which the pronoun refers (its antecedent).

- Steel is one of those *industries* that *are* most affected by energy costs.
  [*That* refers to *industries*.]
- She is one of those *employees* who *are* rarely absent.
  [*Who* refers to *employees*.]

The word *number* sometimes causes confusion. When used to mean a specific number, it is singular.

- *The number* of committee members *was* six.

When used to mean an approximate number, it is plural.

- *A number* of people *were* waiting for the announcement.

Relative pronouns (*who*, *which*, *that*) may take either singular or plural verbs, depending on whether the antecedent is singular or plural. See also who/whom.

- He is a bookkeeper *who takes* work home at night.
- He is one of those bookkeepers *who take* work home at night.

Some abstract nouns are singular in meaning but plural in form: *mathematics*, *news*, *physics*, and *economics*.

- *Mathematics is* essential in computer science.

Some words, such as the plural *jeans* and *scissors*, cause special problems.

- The *scissors were* ordered last week.
  [The subject is the plural *scissors*.]
- *A pair* of scissors *is* on order.
  [The subject is the singular *pair*.]

A book with a plural title requires a singular verb.

- *Basic Medical Laboratory Techniques* is an essential resource.

A collective noun (*committee*, *faculty*, *class*, *jury*) used as a subject takes a singular verb when the group is thought of as a unit and a plural verb when the individuals in the group are thought of separately.

- The *committee is* unanimous in its decision.
- The *committee are* returning to their offices.

A clearer way to emphasize the individuals would be to use a phrase.

- The *committee members are* returning to their offices.

Compound Subjects

A compound subject is composed of two or more elements joined by a conjunction such as *and*, *or*, *nor*, *either . . . or*, or *neither . . . nor*. Usually, when the elements are connected by *and*, the subject is plural and requires a plural verb.

- *Accounting and chemistry are* both prerequisites for this position.

One exception occurs when the elements connected by *and* form a unit or refer to the same thing. In that case, the subject is regarded as singular and takes a singular verb.

- *Bacon and eggs is* a high-cholesterol meal.
- Our greatest *challenge and business opportunity is* the Internet.

A compound subject with a singular element and a plural element joined by *or* or *nor* requires that the verb agree with the closer element.

- Neither the director nor the *project assistants were* there.
- Neither the project assistants nor the *director was* there.

If *each* or *every* modifies the elements of a compound subject, use the singular verb.

- *Each* manager and supervisor *has* a production goal to meet.
- *Every* manager and supervisor *has* a production goal to meet.

Pronoun-Antecedent Agreement

Every pronoun must have an antecedent—a noun to which it refers. See also pronoun reference.

- When *employees* are hired, *they* must review the policy manual. [The pronoun *they* refers to the antecedent *employees*.]

Gender. A pronoun must agree in gender with its antecedent.

- *Mr. Swivet* in the accounting department acknowledges *his* share of the responsibility for the misunderstanding, just as *Ms. Barkley* in the research division must acknowledge *hers*.

Traditionally, a masculine, singular pronoun was used to agree with such indefinite antecedents as *anyone* and *person*.

- *Each* may stay or go as *he* chooses.

Because such usage ignores or excludes women, use alternatives when they are available. One solution is to use the plural. Another is to use both feminine and masculine pronouns, although that combination is clumsy when used too often.

- ~~Every employee~~ *All employees* must sign ~~his~~ *their* time ~~card~~ *cards*.
- Every employee must sign his *or her* time card.

Do not attempt to avoid expressing gender by resorting to a plural pronoun when the antecedent is singular. You may be able to avoid the pronoun entirely.

- Every employee must sign ~~their~~ *a* time card.

Avoid gender-related stereotypes in general references, as in "the nurse . . . *she*" or "the doctor . . . *he*." What if the nurse is male or the doctor female? See also biased language.

Number. A pronoun must agree with its antecedent in number. Many problems of agreement are caused by expressions that are not clear in number.

- Although the typical engine runs well in moderate temperatures, ~~they~~ *it* often ~~stall~~ *stalls* in extreme cold.

Use singular pronouns with the antecedents *everybody* and *everyone* unless to do so would be illogical because the meaning is obviously plural. See also everybody/everyone.

- *Everyone* pulled *his or her* share of the load.
- *Everyone* thought my plan should be revised, and I really couldn't blame *them.*

Collective nouns may use a singular or plural pronoun, depending on the meaning.

- The *committee* arrived at the recommended solutions only after *it* had deliberated for days.
  [*committee* thought of as collective]
- The *committee* quit for the day and went to *their* respective homes.
  [*committee* thought of as plural]

Demonstrative adjectives sometimes cause problems with agreement of number. *This* and *that* are used with singular nouns, and *these* and *those* are used with plural nouns. Demonstrative adjectives often cause problems when they modify the nouns *kind*, *type*, and *sort.* When used with those nouns, demonstrative adjectives should agree with them in number.

- *this* kind, *these* kinds; *that* type, *those* types

Confusion often develops when the preposition *of* is added (*this kind of*, *these kinds of*) and the object of the preposition does not conform in number to the demonstrative adjective and its noun.

- This kind of retirement ~~plans are~~ *plan is* best.

Avoid that error by remembering to make the demonstrative adjective, the noun, and the object of the preposition—all three—agree in number. The agreement makes the sentence not only correct but also more precise. Keep in mind that using demonstrative adjectives with words like *kind*, *type*, and *sort* can easily lead to vagueness. See kind of/sort of.

Compound Antecedents. A compound antecedent joined by *or* or *nor* is singular if both elements are singular and plural if both elements are plural.

- Neither the *engineer* nor the *technician* could do *his* job until *he* understood the new concept.
- Neither the *executives* nor the *directors* were pleased at the performance of *their* company.

When one of the antecedents connected by *or* or *nor* is singular and the other plural, the pronoun agrees with the closer antecedent.

- Either the *computer* or the *printers* should have *their* serial numbers registered.
- Either the *printers* or the *computer* should have *its* serial number registered.

A compound antecedent with its elements joined by *and* requires a plural pronoun.

- *Seon Ju and Juanita* took *their* layout drawings with them.

If both elements refer to the same person, however, use the singular pronoun.

- The noted *scientist and author* departed from *her* prepared speech.

all ready / already

All ready is a two-word phrase meaning "completely prepared." *Already* is an adverb that means "before this time" or "previously."(They were *all ready* to cancel the order; fortunately, we had *already* corrected the shipments.)

all right

All right means "all correct." (The answers were *all right*.) In formal writing, it should not be used to mean "good" or "acceptable." It is always written as two words, with no hyphen; *alright* is nonstandard.

all together / altogether

All together means "all acting together" or "all in one place." (The necessary instruments were *all together* on the tray.) *Altogether* means "entirely" or "completely." (The trip was *altogether* unnecessary.)

allude / elude / refer

Allude means to make an indirect reference to something. (The report simply *alluded* to the problem, rather than stating it explicitly.) *Elude* means to escape notice or detection. (The leak *eluded* the inspectors.) *Refer* is used to indicate a direct reference to something. (She *referred* to the chart three times during her presentation.)

allusion / illusion

An *allusion* is an indirect reference to something not specifically mentioned. (The report made an *allusion* to metal fatigue in the support structures.) An *illusion* is a mistaken perception or a false image. (County officials are under the *illusion* that the landfill will last indefinitely.)

allusions

An allusion is an indirect reference to something from past or current events, literature, or other familiar sources. The use of allusion promotes economical writing because it is a shorthand way of referring to a body of material in a few words or of helping to explain a new and unfamiliar process in terms of one that is familiar. In the following example, the writer sums up a description with an allusion to a well-known story. The allusion, with its implicit reference to "right standing up to might," concisely emphasizes the writer's point.

- As it currently exists, the review process involves the consumer's attorney sitting alone, usually without adequate technical assistance, faced by two or three government attorneys, two or three attorneys from CompuSystems, and large teams of experts who support the government and the corporation. The entire proceeding is reminiscent of David versus Goliath.

Be sure, of course, that your reader is familiar with the material to which you allude. Allusions should be used with restraint, especially in international correspondence. If overdone, allusions can lead to affectation or can be viewed merely as clichés. See also technical writing style.

almost / most

Do not use *most* as a colloquial substitute for *almost* in your writing.

- New shipments arrive ~~most~~ *almost* every day.

also

Also is an adverb that means "additionally." (Two 5,000-gallon tanks are on-site, and several 2,500-gallon tanks are *also* available.) *Also* should not be used as a connective in the sense of "and."

- He brought the reports, the letters, ~~also~~ *and* the section supervisor's recommendations.

Avoid opening sentences with *also*. It is a weak transitional word that suggests an afterthought rather than planned writing.

- ~~Also~~ *In addition*, he brought statistical data to support his proposal.
- ~~Also, he~~ *He also* brought statistical data to support his proposal.

ambiguity

A word or passage is ambiguous when it can be interpreted in two or more ways, yet provides the reader with no certain basis for choosing among the alternatives.

- Mathematics is more valuable to an engineer than a computer.

[Does that mean an engineer is more in need of mathematics than a computer is? Or does it mean that mathematics is more valuable to an engineer than a computer is?]

Ambiguity can take many forms, as in ambiguous pronoun reference.

AMBIGUOUS	Inadequate quality-control procedures have resulted in more equipment failures. This is our most serious problem at present. [Does *this* refer to *inadequate quality-control procedures* or to *equipment failures*?]
SPECIFIC	Inadequate quality-control procedures have resulted in more equipment failures. *These failures* are our most serious problem at present.
SPECIFIC	Inadequate quality-control procedures have resulted in more equipment failures. *Quality control* is our most serious problem at present.

Incomplete comparison and missing or misplaced modifiers (including dangling modifiers) cause ambiguity.

- Ms. Lee values rigid quality-control standards more than Mr. Rosenblum ^does^. [Complete the comparison.]

- He lists his hobby as cooking. He ^also^ is especially fond of cocker spaniels. [Add the missing modifier.]

The placement of some modifiers enables them to be interpreted in either of two ways.

- She volunteered *immediately* to deliver the toxic substance.

By moving the word *immediately*, the meaning can be clarified.

- She *immediately* volunteered to deliver the toxic substance.
- She volunteered to deliver *immediately* the toxic substance.

Imprecise word choice (including faulty idioms) can cause ambiguity.

- The general manager has denied reports that the plant's recent fuel-allocation cut will be ~~restored~~ ^rescinded^. [inappropriate word choice]

Various forms of awkwardness also can cause ambiguity.

amount / number

Amount is used with things that are thought of in bulk and that cannot be counted (mass nouns), as in "the *amount* of electricity." *Number* is used with things that can be counted as individual items (count nouns), as in "the *number* of employees."

ampersands

The ampersand (&) is a symbol sometimes used to represent the word *and*, especially in the names of organizations (Kirkwell & Associates). When you are writing the name of an organization in sentences, addresses, or references, spell out the word *and* unless the ampersand appears in the organization's official name on its letterhead stationery or Web site.

analogies (*see* **figures of speech**)

and / or

And/or means that either both circumstances are possible or only one of two circumstances is possible. This term is awkward and confusing because it makes the reader stop to puzzle over your distinction.

AWKWARD Use A *and/or* B.

IMPROVED Use A or B or both.

antonyms

An antonym is a word with a meaning opposite that of another word (*good/bad, wet/dry, fresh/stale*). Many pairs of words that look as if they are antonyms, such as *limit/delimit*, are not. When in doubt, consult a thesaurus or dictionary. See also synonyms.

apostrophes

An apostrophe (') is used to show possession or to indicate the omission of letters. Sometimes it can be used to avoid confusion for certain plurals of words, letters, and abbreviations.

Showing Possession

An apostrophe is used with an *s* to form the possessive case of some nouns (the *report's* title). For further advice on using apostrophes to show possession, see possessive case.

Indicating Omission

An apostrophe is used to mark the omission of letters or numbers in a contraction or a date (*can't, I'm, I'll*; the class of *'06*).

Forming Plurals

An apostrophe can be used in forming the plurals of letters, words, or lowercase abbreviations if confusion might result from using *s* alone.

- The search program does not find *a*'s and *i*'s.
- Do not replace all *of which*'s in the document.
- *I*'s need to be distinguished from the number 1.
- Check for any c.o.d.'s.

Generally, however, add only *s* in roman type when referring to words as words or capital letters.

- Five *and*s appear in the first sentence.
  [The word itself is in italics.]
- The applicants received *A*s and *B*s in their courses.

Do not use an apostrophe for plurals of abbreviations with all capital letters (DVDs) or a final capital letter (ten Ph.D.s) or for plurals of numbers (7s, the late 1990s).

appendixes

An appendix, located at the end of a formal report, proposal, or other major document, supplements or clarifies the information in the body of the document. Appendixes (or *appendices*) can provide information that is too detailed or lengthy for the primary audience of the document. For example, an appendix could contain material of interest to secondary readers, such as maps, statistical analysis, or résumés of key personnel involved in a proposed project.

A document may have more than one appendix, with each providing only one type of information. When the document contains more than one appendix, arrange them in the order they are mentioned in the text. Begin each appendix on a new page, and identify each with a letter, starting with the letter *A* (Appendix A: Sample Questionnaire). If you have only one appendix, title it simply "Appendix." List the titles and beginning page numbers of the appendixes in the table of contents.

application letters

When applying for a job, you usually need to submit both an application letter (also referred to as a *cover letter*) and a résumé. Employers may ask you to submit them by standard mail, fax, or e-mail. See also job search.

The application letter is essentially a sales letter in which you market your skills, abilities, and knowledge. Therefore, your application letter must be persuasive. The successful application letter accomplishes four tasks: (1) it catches the reader's attention favorably by describing how your skills will contribute to the organization, (2) it explains which particular job interests you and why, (3) it convinces the reader that you are qualified for the job by drawing attention to particular elements in your résumé, and (4) it requests an interview. See also correspondence, interviewing for a job, persuasion, and salary negotiations.

The sample application letters shown in Figures A–10 through A–12 follow the application-letter structure described in this entry. Each sample's emphasis, tone, and style are tailored to fit the applicant's experience and the particular audience.

Opening

In the opening paragraph, provide context and show your enthusiasm. Indicate how you heard about the position and name the specific job title or area. If you have been referred to a company by an employee, a career counselor, a professor, or someone else, be sure to say so. ("I recently learned from Jerry Gill, a mechanical engineer at Blake Heavy Equipment and a former colleague, that you are recruiting for mechanical engineers.") Explain why you are interested in the job and demonstrate your initiative as well as knowledge of the organization by relating your interest to some facet of the organization. ("Your position interests

From: Molly Sennett <sennett@execpr.com>
To: Bob Lupert <lupert@appliedsciences.com>
Sent: Monday, February 28, 2006 10:47 AM
Subject: Application for Summer Internship
Attachment: Sennett_Resume.pdf

Dear Mr. Lupert:

In the February 24 issue of the *Butler Gazette,* I learned that you have summer technical training internships available. This opportunity interests me because I have learned that Applied Sciences has one of the leading user-education centers in the country.

From your Web site, I have also learned that you value collaborative and leadership skills. My current project at Penn State University—a computer tutoring system that teaches LISP—has helped me to develop these skills. This tutoring system, which I helped develop as part of a student engineering team, is the first of its kind and is now being sold across the country to various corporations and universities. As part of the team, I solved specific problems and then helped test and revise our work to improve efficiency.

Pursuing degrees in Industrial Management and Computer Science has prepared me well to make valuable contributions to your goal of successful implementation of new software. Through my varied courses, which are described on my résumé, I have developed the ability to learn new skills quickly and independently and to interact effectively in a technical environment. I would look forward to applying these abilities to user education at Applied Sciences, Inc.

I would appreciate the chance to interview with you at your earliest convenience. If you would like additional information, please contact me at (435) 228-3490 any Tuesday or Thursday after 10 a.m. or e-mail me at <sennett@execpr.com>.

Thank you for your consideration.

Sincerely,

Molly Sennett

FIGURE A–10. Application Letter Sent as E-mail (College Student Applying for an Internship)

237 Darnell Drive
Sagle, ID 83860
July 3, 2006

James Jody
Employee Development
Blake Heavy Equipment Corporation
796 Pike Street
Baltimore, MD 21230

Dear Mr. Jody:

I recently learned from Jerry Gill, a mechanical engineer at Blake Heavy Equipment and a former colleague, that you are recruiting for mechanical engineers. Your position interests me greatly not only because your firm is number one in the region but also because I am impressed with Blake's Employee Development Program.

I understand that you especially need engineers who can also write well because your engineers write most of your manuals. As noted in my enclosed résumé, I carried straight *A*s in my college writing courses and have published an article in the trade journal *American Engineering* (Spring 2005 issue). I have the education and professional experience to solve both simple and complex problems. I work well both individually and in team efforts.

Could we schedule a meeting at your convenience to discuss this career opportunity further? Feel free to call me any weekday morning at 866-458-3215 or e-mail me at <jeb@pcexec.com> if you have questions or need further information. I also encourage you to contact one of my many references. Thank you for your consideration.

Sincerely,

Harry Jeb Stinson

Harry Jeb Stinson

Enclosures: Résumé
American Engineering article

FIGURE A–11. Application Letter (Recent College Graduate Applying for a Job)

2116 Billingham Drive
Atlanta, GA 30302
July 5, 2006

Ms. Cecilia Wysocki
Manager, Medical Products Department
HiTech Corporation
2094 Pleasant Avenue
Tempe, AZ 85281

Dear Ms. Wysocki:

During the recent AMA convention in Seattle, one of your engineers, Bob Griffey, told me of a possible opening for a calibration technician in your department. My extensive background in the calibration field, as well as my familiarity with the systems used at HiTech, make me an ideal candidate for the position.

I was with Medical Division Technology, Inc., from its formation in 1997 until its closing this year. During that period, I was involved in all areas of medical products calibration, receiving excellent performance reviews and promotions four times during my employment. I have experience with your HT-3 molecular scanner, so I would be able to work on that system with a minimum of training.

I would be pleased to discuss my qualifications in an interview at your convenience. If you have any questions, you can write to me at the address above, telephone me at (896) 539-9935, or e-mail me at <aald@cs.com>.

Sincerely,

Alice Arnold

Alice Arnold

Enclosure: Résumé

FIGURE A–12. Application Letter (Applicant with Years of Experience)

me greatly not only because your firm is number one in the region but also because I am impressed with Blake's Employee Development Program.")

Body

In the middle paragraphs, show through examples that you are qualified for the job. Limit each of these paragraphs to just one basic point that is clearly stated in the topic sentence. For example, your second paragraph might focus on work experience and your third paragraph on educational achievements. Do not just *tell* readers that you are qualified—*show* them by including examples and details. Come across as proud of your achievements and refer to your enclosed résumé, but do not simply summarize your résumé. Indicate how you with your talents can make valuable contributions to their company. ("I have experience with your HT-3 molecular scanner, so I would be able to work on that system with a minimum of training.")

Closing

In the final paragraph, request an interview. Let the reader know how to reach you by including your phone number or e-mail address. End with a statement of goodwill, even if only a "thank you."

Proofread your letter *carefully*. Research indicates that if employers notice even one spelling, grammatical, or mechanical error, they often eliminate candidates from consideration immediately. Such errors will give employers the impression that you lack writing skills or that you are careless in the way you present yourself professionally. See also proofreading.

appositives

An appositive is a noun or noun phrase that follows and amplifies another noun or noun phrase. It has the same grammatical function as the noun it complements.

- Dennis Gabor, *the famous British scientist*, experimented with coherent light in the 1940s.
- The famous British scientist *Dennis Gabor* experimented with coherent light in the 1940s.

For detailed information on the use of commas with appositives, see restrictive and nonrestrictive elements.

If you are in doubt about the case of an appositive, check it by substituting the appositive for the noun it modifies.

me
- My boss gave the two of us, Jim and I, the day off.

  [You would not say, "My boss gave *I* the day off."]

articles

Articles (*a*, *an*, *the*) function as adjectives because they modify the items they designate by either limiting them or making them more specific. The two kinds of articles are indefinite and definite.

TIPS FOR USING ARTICLES

Whether to use a definite or an indefinite article is determined by what you can safely assume about your audience's knowledge. In each of these sentences, you can safely assume that the reader can clearly identify the noun. Therefore, use a definite article.

- *The* sun rises in the east.
  [The Earth has only one *sun*.]
- Did you know that yesterday was *the* coldest day of the year so far?
  [The modified noun refers to *yesterday*.]
- *The* man who left his briefcase in the conference room was in a hurry.
  [The relative phrase *who left his briefcase in the conference room* restricts and, therefore, identifies the meaning of *man*.]

In the following sentence, however, you cannot assume that the reader can clearly identify the noun.

- *A* package is on the way.
  [It is impossible to identify specifically what package is meant.]

A more important question for some nonnative speakers of English is when *not* to use articles. These generalizations will help. Do not use articles with the following:

singular proper nouns

- Utah, Main Street, Harvard University, Mount Hood

plural nonspecific countable nouns (when making generalizations)

- Helicopters are the new choice of transportation for the rich and famous.

singular uncountable nouns

- She loves chocolate.

plural countable nouns used as complements

- Those women are physicians.

The indefinite articles, *a* and *an*, denote an unspecified item.

- *A* package was delivered yesterday.
  [This is not a specific package but an unspecified package.]

The choice between *a* and *an* depends on the sound rather than on the letter following the article, as described in the entry a/an.

The definite article, *the*, denotes a particular item.

- *The* package was delivered yesterday.
  [This is not just any package but one specific package.]

Do not omit all articles from your writing. Including articles costs nothing; eliminating them makes reading more difficult. (See also telegraphic style.) However, do not overdo it. An article can be superfluous.

- Fill with *a* half *a* pint of fluid.
  [Choose one article and eliminate the other.]

Do not capitalize articles in titles except when they are the first word. (*The Scientist* reviewed *Saving the Humboldt Penguin*.) See also capitalization and English as a second language.

as / because / since

As, *because*, and *since* are commonly used to mean "because." To express cause, *because* is the strongest and most specific connective; *because* is unequivocal in stating a causal relationship. (*Because* she did not have an engineering degree, she was not offered the job.)

Since is a weak substitute for *because* as a connective to express cause. However, *since* is an appropriate connective when the emphasis is on circumstance, condition, or time rather than on cause and effect. (*Since* it went public, the company has earned a profit every year.)

As is the least definite connective to indicate cause; its use for that purpose is best avoided. (See also subordination.) The word *as* can also contribute to awkwardness by appearing too many times in a sentence.

- ~~As we~~ *We* realized ~~as soon as~~ *the moment* we began the project~~,~~ *that* the problem needed a solution.

Avoid colloquial, nonstandard, or wordy phrases sometimes used instead of *as*, *because*, or *since*. See also affectation, as such, and due to/because of.

PHRASE	REPLACE WITH
being as, being that	because, since
inasmuch as	because
insofar as	since
on account of	because
on the grounds of/that	because
due to the fact that	because

as much as / more than

The phrases *as much as* and *more than* are sometimes incorrectly combined, especially when intervening phrases delay the completion of the phrase.

- The engineers had as much~~, if not more,~~ influence in planning the program ~~than~~ *as* the accountants*, if not more*.

as such

The phrase *as such* is seldom useful and should be omitted.

- Style sheets~~, as such,~~ are useful in designing Web pages.

as well as

Do not use *as well as* with *both*. The two expressions have similar meanings; use one or the other and adjust the verb as needed.

- Both General Motors ~~as well as~~ *and* Ford ~~is~~ *are* marketing hybrid vehicles.
- ~~Both~~ General Motors as well as Ford is marketing hybrid vehicles.

audience

Although the word *audience* can have a slightly different meaning for a writer than for a speaker, it is crucial to both. The writer and the speaker must know as much as possible about the people they are trying to reach with their message—readers or listeners—regardless of its method of delivery. For meeting the needs of some particular audiences, see global communication, international correspondence, and presentations.

augment / supplement

Augment means to increase or magnify in size, degree, or effect. (Many employees *augment* their incomes by freelancing.) *Supplement* means to add something to make up for a deficiency. (He will *supplement* his diet with vitamins.)

average / median / mean

The *average* (or arithmetic *mean*) is determined by adding two or more quantities and dividing the sum by the number of items totaled. For example, if one report is 10 pages, another is 30 pages, and a third is 20 pages, their *average* length is 20 pages. It is incorrect to say that "each report averages 20 pages" because each report is a specific length.

- *The three reports average*
~~Each report averages~~ 20 pages.

The *median* is the middle number in a sequence of numbers. For example, the *median* of the series 1, 3, 4, 7, 8 is 4.

awhile / a while

The adverb *awhile* means "for a short time." The preposition *for* should not precede *awhile* because *for* is inherent in the meaning of *awhile*. The two-word noun phrase *a while* means "a period of time."

- Wait ~~for~~ awhile before testing the sample.
- *a while*
Wait for ~~awhile~~ before testing the sample.

awkwardness

Any writing that strikes the readers as awkward—that is, as forced or unnatural—impedes the readers' understanding. The following checklist and the entries indicated will help you smooth out most awkward passages.

Writer's Checklist: Eliminating Awkwardness

- ☑ Strive for clarity and coherence during revision.
- ☑ Check for organization to ensure your writing develops logically.
- ☑ Keep sentence construction as direct and simple as possible.
- ☑ Use subordination appropriately and avoid needless repetition.
- ☑ Correct any logic errors within your sentences.
- ☑ Revise for conciseness and avoid expletives where possible.
- ☑ Use the active voice unless you have a justifiable reason to use the passive voice.
- ☑ Eliminate jammed or misplaced modifiers and, for particularly awkward constructions, apply the tactics in garbled sentences.

B

B

bad / badly

Bad is the adjective form that follows such linking verbs as *feel* and *look*. (We don't want to look *bad* at the meeting.) *Badly* is an adverb. (The new model performed *badly* during the test.) To say "I feel *badly*" would mean, literally, that your sense of touch is impaired. See also good/well.

be sure to

The phrase *be sure and* is colloquial and unidiomatic when used for *be sure to.* See also idioms.

- When you sign the contract, be sure ~~and~~ *to* keep a copy.

beside / besides

Besides, meaning "in addition to" or "other than," should be carefully distinguished from *beside,* meaning "next to" or "apart from." (*Besides* two of us from the Information Systems Department, three people from Production stood *beside* the president during the ceremony.)

between / among

Between is normally used to relate two items or persons. (Preferred stock offers a middle ground *between* bonds and common stock.) *Among* is used to relate more than two. (The subcontracting was distributed *among* three firms.)

B

between you and me

The expression *between you and I* is incorrect. Because the pronouns are objects of the preposition *between*, the objective form of the personal pronoun (*me*) must be used. See also case.

- Between you and ~~I~~ *me*, Joan should be promoted.

bi- / semi-

When used with periods of time, *bi-* means "two" or "every two," as in *bimonthly*, which means "once in two months." When used with periods of time, *semi-* means "half of" or "occurring twice within a period of time." *Semimonthly* means "twice a month." Both *bi-* and *semi-* normally are joined with the following element without a space or hyphen.

biannual / biennial

In conventional usage, *biannual* means "twice during the year," and *biennial* means "every other year." See also bi-/semi-.

biased language

Biased language refers to words and expressions that offend because they make inappropriate assumptions or stereotypes about gender, ethnicity, physical or mental disability, age, or sexual orientation.

ETHICS NOTE The easiest way to avoid bias is simply not to mention differences among people unless the differences are relevant to the discussion. Keep current with accepted usage and, if you are unsure of the appropriateness of an expression or the tone of a passage, have several colleagues review the material and give you their assessments. ✦

Sexist Language

Sexist language can be an outgrowth of sexism, the arbitrary stereotyping of men and women—it can breed and reinforce inequality. To avoid sexism in your writing, treat men and women equally, and do not make assumptions about traditional or occupational roles. Accordingly, use nonsexist occupational descriptions in your writing.

INSTEAD OF	CONSIDER
chairman, chairwoman	chair, chairperson
foreman	supervisor, manager
man-hours	staff hours, worker hours
policeman, policewoman	police officer
salesman, saleswoman	salesperson

Use parallel terms to describe men and women.

INSTEAD OF	USE
man and wife	husband and wife
Ms. Jones and Bernard Weiss	Ms. Jones and Mr. Weiss; Mary Jones and Bernard Weiss
ladies and men	ladies and gentlemen; women and men

Sexism can creep into your writing by the unthinking use of male or female pronouns where a reference could apply equally to a man or a woman. One way to avoid such usage is to rewrite the sentence in the plural.

All employees *their managers* *their travel vouchers.*
- ~~Every employee~~ will have ~~his manager~~ sign ~~his travel voucher.~~

Other possible solutions are to use *his or her* instead of *his* alone or to omit the pronoun completely if it is not essential to the meaning of the sentence. See also pronoun reference.

an
- Everyone must submit ~~his~~ expense report by Monday.

He or she can become monotonous when used repeatedly, and a pronoun cannot always be omitted without changing the meaning of a sentence. Another solution is to omit troublesome pronouns by using the imperative mood whenever possible.

Submit all *reports*
- ~~Everyone must submit his or her~~ expense ~~report~~ by Monday.

Other Types of Biased Language

Identifying people by racial, ethnic, or religious categories is simply not relevant in most workplace writing. Telling readers that an engineer is Native American or a mathematician is Asian American almost never conveys useful information. Linking a profession or characteristic to race or ethnicity reinforces stereotypes, implying that it is rare or expected for a person of a particular background to have achieved a certain position.

Consider how you refer to people with disabilities. If you refer to "a disabled employee," you imply that the part (*disabled*) is as significant as the whole (*employee*). Use "an employee with a disability" instead. Similarly, the preferred usage is "a person who uses a wheelchair" rather than "a wheelchair-bound person," an expression that inappropriately equates the wheelchair with the person. Likewise, references to a person's age can also be inappropriate, as in expressions like "middle-aged manager" or "young Web designer."

In most workplace writing, such issues are simply not relevant. Of course, in some contexts, race, ethnicity, religion, disability, or age should be identified. For example, if you are writing an Equal Employment Opportunity Commission report about your firm's hiring practices, the racial composition of the workforce is relevant. In such cases, you need to present the issues in ways that respect and do not demean the individuals or groups to which you refer. See also <u>ethics in writing</u>.

bibliographies

A bibliography is an alphabetical list of books, articles, Web sources, and other works that have been consulted in preparing a document or that are useful for reference purposes. In a <u>formal report</u> or other document, a bibliography can provide a convenient alphabetical listing of sources in a standardized form for readers interested in getting further information on the topic or in assessing the scope of the research.

While a list of references, footnotes/endnotes, and a works-cited list refer to works actually cited in the text, a bibliography includes works consulted for general background information. For information on using various citation styles, see <u>documenting sources</u>.

Entries in a bibliography are listed alphabetically by the author's last name. If an author is unknown, the entry is alphabetized by the first word in the title (other than *A*, *An*, or *The*). Entries also can be arranged by subject and then ordered alphabetically within those categories.

An annotated bibliography includes complete bibliographic information about a work (author, title, publisher, and date) followed by a brief description or evaluation of what the work contains. The following is an annotation of a book on style and format:

Sabin, William A. 2005. The *Gregg Reference Manual: A Manual of Style, Grammar, Usage, and Formatting.* 10th ed. New York: McGraw Hill/Irwin. 688 pp.

> The author states that the tenth edition of this standard office guide "is intended for anyone who writes, edits, or prepares material for distribution or publication" (p. viii). The book begins with a section titled "How to Look Things Up," which includes a reference to an online index at <www.gregg.com>. The book is then divided into three parts. Part 1: "Grammar, Usage, and Style" contains 11 sections on topics such as punctuation, abbreviations, spelling, and usage. Part 2: "Techniques and Formats" contains seven sections on such topics as proofreading, correspondence (letters, memos, and e-mail), and other elements of manuscript preparation. Part 3: "References" includes four appendixes: several essays on style, rules for alphabetic filing, a glossary of grammatical terms, and an alphabetical list offering appropriate pronunciation for words used in business, legal, and professional contexts.

both . . . and

Statements using the *both . . . and* construction should always be balanced grammatically and logically. See also <u>parallel structure</u>.

- For success in engineering, it is necessary *both* to develop writing skills *and* ~~mastering~~ *to master* mathematics.

brackets

The primary use of brackets ([ ]) is to enclose a word or words inserted by the writer or editor into a quotation.

- The text stated, "Hypertext systems can be categorized as either modest [not modifiable] or robust [modifiable]."

Brackets are used to set off a parenthetical item within parentheses.

- We should be sure to give Emanuel Foose (and his brother Emilio [1812–1882]) credit for his role in founding the institute.

Brackets are also used to insert the Latin word *sic*, which is a scholarly <u>abbreviation</u> that indicates a writer has quoted material exactly as it appears in the original, even though it contains a misspelled or wrongly used word. See also <u>quotations</u>.

- Dr. Smith wrote that "the earth does not revolve around the son [*sic*] at a constant rate."

brainstorming

Brainstorming, a form of free association used to generate ideas, can be done individually or in groups. Brainstorming can stimulate creative thinking about a topic and reveal fresh perspectives and new connections. When preparing to write, jot down as many random ideas as you can think of about the topic. When a group brainstorms, a designated person writes down the words or phrases as the other group members suggest them. Do not stop to analyze ideas or hold back looking for only the "best" ideas; just write down everything that comes to mind. After compiling a list of initial ideas, ask *what, when, who, where, how,* and *why* for each idea, then list additional details that those questions bring to mind. When you run out of ideas, analyze each one you recorded, discarding those that are redundant. Then group the items in the most logical order, based on the purpose of the document and the needs of the readers; that will result in a tentative outline of the document. Although the outline will be sketchy and incomplete, it will show where further brainstorming or research is needed and provide a framework for any new details that additional research yields.

Many writers find a technique called *clustering* (also called *mind mapping*), as shown in Figure B–1, helpful in recording and organizing ideas created during a brainstorming session. To cluster, begin with a blank sheet of paper or a flip chart. Think of a keyword or key phrase that best characterizes your topic and put it in a circle at the center of the paper. Figure B–1 shows brainstorming about the best way to communicate with customers, so that topic could be "Customer communication." Then, think of subtopics most closely related to the main topic. In Figure B–1, the main topic ("Customer communication") might lead to such subtopics as "Advertising," "Internet," "Direct mail," and "Conferences and trade shows." Draw circles or boxes around the subtopics, connecting each to the center circle, like adding spokes to a wheel hub. Repeat the exercise for each subtopic. Each subtopic should stimulate additional subtopics, which you add in circles connected to the parent subtopic, as you did with the main topic at the center of the page. In Figure B–1, for example, one subtopic is "Internet," which could stimulate additional subtopics like "E-mail" and "Web page." Continue the process until you exhaust all possible ideas. The resulting "map" will show clusters of terms grouped around the central concept.

B

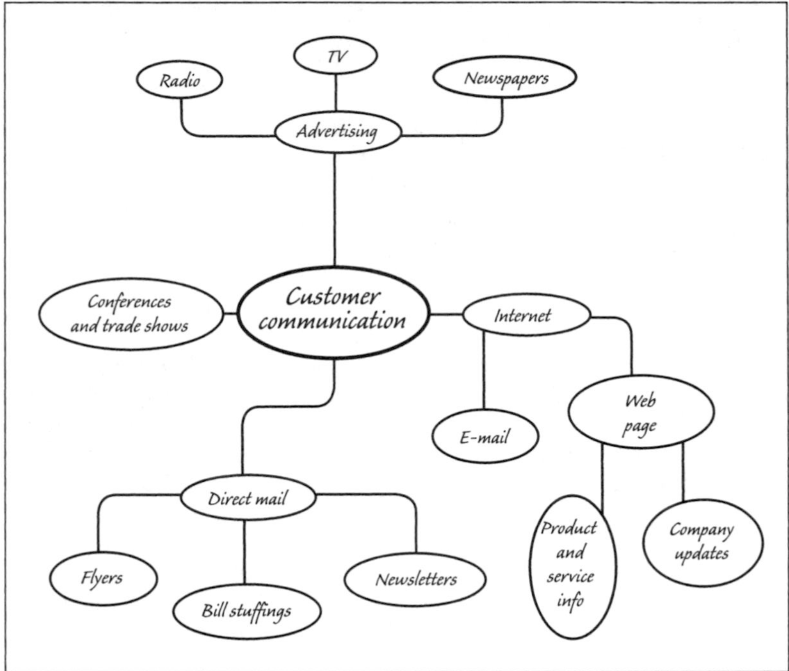

FIGURE B–1. Cluster Map from a Brainstorming Session

brochures

Brochures are printed publications that promote the products and services offered by a business or that promote the image of a business or an organization by providing general or specific technical information important to a target audience. The goal of a brochure is to inform or to persuade or both. See also clarity and persuasion.

Types of Brochures

The two major types of brochures are sales brochures and informational brochures. *Sales brochures* are created specifically to sell a company's products and services. A technically oriented sales brochure can be created to promote the performance data of various models of a product or to list the benefits or capabilities of different types of equipment available for specific jobs. For example, the brochure of a microwave oven manufacturer might describe the various models and their accessories and how the ovens are to be installed over stovetops.

Informational brochures are created to inform and educate the reader as well as to promote goodwill and raise the profile of an organization. For example, an informational brochure for a pesticide company might show the reader how to identify various types of pests like silverfish and termites, detail the damage they can do, and explain how to protect your home from them.

Designing the Brochure

Before you begin to write, determine the specific purpose of the brochure—to provide information about a service? to sell a product? to educate readers about a public health issue? You must also identify your target audience—general reader? expert? potential or existing client? Understanding your purpose, audience, and context is essential to creating content and design that will be both rhetorically appropriate and persuasive to your target audience. Once you make these decisions, gather samples of brochures for similar products or services to stimulate your thinking.

Cover Panel. The main goal of the cover panel of an informational brochure is to gain the audience's attention. The cover panel should clearly identify the organization or service being promoted as well as provide a carefully selected visual image or headline geared toward the interest of the audience. Accompany the image with a minimal amount of text—for example, a statement about the value of the information or a brief promotional quotation from a satisfied customer. Figure B–2 shows a brochure produced by the National Cancer Institute, which hopes to attract health-care professionals to a course on conducting clinical trials in their practices. The cover photo fea-

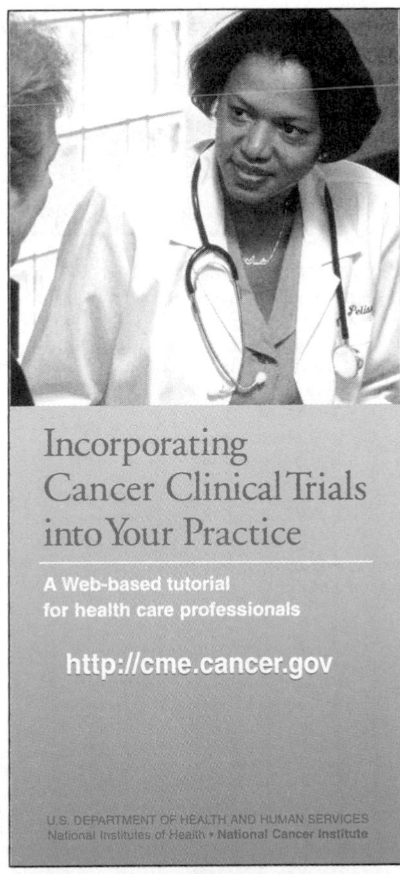

FIGURE B–2. Brochure (Cover Panel)

B

tures a health-care professional wearing a stethoscope and apparently counseling a patient.

First Inside Panel. The first inside panel of a brochure should again identify the organization or service with headlines and brief, readable content, such as that used in advertising. The first inside panel of the National Cancer Institute brochure uses boldface headings to highlight brief descriptions of the course and those participants who may benefit.

Subsequent Panels. Subsequent panels should describe the product or service with your readers' needs in mind, clearly stating the benefits and solutions to problems that your product or service offers. Installation options could be included here, as well as relevant and

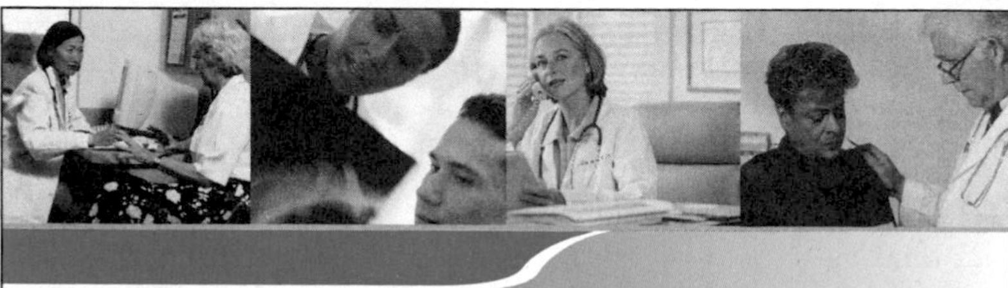

What is "Incorporating Cancer Clinical Trials Into Your Practice?"

It's a free Web-based tutorial for health care professionals that emphasizes the importance of participating in cancer clinical trials, either through patient referral or becoming a clinical trials investigator.

The course begins with a brief overview of cancer clinical trials and explains ways one can become involved in clinical trials. It continues with practical information and guidance for professionals interested in referring patients to, or conducting, clinical trials.

Participants will hear from experienced clinicians about some of the issues they have faced and how these issues were addressed.

Who should take this course?

The course is designed for health care professionals who are new to the clinical trials process, including oncologists, family physicians, internists, nurses, and other clinical staff. Specifically, it is for those clinicians that are referring patients to cancer clinical trials for the first time, and clinical research staff who are new to conducting clinical trials in their practices.

What are the course objectives?

At the end of the course, participants should be able to:

- Describe the role of clinical research in advancing cancer care
- Explain how clinical trials are conducted
- Describe the sponsorship of clinical trials
- Demonstrate how to locate clinical trials
- Understand how clinical trial sites are set up, staffed, and funded
- Make successful patient referrals to clinical trials
- Know how a physician may become a cancer clinical trials investigator
- Understand regulatory, record keeping, and reporting requirements in conducting clinical trials
- Identify key issues in communication among the patient, referring clinician, and clinical trials investigator

To register for this course, go to **http://cme.cancer.gov**, where you will be given further instructions.

Participants may receive continuing medical education (CME) credits or contact hours for course completion.

FIGURE B–2. Brochure (*continued*) (Inside Panels)

accurate supporting facts and visuals. You might further establish credibility with a company history, a product history, and a list of current clients or testimonials. Use subheadings and bulleted lists to break up the text and make the brochure easy to read. In the final panel, be clear about the action you want the reader to take, such as calling for an appointment, e-mailing for technical support, or visiting a Web site for more information. In Figure B–2, the second inside panel describes the course objectives and tells readers how to register.

Layout-and-Design Features. As you plan the design for a brochure, create a rough planning sketch. To find the best way to present your content, experiment with margins, spacing, and the arrangement and amount of text on each panel (you may want to make changes to content or length based on concerns of layout). Allow adequate white space for readability. Experiment with various fonts and formatting. Consider, for example, enlarging the first letter of the first word in a paragraph, but do not overdo the use of unusual fonts or alternative styles (such as running type vertically). Choose a style that unifies your brochure, and use it consistently throughout. Be judicious about color choices. For some brochures, black and white, which is cheaper, is more effective; in other circumstances (such as brochures for medical services) color, although more expensive, is a must. Depending on your budget and the scale of your project, consider using a professional printer. See also layout and design.

Writer's Checklist: Designing Brochures

- ☑ Collect other brochures to stimulate your thinking as you consider the goals and content for your brochure and each panel.
- ☑ Create a rough sketch that maps out the content of each panel to help you select visuals, color schemes, and the number of panels.
- ☑ Experiment with margins, spacing, and the arrangement and amount of text on each panel; make appropriate changes to content or length and allow adequate white space for readability.
- ☑ Experiment with fonts and formatting, such as enlarging the first letter of the first word in a paragraph, but do not overuse unusual fonts or alternative styles, such as running type vertically.
- ☑ Consider color choices: black and white, which is cheaper, can be effective; in some circumstances color, although more expensive, is a must.
- ☑ Evaluate the impact of the design with your content. Does the design complement the content and make the brochure more persuasive? Have you adequately considered the needs of your readers? When your brochure is distributed, will it fulfill its purposes?

B

Writer's Checklist: Designing Brochures (continued)

☑ Depending on your budget and the scale of your project, consider using a professional printer.

WEB LINK **SAMPLE BROCHURES**

For links to Web sites with sample brochures, see <*bedfordstmartins.com/alredtech*> and select *Links for Handbook Entries*.

buzzwords

Buzzwords are words or phrases that suddenly become popular and, because of an intense period of overuse, lose their freshness and preciseness. They may become popular through their association with technology, popular culture, or even sports. We include them in our vocabulary because they *seem* to give force and vitality to our language. Actually, buzzwords sound pretentious in workplace writing. See also word choice.

interface [as a verb]	deliverables	dialogue [as a verb]
impact [as a verb]	24/7	bleeding edge
skill sets	dot-com	cash cow

Obviously, some buzzwords are appropriate in the right context.

- We must establish an *interface* between the computer and the satellite hardware.
  [*Interface* is appropriately used as a noun.]

When writers needlessly shift the normal function of a word, they often create a buzzword that is imprecise.

- We must ~~interface~~ *cooperate* with the Human Resources Department.

  [*Interface* is inappropriately used as a verb; *cooperate* is more precise.]

WEB LINK **BUZZWORDS**

Former newspaper editor John Walston offers BuzzWhack, a lighthearted site dedicated to the meaningful use of buzzwords. See <*bedfordstmartins.com/alredtech*> and select *Links for Handbook Entries*.

C

can / may

In writing, *can* refers to capability (I *can* have the project finished today). *May* refers to possibility (I *may* be in Boston on Monday) or permission (*May* I leave early?).

cannot

Cannot is one word.

- We ~~can not~~ *cannot* meet the deadline specified in the contract.

capitalization

DIRECTORY

The use of capital, or uppercase, letters is determined by custom. Capital letters are used to call attention to certain words, such as proper nouns and the first word of a sentence. Use them carefully, especially when they affect a word's meaning (march/March, china/China, turkey/Turkey) and because a spell checker would fail to identify an error.

Proper Nouns

Proper nouns name specific persons, places, things, concepts, or qualities and are capitalized (Technical Writing 206, Microsoft, Pat Wilde, Peru).

Common Nouns

Common nouns name general classes or categories of people, places, things, concepts, or qualities rather than specific ones and are not capitalized (technical writing class, company, person, country).

First Words

The first letter of the first word in a sentence is always capitalized. (Of the plans submitted, ours is best.) The first word after a colon is capitalized when the colon introduces two or more complete sentences or if the colon precedes a statement requiring special emphasis.

- The meeting will address only one issue: What is the firm's role in environmental protection?

If a subordinate element follows the colon or if the thought is closely related, use a lowercase letter following the colon.

- We kept working for one reason: the approaching deadline.

The first word of a complete sentence in quotation marks is capitalized.

- Albert Einstein stated, "Imagination is more important than knowledge."

The first word in the salutation (Dear Mr. Smith:) and complimentary close (Sincerely yours,) is capitalized. See also correspondence and e-mail.

Specific Groups

Capitalize the names of ethnic groups, religions, and nationalities (Native American, Christianity, Mongolian). Do not capitalize the names of social and economic groups (middle class, working class, unemployed).

Specific Places

Capitalize the names of all political divisions (Ward Six, Chicago, Cook County, Illinois). Capitalize the names of geographical divisions (Europe, Asia, North America, the Middle East). Do not capitalize geographic features unless they are part of a proper name.

- The mountains in some areas, such as the *Great Smoky Mountains*, make television transmission difficult.

The words *north*, *south*, *east*, and *west* are capitalized when they refer to sections of the country. They are not capitalized when they refer to directions.

- I may travel *north* when I relocate, but my family will remain in the *South*.

Capitalize the names of stars, constellations, and planets.

- North Star, Andromeda, Saturn

Do not capitalize *earth*, *sun*, and *moon* except when they are referred to formally as astronomical bodies.

- My workday was so long that I saw the *sun* rise over the lake and the *moon* appear as darkness settled over the *earth*.
- The various effects of the *Sun* on *Earth* and the *Moon* were discussed at the symposium.

Specific Institutions, Events, Concepts

Capitalize the names of institutions, organizations, and associations (Society for Technical Communication). An organization usually capitalizes the names of its internal divisions and departments (Aeronautics Division, Human Resources Department). Types of organizations are not capitalized unless they are part of an official name (a medical association; American Medical Association). Capitalize historical events (the Great Depression of the 1930s). Capitalize words that designate holidays, specific periods of time, months, or days of the week (Labor Day, the Renaissance, January, Monday). Do not capitalize seasons of the year (spring, autumn, winter, summer).

Capitalize the scientific names of classes, families, and orders but not the names of species or English derivatives of scientific names (Mammalia/mammal, Carnivora/carnivorous).

Titles of Works

Capitalize the initial letters of the first and last words of the title of a book, an article, a play, or a film, as well as all major words in the title. Do not capitalize articles (*a, an, the*), coordinating conjunctions (*and, but*), or prepositions unless they begin or end the title (*The Lives of a Cell*). Capitalize prepositions within titles when they contain more than four letters (*Between, Within, Until, After*), unless you are following a style that recommends otherwise. The same rules apply to the subject line of an e-mail or a memo.

Professional and Personal Titles

Titles preceding proper names are capitalized (Ms. Berger, Senator Lieberman). Appositives following proper names normally are not capitalized (Joseph Lieberman, *senator* from Connecticut). However, the word *President* is capitalized when it refers to the chief executive of a national government.

Job titles used with personal names are capitalized (Ho-shik Kim, *Laboratory Director*). Job titles used without personal names are not capitalized.

- The *laboratory director* will meet us tomorrow.

Use capital letters to designate family relationships only when they occur before a name (my uncle; Uncle Fred).

Abbreviations

Capitalize abbreviations if the words they stand for would be capitalized, such as B.S.Ch.E. (Bachelor of Science in Chemical Engineering).

Letters

Capitalize letters that serve as names or indicate shapes (vitamin B, T-square, U-turn, I-beam).

Miscellaneous Capitalizations

The first word of a complete sentence enclosed in dashes, brackets, or parentheses is not capitalized when it appears as part of another sentence.

- We must make an extra effort in safety this year (accidents last year were up 10 percent).

Certain units, such as parts and chapters of books and rooms in buildings, when specifically identified by number, are capitalized (Chapter 5, Ch. 5; Room 72, Rm. 72). Minor divisions within such units are not capitalized unless they begin a sentence (page 11, verse 14, seat 12).

case

DIRECTORY

C

Grammatical case indicates the functional relationship of a noun or a pronoun to the other words in a sentence. Nouns change form only in the possessive case; pronouns may show change for the subjective, the objective, or the possessive case.

The case of a noun or pronoun is always determined by its function in a phrase, clause, or sentence. If it is the subject of a phrase, clause, or sentence, it is in the subjective case; if it is an object in a phrase, clause, or sentence, it is in the objective case; if it reflects possession or ownership and modifies a noun, it is in the possessive case. Figure C–1 is a table of pronouns in the subjective, objective, and possessive cases.

SINGULAR	**SUBJECTIVE**	**OBJECTIVE**	**POSSESSIVE**
First person	I	me	my, mine
Second person	you	you	your, yours
Third person	he, she, it	him, her, it	his, her, hers, its
PLURAL	**SUBJECTIVE**	**OBJECTIVE**	**POSSESSIVE**
First person	we	us	our, ours
Second person	you	you	your, yours
Third person	they	them	their, theirs

FIGURE C–1. Pronoun Case Chart

The subjective case can indicate the person or thing acting (*He* sued the vendor), the person or thing acted upon (*He* was sued by the vendor), or the topic of description (*He* is the vendor). The objective case can indicate the thing acted on (The vendor sued *him*) or the person or thing acting but in the objective position (The vendor was sued by *him*). (See also voice.) The possessive case indicates the person or thing owning or possessing something (It was *his* company). See also modifiers.

Subjective Case

C

A pronoun is in the subjective case (also called the *nominative case*) when it represents the person or thing acting or is the receiver of the action but in the subject position.

- *I* wrote a proposal.
- *I* was praised for my proposal writing.

A linking verb links a pronoun to its antecedent to show that they identify the same thing. Because they represent the same thing, the pronoun is in the subjective case even when it follows the verb, which makes it a subjective complement.

- *He* is the head of the Quality Control Group. [subject]
- The head of the Quality Control Group is *he*. [subjective complement]

The subjective case is used after the words *than* and *as* because of the understood (although unstated) portion of the clauses in which those words appear.

- George is as good a designer as *I* [am].
- Our subsidiary can do the job better than *we* [can].

Objective Case

A pronoun is in the objective case (also called the *accusative case*) when it indicates the person or thing receiving the action expressed by a verb in the active voice.

- They informed *me* by letter that they had received my résumé.

Pronouns that follow action verbs (which excludes all forms of the verb *be*) must be in the objective case. Do not be confused by an additional name.

- The company promoted *me* in July.
- The company promoted John and *me* in July.

A pronoun is in the objective case when it is the object of a gerund or a preposition or the subject of an infinitive.

- Between *you* and *me,* his facts are questionable. [objects of a preposition]
- Many of *us* attended the conference. [object of a preposition]

- Training *him* was the best thing I could have done. [object of a gerund]
- We asked *them* to return the equipment. [subject of an infinitive]

English does not differentiate between direct objects and indirect objects; both require the objective form of the pronoun. See also complements.

- The interviewer seemed to like *me.* [direct object]
- They wrote *me* a letter. [indirect object]

Possessive Case

A noun or pronoun is in the possessive case when it represents a person, place, or thing that possesses something. To make a singular noun possessive, add *'s* (the *manufacturer's* robotic inventory system). With plural nouns that end in *s*, show the possessive by placing an apostrophe after the *s* that forms the plural (a *managers'* meeting). For other guidelines, see possessive case.

Appositives

An appositive is a noun or noun phrase that follows and amplifies another noun or noun phrase. Because it has the same grammatical function as the noun it complements, an appositive should be in the same case as the noun with which it is in apposition.

- Two nurses, Jim Knight and *I,* were asked to review the patient's chart. [subjective case]
- The group leader selected two members to represent the lab—Mohan Pathak and *me.* [objective case]

Determining the Case of Pronouns

One test to determine the proper case of a pronoun is to try it with some transitive verb such as *resembled* or *hit.* If the pronoun would logically precede the verb, use the subjective case; if it would logically follow the verb, use the objective case.

- *She* [*He, They*] resembled her father. [subjective case]
- Angela resembled *him* [*her, them*]. [objective case]

In the following example, try omitting the noun to determine the case of the pronoun.

SENTENCE	(*We*/*Us*) pilots fly our own airplanes.
INCORRECT	*Us* fly our own airplanes. [This incorrect usage is obviously wrong.]
CORRECT	*We* fly our own airplanes. [This correct usage sounds right.]

To determine the case of a pronoun that follows *as* or *than*, mentally add the words that are omitted but understood.

- The other technical writer is not paid as well as *she* [is paid].
  [You would not write, "*Her* is paid."]
- His partner was better informed than *he* [was informed].
  [You would not write, "*Him* was informed."]

If pronouns in compound constructions cause problems, try using them singly to determine the proper case.

SENTENCE	(*We*/*Us*) and the Johnsons are going to the Grand Canyon.
CORRECT	*We* are going to the Grand Canyon. [You would not write, "*Us* are going to the Grand Canyon."]

For advice on when to use *who* and *whom*, see who/whom.

cause-and-effect method of development

The cause-and-effect method of development is a common strategy to explain why something happened or why you think something will happen. The goal of the cause-and-effect method of development is to make as plausible as possible the relationship between a situation and either its cause or its effect. The conclusions you draw about the relationships should be based on evidence you have gathered. Like all methods, this one is often used in combination with others. If you were examining a problem with multiple causes, for example, you might combine cause-and-effect with order-of-importance as you examine each cause and its effect.

Evaluating Evidence

Because not all evidence will be of equal value, keep the following guidelines in mind as you gather evidence:

- *Your facts and arguments should be relevant to your topic.* Be careful not to draw a conclusion that your evidence does not lead to or support. You may have researched some statistics, for example, showing that an increasing number of Americans are licensed to fly small airplanes. You cannot use that information as evidence for a decrease in new car sales in the United States—the evidence does not lead to that conclusion.

- *Your evidence should be adequate.* Incomplete evidence can lead to false conclusions.

  - Driver training classes do not help prevent auto accidents. Two people I know who completed driver training classes were involved in accidents.

  A thorough investigation of the usefulness of driver training classes in keeping down the accident rate would require more than one or two examples. It would require a systematic comparison of the driving records for a representative sample of drivers who had completed driver training and those who had not.

- *Your evidence should be representative.* If you conduct a survey to obtain your evidence, do not solicit responses only from individuals or groups whose views are identical to yours; be sure you obtain a representative sampling.
- *Your evidence should also be demonstrable.* Two events that occur close to each other in time or place may or may not be causally related. For example, that new traffic signs were placed at an intersection and the next day an accident occurred does not necessarily prove that the signs caused the accident. You must demonstrate the relationship with pertinent facts and arguments. See also logic errors.

Linking Causes to Effects

To show a true relationship between a cause and an effect, you must demonstrate that the existence of the one *requires* the existence of the other. It is often difficult to establish beyond any doubt that one event was the cause of another event. More often, a result will have more than one cause. As you research a subject, your task is to determine which cause or causes are most plausible.

When several probable causes are equally valid, report your findings accordingly, as in the following excerpt from an article on the use of an energy-saving device called a furnace-vent damper. The damper is a metal plate that fits inside the flue or vent pipe of a natural-gas or fuel-oil furnace. When the furnace is on, the damper opens to allow gases to escape up the flue. When the furnace shuts off, the damper closes, thus preventing warm air from escaping up the flue stack. The dampers are potentially dangerous, however. If a damper fails to open at the proper time, poisonous furnace gases could back up into the house and asphyxiate the occupants. Tests run on several dampers showed a number of probable causes for their malfunctioning.

C

- One damper was sold without proper installation instructions, and another was wired incorrectly. Two of the units had slow-opening dampers (15 seconds) that prevented the [furnace] burner from firing. And one damper jammed when exposed to a simulated fuel temperature of more than 700 degrees.

—Don DeBat, "Save Energy but Save Your Life, Too," *Family Safety*

The investigator located more than one cause of damper malfunctions and reported on them. Without such a thorough account, recommendations to prevent malfunctions would be based on incomplete evidence.

center on

Use the phrase *center on* in writing, not *center around.* (The experiments *center on* the new discovery.) Often the idea intended by *center on* is better expressed by other words.

- The hearings on computer security ~~centered on~~ *dealt with* access codes.

chronological method of development

The chronological method of development arranges the events under discussion in sequential order, emphasizing time as it begins with the first event and continues chronologically to the last. Trip reports, laboratory reports, work schedules, some minutes of meetings, and certain trouble reports are among the types of writing in which information is organized chronologically. Chronological order is typically used in narration.

Figure C–2 is a description of the development of a fire, as reported in a trouble report. After providing important background information, the writer presents the events in chronological order.

cite / sight / site

Cite means "acknowledge" or "quote an authority." (The speaker *cited* several ecology experts.) *Sight* is the ability to see. (He feared that he might lose his *sight.*) *Site* is a plot of land (a construction *site*) or the place where something is located (a Web *site*).

C

Memorandum

To: Charles Artmier, Chairman, Safety Committee
From: Willard Ricke, Plant Superintendent, Sequoia Plant *WR*
Date: November 17, 2006
Subject: Fire at our Sequoia Plant in Ardville

The following description provides an account of the conditions and development of the fire at our Sequoia Plant in Ardville. Because this description will be incorporated in the report that the Safety Committee will prepare, please let me know if any details should be added or clarified.

Three piles of scrap paper were located about 100 feet south of Building A and 150 feet west of Building B at our Sequoia Chemical Plant in Ardville. Building A, which consisted of one story and a partial attic, was used in part as an electric shop and in part for equipment storage. Six pallets of Class I flammable liquids in 55-gallon drums were temporarily stored just outside of Building A, along its west wall.

A spark created in the electrical shop in Building A started a fire. The night watchman first noticed the fire in the electric shop at about 6:00 a.m. and immediately called the fire department and the plant superintendent. Before the fire department arrived, the fire spread to the drums of flammable liquid just outside the west wall and quickly engulfed Building A in flames. The fire then spread to the scrap paper outside, with a 40 mph wind blowing it in the direction of Building B.

During this time, the watchman began to hose down the piles of scrap paper between the buildings to try to keep the fire from reaching Building B. He also set up lawn sprinklers between the fire and Building B as part of his attempt to protect Building B from the fire. However, he saw smoke coming from Building B at 6:15 a.m., in spite of his best efforts to protect it.

The volunteer fire department, which was 20 miles away, reached the scene at 6:57 a.m., nearly an hour after the fire was discovered. It is estimated that the fire department's pumps were started after the fire had been burning in Building B for at least 20 minutes.

Building A was destroyed, along with its contents, and the fire burned into the hollow joisted roof of Building B, which sustained a $250,000 loss. . . .

FIGURE C–2. Chronological Method of Development

clarity

Clarity is essential to effective communication with your readers. You cannot achieve your purpose or a goal like persuasion without clarity. Many factors contribute to clarity, just as many other elements can defeat it.

A logical method of development and an outline will help you avoid presenting your reader with a jumble of isolated thoughts. A method of development and an outline that puts your thoughts in a logical, meaningful sequence brings coherence as well as unity to your writing. Clear transition contributes to clarity by providing the smooth flow that enables the reader to connect your thoughts with one another without conscious effort. See also outlining.

Proper emphasis and subordination are mandatory if you want to achieve clarity. If you do not use those two complementary techniques wisely, all your clauses and sentences will appear to be of equal importance. Your reader will only be able to guess which are most important, which are least important, and which fall somewhere in between. The pace at which you present your ideas is also important to clarity; if the pace is not carefully adjusted to both the topic and the reader, your writing will appear cluttered and unclear.

Point of view establishes through whose eyes or from what vantage point the reader views the subject. A consistent point of view is essential to clarity; if you inappropriately switch from the first person to the third person in midsentence, you are certain to confuse your reader.

Precise word choice contributes to clarity and helps eliminate ambiguity and awkwardness. Vague words, clichés, poor use of idiom, and inappropriate usage detract from clarity.

That conciseness is a requirement of clearly written communication should be evident to anyone who has ever attempted to decipher an insurance policy or a legal contract. Although words are our chief means of communication, more words than necessary can impede clear communication, just as too many cars on a highway can impede traffic. For the sake of clarity, remove unnecessary words from your writing.

clauses

A clause is a group of words that contains a subject and a predicate and that functions as a sentence or as part of a sentence. (See sentence construction.) Every subject-predicate word group in a sentence is a clause, and every sentence must contain at least one independent clause; otherwise, it is a sentence fragment.

A clause that could stand alone as a simple sentence is an *independent clause.*

- *The scaffolding fell* when the rope broke.

A clause that could not stand alone if the rest of the sentence were deleted is a *dependent* (or *subordinate*) *clause.*

- I was at the St. Louis branch *when the decision was made.*

Dependent clauses are useful in making the relationship between thoughts clearer and more succinct than if the ideas were presented in a series of simple sentences or compound sentences.

FRAGMENTED	The recycling facility is located between Millville and Darrtown. Both villages use it. [The two thoughts are of approximately equal importance.]
SUBORDINATED	The recycling facility, *which is located between Millville and Darrtown*, is used by both villages. [One thought is subordinated to the other.]

Subordinate clauses are especially effective for expressing thoughts that describe or explain another statement. Too much subordination, however, can be confusing and produce wordiness. See also conciseness.

- He selected instructors whose classes ~~had a slant that was~~ *were* specifically directed ~~toward students who intended to go into engineering.~~ *at engineering students.*

A clause can be connected with the rest of its sentence by a coordinating conjunction, a subordinating conjunction, a relative pronoun, or a conjunctive adverb.

- It was 500 miles to the facility, *so* we made arrangements to fly. [coordinating conjunction]
- Mission control will have to be alert *because* at launch the space shuttle will contain a highly flammable fuel. [subordinating conjunction]
- Robert M. Fano was the scientist *who* developed the earliest multiple-access computer system at MIT. [relative pronoun]
- We arrived in the evening; *nevertheless*, we began the tour of the facility. [conjunctive adverb]

clichés

C

Clichés are expressions that have been used for so long that they are no longer fresh but come to mind easily because they are so familiar. Clichés are often wordy as well as vague and can be confusing, especially to nonnative speakers of English. A better, more direct word or phrase is given for each of the following clichés.

INSTEAD OF	USE
all over the map	scattered, unfocused
run it up the flagpole	see what others think
last but not least	last, finally

Some writers use clichés in a misguided attempt to appear casual or spontaneous, just as other writers try to impress readers with buzzwords. Although clichés may come to mind easily while you are writing a draft, eliminate them during revision. See also affectation, conciseness, and international correspondence.

coherence

Writing is coherent when the relationships among ideas are clear to readers. The major components of coherent writing are a logical sequence of related ideas and clear transitions between these ideas. See also clarity and organization.

Presenting ideas in a logical sequence is the most important requirement in achieving coherence. The key to achieving a logical sequence is the use of a good outline (see outlining). An outline forces you to establish an introduction, a middle (body), and a conclusion. That structure contributes greatly to coherence by enabling you to experiment with sequences and to lay out the most direct route to your purpose without digressing.

Thoughtful transition is also essential; without it, your writing cannot achieve the smooth flow from sentence to sentence and from paragraph to paragraph that results in coherence.

Check your draft carefully for coherence during revision. If possible, have someone else review your draft for how well it expresses the relationships between ideas. Such help is especially important in collaborative writing where team member contributions need to be blended into a coherent document. See also unity.

collaborative writing

Collaborative writing occurs when two or more writers work together to produce a single document for which they share responsibility and decision-making authority. Collaborative writing teams are formed when (1) the size of a project or the time constraints imposed on it require collaboration, (2) the project involves multiple areas of expertise, or (3) the project requires the melding of divergent views into a single perspective that is acceptable to the whole team or to another group.

Tasks of the Collaborative Writing Team

The collaborating team strives to achieve a compatible working relationship by dividing the work in a way that uses each writer's expertise and experience to their collective advantage. The team should also designate a coordinator who will guide the team members' activities, organize the project, and ensure coherence within the document. The coordinator's duties can be determined by mutual agreement or, if the team often works together, assigned on a rotating basis.

Planning. The team collectively identifies the readers, purpose, and scope of the project. The team conceptualizes the document to be produced, creates a broad outline of the document (see outlining), divides it into segments, and assigns each segment to individual team members, often on the basis of expertise. See also meetings and "Five Steps to Successful Writing."

In the planning stage, the team establishes a schedule and sets any writing style standards that the team is expected to follow. The schedule includes due dates for drafts, reviews of the drafts, revisions, and the final document. Milestone deadlines must be met, even if the drafts are not as polished as the individual writers would like: One missed deadline can delay the entire project.

Research and Writing. Planning is followed by research and writing, a period of intense independent activity by members of the team. Each team member researches his or her assigned segment of the document, expands and develops the broad outline, and produces a draft from the detailed outline. See also writing a draft.

Reviewing. Keeping the readers' needs and the document's purpose in mind, each member critically yet diplomatically reviews the other team members' work, from the overall organization to the clarity of each paragraph, and offers advice to help improve the writer's segment.

DIGITAL TIP **REVIEWING COLLABORATIVE DOCUMENTS**

Adobe Acrobat™ and many word-processing packages have options for providing feedback on the drafts of collaborators' documents. You can add text or voice annotations within the text, allowing your readers to read or hear your comments, as well as track changes in your text and allow readers to accept or reject each change. In Acrobat, you can also use a drawing tool to input traditional proofreaders' marks. For more on this topic, see <*bedfordstmartins.com/alredtech*> and select *Digital Tips,* "Reviewing Collaborative Documents."

Team members can easily solicit feedback by sharing electronic files. Tracking and commenting features allow the reviewer to show the suggested changes without deleting the original text. The author can then easily accept or reject the proposed changes.

Revising. In this stage, individual writers evaluate their colleagues' reviews and accept, reject, or build on their suggestions. Then, all drafts can be consolidated into a final master copy maintained by the team coordinator. See also revision.

Conflict

As you collaborate, be ready to tolerate some disharmony, but temper it with mutual respect. Team members may not agree on every subject, and differing perspectives can easily lead to conflict, ranging from mild differences over minor points to major showdowns. However, creative differences resolved respectfully can energize the team and, in fact, strengthen a finished document by compelling writers to reexamine assumptions and issues in unanticipated ways. See also listening.

Writer's Checklist: Writing Collaboratively

- ☑ Designate one person as the team coordinator.
- ☑ Identify the audience, purpose, and scope of the project.
- ☑ Create a working outline of the document.
- ☑ Assign segments or tasks to each team member.
- ☑ Establish a schedule: due dates for drafts, revisions, and final versions.
- ☑ Agree on a standard reference guide for style and format.
- ☑ Research and write drafts of document segments.

Writer's Checklist: Writing Collaboratively (continued)

☑ Exchange segments for team member reviews.
☑ Revise segments as needed.
☑ Meet your established deadlines.

C

colons

The colon (:) is a mark of introduction that alerts readers to the close connection between the preceding statement and what follows.

Colons in Sentences

A colon links independent clauses to words, phrases, clauses, or lists that identify, rename, explain, emphasize, amplify, or illustrate the sentence that precedes the colon.

- Two topics will be discussed: *the new lab design and the revised safety procedures.* [phrases that identify]
- Only one thing will satisfy Mr. Sturgess: *our finished proposal.* [appositive (renaming) phrase for emphasis]
- Any organization is confronted with two separate, though related, information problems: *It must maintain an effective internal communication system and an effective external communication system.* [clause to amplify and explain]
- Heart patients should make key lifestyle changes: *stop smoking, exercise regularly, eat a low-fat diet, and reduce stress.* [list to identify and illustrate]

Colons with Salutations, Titles, Citations, and Numbers

A colon follows the salutation in correspondence, even when the salutation refers to a person by first name.

- Dear Professor Jeffers: *or* Dear Georgia:

Colons separate titles from subtitles and separate references to sections of works in citations. See also documenting sources.

- *Writing That Works: Communicating Effectively on the Job*
- Genesis 10:16 [chapter 10, verse 16]

Colons separate numbers in time references and indicate numerical ratios.

- 9:30 a.m. [9 hours and 30 minutes]
- The cement is mixed with water and sand at 5:3:1. [The colon is read as the word *to*.]

Punctuation and Capitalization with Colons

A colon always goes outside quotation marks.

- This was the real meaning of the manager's "suggestion": Cooperation within our department must improve.

As this example shows, the first word after a colon may be capitalized if the statement following the colon is a complete sentence and functions as a formal resolution or question. If the element following the colon is subordinate, however, use a lowercase letter to begin that element. See also capitalization.

- We have only one way to stay within our present budget: to reduce expenditures for research and development.

Unnecessary Colons

Do not place a colon between a verb and its objects.

- Three fluids that clean pipettes are~~:~~ water, alcohol, and acetone.

Likewise, do not use a colon between a preposition and its object.

- I may be transferred to~~:~~ Tucson, Boston, or Miami.

Do not insert a colon after *including, such as,* or *for example* to introduce a simple list.

- Office computers should not be used for activities such as~~:~~ personal e-mail, Web surfing, Internet shopping, and playing computer games.

One common exception is made when a verb or preposition is followed by a stacked list; however, it may be possible to introduce the list with a complete sentence instead.

- *The following corporations* ~~Corporations that~~ manufacture computers*:* ~~include:~~

Apple	Compaq	Dell
Gateway	IBM	Micron

C

comma splice

A comma splice is a grammatical error in which two independent clauses are joined by only a comma.

INCORRECT It was 500 miles to the facility, we arranged to fly.

A comma splice can be corrected in several ways.

1. Substitute a semicolon, a semicolon and a conjunctive adverb, or a comma and a coordinating conjunction.
   - It was 500 miles to the facility*;* we arranged to fly.
   - It was 500 miles to the facility*; therefore,* we arranged to fly.
   - It was 500 miles to the facility*, so* we arranged to fly.
2. Create two sentences.
   - It was 500 miles to the facility*. We* arranged to fly.
3. Subordinate one clause to the other. (See subordination.)
   - *Because it was 500 miles to the facility,* we arranged to fly.

See also sentence construction and sentence faults.

commas

DIRECTORY

Like all punctuation, the comma (,) helps readers understand the writer's meaning and prevents ambiguity. Notice how the comma helps make the meaning clear in the second example.

AMBIGUOUS To be successful nurses with M.S. degrees must continue to learn.

CLEAR To be successful, nurses with M.S. degrees must continue to learn. [The comma makes clear where the main part of the sentence begins.]

Do not follow the old myth that you should insert a comma wherever you would pause if you were speaking. Although you would pause wherever you encounter a comma, you should not insert a comma wherever you might pause. Effective use of commas depends on an understanding of sentence construction.

Linking Independent Clauses

Use a comma before a coordinating conjunction (*and*, *but*, *or*, *nor*, and sometimes *so*, *yet*, and *for*) that links independent clauses.

- The new microwave disinfection system was delivered, *but* the installation will require an additional week.

However, if two independent clauses are short and closely related—and there is no danger of confusing the reader—the comma may be omitted. Both of the following examples are correct.

- The cable snapped and the power failed.
- The cable snapped, and the power failed.

Enclosing Elements

Commas are used to enclose nonessential information in nonrestrictive clauses, phrases, and parenthetical elements. See also restrictive and nonrestrictive elements.

- Our new factory, *which began operations last month,* should add 25 percent to total output. [nonrestrictive clause]
- The technician, *working quickly and efficiently,* finished early. [nonrestrictive phrase]
- We can, *of course,* expect their lawyer to call us. [parenthetical element]

Yes and *no* are set off by commas in such uses as the following:

- I agree with you, *yes.*
- *No,* I do not think we can finish by the deadline.

A direct address should be enclosed in commas.

- You will note, *Mark,* that the surface of the brake shoe complies with the specification.

An appositive phrase (which re-identifies another expression in the sentence) is enclosed in commas.

- Our company, *Envirex Medical Systems,* won several awards last year.

Interrupting parenthetical and transitional words or phrases are usually set off with commas. See also transition.

C

- The report, *therefore,* needs to be revised.

Commas are omitted when the word or phrase does not interrupt the continuity of thought.

- I *therefore* suggest that we begin construction.

For other means of punctuating parenthetical elements, see dashes and parentheses.

Introducing Elements

Clauses and Phrases. Generally, place a comma after an introductory clause or phrase, especially if it is long, to identify where the introductory element ends and the main part of the sentence begins.

- *Because we have not yet contained the new strain of influenza,* we recommend vaccination for high-risk patients.

A long modifying phrase that precedes the main clause should always be followed by a comma.

- *During the first series of field-performance tests at our Colorado proving ground,* the new engine failed to meet our expectations.

When an introductory phrase is short and closely related to the main clause, the comma may be omitted.

- *In two seconds* a 5°C temperature rise occurs in the test tube.

A comma should always follow an absolute phrase, which modifies the whole sentence.

- *The tests completed,* we organized the data for the final report.

Words and Quotations. Certain types of introductory words are followed by a comma. One example is a transitional word or phrase (*however, in addition*) that connects the preceding clause or sentence with the thought that follows.

- *Furthermore,* steel can withstand a humidity of 99 percent, provided that there is no chloride or sulfur dioxide in the atmosphere.
- *For example,* this change will make us more competitive in the global marketplace.

When an adverb closely modifies the verb or the entire sentence, it should not be followed by a comma.

C

- *Perhaps* we can still solve the turnover problem. *Certainly* we should try.
  [*Perhaps* and *certainly* closely modify each statement.]

A proper noun used in an introductory direct address is followed by a comma, as is an interjection (such as *oh*, *well*, *why*, *indeed*, *yes*, and *no*).

- *Nancy,* enclosed is the article you asked me to review.
  [direct address]
- *Indeed,* I will ensure that your request is forwarded. [interjection]

Use a comma to separate a direct quotation from its introduction.

- Morton and Lucia White *said,* "People live in cities but dream of the countryside."

Do not use a comma when giving an indirect quotation.

- Morton and Lucia White *said that* people dream of the countryside, even though they live in cities.

Separating Items in a Series

Although the comma before the last item in a series is sometimes omitted, it is generally clearer to include it.

- Random House, Bantam, Doubleday*,* and Dell were individual publishing companies.
  [Without the final comma, "Doubleday and Dell" might refer to one company or two.]

Phrases and clauses in coordinate series are also punctuated with commas.

- Plants absorb noxious gases, act as receptors of dirt particles, and cleanse the air of other impurities.

When phrases or clauses in a series contain commas, use semicolons rather than commas to separate the items.

- Our new products include amitriptyline, which has sold very well*;* dipyridamole, which has not sold well*;* and cholestyramine, which was just introduced.

When adjectives modifying the same noun can be reversed and make sense, or when they can be separated by *and* or *or*, they should be separated by commas.

- The aircraft featured a *modern, sleek, swept-wing* design.

When an adjective modifies a phrase, no comma is required.

- She was investigating the *damaged inventory-control system.* [The adjective *damaged* modifies the phrase *inventory-control system.*]

Never separate a final adjective from its noun.

- He is a conscientious, honest, reliable,/ worker.

Clarifying and Contrasting

Use a comma to separate two contrasting thoughts or ideas.

- The project was finished on time, but not within the budget.

Use a comma after an independent clause that is only loosely related to the dependent clause that follows it or that could be misread without the comma.

- I should be able to finish the plan by July, even though I lost time because of illness.

Showing Omissions

A comma sometimes replaces a verb in certain elliptical constructions.

- Some were punctual; *others, late.* [The comma replaces *were.*]

It is better, however, to avoid such constructions in technical writing.

Using with Numbers and Names

Commas are conventionally used to separate distinct items. Use commas between the elements of an address written on the same line (but not between the state and the zip code).

- Kristen James, 4119 Mill Road, Dayton, Ohio 45401

A full date that is written in month-day-year format uses a comma preceding and following the year.

- November 30, 2015, is the payoff date.

Do not use commas for dates in the day-month-year format, which is used in many parts of the world and by the U.S. military. See also international correspondence.

- Note that 30 November 2015 is the payoff date.

C

No commas are used if only the day or year is included. See also dates.

- The target date of May 2007 is optimistic, so I would like to meet on March 4 to discuss our options.

Use commas to separate the elements of Arabic numbers.

- 1,528,200 feet

However, because many countries use the comma as the decimal marker, use spaces or periods rather than commas in international documents.

- 1.528.200 meters *or* 1 528 200 meters

A comma may be substituted for the colon in the salutation of a personal letter or e-mail. Do not, however, use a comma in the salutation of a business letter or e-mail, even if you use the person's first name.

- Dear Marie, [personal letter or e-mail]
- Dear Marie: [business letter or e-mail]

Use commas to separate the elements of geographical names.

- Toronto, Ontario, Canada

Use a comma to separate names that are reversed or that are followed by an abbreviation.

- Smith, Alvin
- Jane Rogers, Ph.D.

Using with Other Punctuation

Conjunctive adverbs (*however, nevertheless, consequently, for example, on the other hand*) that join independent clauses are preceded by a semicolon and followed by a comma. Such adverbs function both as modifiers and as connectives.

- The idea is good; *however,* our budget is not sufficient.

As shown earlier in this entry, use semicolons rather than commas to separate items in a series when the items themselves contain commas.

When a comma should follow a phrase or clause that ends with words in parentheses, the comma always appears outside the closing parenthesis.

- Although we left late (at 7:30 p.m.), we arrived in time for the keynote address.

Commas always go inside quotation marks.

- The operator placed the discharge bypass switch at *"normal,"* which triggered a second discharge.

Except with abbreviations, a comma should not be used with a dash, an exclamation mark, a period, or a question mark.

- "Have you finished the project?,/" I asked.

Avoiding Unnecessary Commas

A number of common writing errors involve placing commas where they do not belong. As stated earlier, such errors often occur because writers assume that a pause in a sentence should be indicated by a comma.

Do not place a comma between a subject and verb or between a verb and its object.

- The conditions at the test site in the Arctic,/ made accurate readings difficult.
- She has often said,/ that one company's failure is another's opportunity.

Do not use a comma between the elements of a compound subject or compound predicate consisting of only two elements.

- The director of the design department,/ and the supervisor of the quality-control section were opposed to the new schedules.
- The design director listed five major objections,/ and asked that the new schedule be reconsidered.

Do not include a comma after a coordinating conjunction such as *and* or *but.*

- The chairperson formally adjourned the meeting, but,/ the members of the committee continued to argue.

Do not place a comma before the first item or after the last item of a series.

- The new products we are considering include,/ calculators, scanners, and cameras.
- It was a fast, simple, inexpensive,/ process.

Do not use a comma to separate a prepositional phrase from the rest of the sentence unnecessarily.

- We discussed the final report,/ on the new project.

compare / contrast

When you *compare* things, you point out similarities or both similarities and differences. (He *compared* the two brands before making his choice.) When you *contrast* things, you point out only the differences. (Their speaking styles *contrasted* sharply.) In either case, you compare or contrast only things that are part of a common category.

When *compare* is used to establish a general similarity, it is followed by *to.* (*Compared to* our old copier, the current one is fast.) When *compare* is used to indicate a close examination of similarities or differences, it is followed by *with.* (We *compared* the features of the new copier *with* those of the current one.)

Contrast is normally followed by *with.* (The new policy *contrasts* sharply *with* the earlier one.) When the noun form of *contrast* is used, one speaks of the *contrast between* two things or of one thing being *in contrast to* the other.

- There is a sharp *contrast between* the old and new policies.
- The new policy is *in* sharp *contrast to* the earlier one.

comparison

When you are making a comparison, be sure that both or all of the elements being compared are clearly evident to your reader.

than the Nicom 2 software

- The Nicom 3 software is better ^.

The things being compared must be of the same kind.

a *briefcase*

- A hard-side briefcase offers more protection than ^ fabric ^.

Be sure to point out the parallels or differences between the things being compared. Do not assume your reader will know what you mean.

it is from

- Washington is farther from Boston than ^ Philadelphia.

A double comparison in the same sentence requires that the first comparison be completed before the second one is stated.

- The discovery of electricity was one of the great ~~if not the greatest~~
  , if not the greatest
  scientific discoveries in history ^.

Do not attempt to compare things that are not comparable.

- Agricultural experts advise that ~~storage space is reduced by 40 percent compared with~~ baled hay *requires 40 percent less storage space than loose hay requires*.

  [*Storage space* is not comparable to *baled hay.*]

comparison method of development

As a method of development, comparison points out similarities and differences between the elements of your subject. The comparison method of development can be used effectively, for example, to explain a difficult or an unfamiliar subject by relating it to a simpler or more familiar one.

You must first determine the basis for the comparison. For example, if you were comparing bids from contractors for a remodeling project at your company, you most likely would compare such factors as price, previous experience, personnel qualifications, availability at a time convenient for you, and completion date. Once you have determined the basis or bases for comparison, you can determine the most effective way to structure your comparison: whole by whole or part by part.

In the *whole-by-whole method,* all the relevant characteristics of one item are examined before all the relevant characteristics of the next item. The description of typical woodworking glues in Figure C–3 is organized according to the whole-by-whole method. It describes each type of glue and all its characteristics before moving to the next one. This description would be useful for those readers who wish to learn about all types of wood glues.

If your purpose were to help readers consider the various characteristics of all the glues, the information might be arranged according to the *part-by-part method of comparison,* in which the relevant features of the items are compared one by one (as shown in Figure C–4). The part-by-part method could accommodate further comparison—such as temperature ranges, special warnings, and common use, in this case.

Comparisons can also be made effectively with the use of tables, as shown in Figure C–5. The advantage of a table is that it provides a quick reference, allowing readers to see and compare all the information at once. The disadvantage is that a table cannot convey as much related detailed information as a narrative description.

C

White glue is the most useful all-purpose adhesive for light construction, but it cannot be used on projects that will be exposed to moisture, high temperature, or great stress. Wood that is being joined with white glue must remain in a clamp until the glue dries, which takes about 30 minutes.

Aliphatic resin glue has a stronger and more moisture-resistant bond than white glue. It must be used at temperatures above 50°F. The wood should be clamped for about 30 minutes. . . .

Plastic resin glue is the strongest of the common wood adhesives. It is highly moisture resistant, though not completely waterproof. Sold in powdered form, this glue must be mixed with water and used at temperatures above 70°F. It is slow setting, and the joint should be clamped for four to six hours. . . .

Contact cement is a very strong adhesive that bonds so quickly it must be used with great care. It is ideal for mounting sheets of plastic laminate on wood. It is also useful for attaching strips of veneer to the edges of plywood. Because this adhesive bonds immediately when two pieces are pressed together, clamping is not necessary, but the parts to be joined must be carefully aligned before being placed together. Most brands are flammable, and the fumes can be harmful if inhaled. To meet current safety standards, this type of glue must be used in a well-ventilated area, away from flames or heat.

FIGURE C–3. Whole-by-Whole Method of Comparison

Woodworking adhesives are rated primarily according to their bonding strength, moisture resistance, and setting time.

Bonding strength is categorized as very strong, moderately strong, or adequate for use with little stress. Contact cement and plastic resin glue bond very strongly, while aliphatic resin glue bonds moderately strongly. White glue provides a bond least resistant to stress.

The *moisture resistance* of woodworking glues is rated as high, moderate, or low. Plastic resin glue and contact cement are highly moisture resistant, aliphatic resin glue is moderately moisture resistant, and white glue is least moisture resistant.

Setting time for the glues varies from an immediate bond to a four-to-six-hour bond. Contact cement bonds immediately and requires no clamping. Because the bond is immediate, surfaces being joined must be carefully aligned before being placed together. White glue and aliphatic resin glue set in 30 minutes; both require clamping to secure the bond. Plastic resin, the strongest wood glue, sets in four to six hours and also requires clamping.

FIGURE C–4. Part-by-Part Method of Comparison

	White Glue	Aliphatic Resin Glue	Plastic Resin Glue	Contact Cement
Bonding Strength	Low	Moderate	High	High
Moisture Resistance	Low	Moderate	High	High
Setting Time	Thirty minutes	Thirty minutes	Four to six hours	Bonds on contact
Common Uses	Light construction	General purpose	General purpose	Laminate and veneer to wood

FIGURE C–5. Comparison Using a Table to Illustrate Key Differences

complaint letters

C

The tone of a complaint letter is important; the most effective ones do not sound complaining. If your letter reflects only your annoyance and anger, you may not be taken seriously. Assume that the recipient will be conscientious in correcting the problem. However, anticipate reader reactions or rebuttals.

- I reviewed carefully the "safe operating guidelines" in my user manual before I installed the device.

Without such explanations, readers may be tempted to dismiss your complaint. Figure C–6 on page 88 shows a typical complaint letter.

Although the circumstances and severity of the problem may vary, effective complaint letters generally follow this pattern:

1. Identify the problem or faulty item(s) and include relevant invoice numbers, part names, and dates. Include a copy of the receipt, bill, or contract, and keep the original for your records.
2. Explain logically, clearly, and specifically what went wrong, especially for a problem with a service. (Avoid guessing why you *think* some problem occurred.)
3. State what you expect the reader to do to solve the problem.

Begin by checking to see if the company's Web site provides instructions for submitting a complaint. Otherwise, for large organizations, you may address your complaint to Customer Service. In smaller organizations, you might write to a vice president in charge of sales or service, or directly to the owner. As a last resort, you may find that sending copies of a complaint letter to more than one person in the company will get faster results. See also adjustment letters and refusal letters.

complement / compliment

Complement means "anything that completes a whole" (see also complements). It is used as either a noun or a verb.

- A *complement* of four employees would bring our staff up to its normal strength. [noun]
- The two programs *complement* one another perfectly. [verb]

Compliment means "praise." It too is used as either a noun or a verb.

- The manager's *compliment* boosted staff morale. [noun]
- The manager *complimented* the staff on its efficient job. [verb]

C

BAKER HEALTHCARE SYSTEMS

Purchasing Department
501 Main Street
Springfield, OH 45321
(513) 683-8100
Fax (513) 683-8000

September 22, 2006

Manager, Customer Relations
Medical Solutions, Inc.
521 West 23rd Street
New York, NY 10011

Subject: ST3 Diagnostic Scanners

On July 10, I ordered nine ST3 Diagnostic Scanners (order # ST3-1179R). The scanners were ordered from your customer Web site.

On August 2, I received from your Newark, New Jersey, parts warehouse seven HL monitors. I immediately returned these monitors with a note indicating the mistake that had been made. However, not only have I failed to receive the ST3 scanners I ordered, but I have also been billed repeatedly.

I have enclosed a copy of my confirmation e-mail, the shipping form, and the most recent bill. If you cannot send me the scanners I ordered by November 1, please cancel my order.

Sincerely,

Paul Denlinger

Paul Denlinger
Manager
pld@baker.org

Enclosures

www.baker.org

FIGURE C–6. Complaint Letter

complements

A *complement* is a word, phrase, or clause used in the predicate of a sentence to complete the meaning of the sentence.

- Pilots fly *airplanes.* [word]
- To live is *to risk death.* [phrase]
- John knew *that he would be late.* [clause]

Four kinds of complements are generally recognized: direct object, indirect object, objective complement, and subjective complement.

A *direct object* is a noun or noun equivalent that receives the action of a transitive verb; it answers the question *What?* or *Whom?* after the verb.

- I designed *a Web page.* [noun phrase]
- I like *to work.* [verbal]
- I like *it.* [pronoun]
- I like *what I saw.* [noun clause]

An *indirect object* is a noun or noun equivalent that occurs with a direct object after certain kinds of transitive verbs such as *give, wish, cause,* and *tell.* It answers the question *To whom or what?* or *For whom or what?*

- We should buy the *office* a *scanner.*
  [*Scanner* is the direct object, and *office* is the indirect object.]

An *objective complement* completes the meaning of a sentence by revealing something about the object of its transitive verb. An objective complement may be either a noun or an adjective.

- They call him *a genius.* [noun phrase]
- We painted the building *white.* [adjective]

A *subjective complement,* which follows a linking verb rather than a transitive verb, describes the subject. A subjective complement may be either a noun or an adjective.

- His sister is *a consultant.* [noun phrase]
- His brother is *ill.* [adjective]

See also sentence construction.

C

compose / constitute / comprise

Compose and *constitute* both mean "make up the whole." The parts *compose* or *constitute* the whole.

- The nine offices *compose* the division.
- Unethical activities *constitute* cause for dismissal.

Comprise means "include," "contain," or "consist of." The whole *comprises* the parts. (The division *comprises* nine offices.)

compound words

A compound word is made from two or more words that function as a single concept. A compound may be hyphenated, written as one word, or written as separate words.

• editor in chief	courthouse	home page
high-energy	nevertheless	post office
low-level	online	Web site

If you are not certain whether a compound word should use a hyphen, check a dictionary.

Be careful to distinguish between compound words (*greenhouse*) and words that simply appear together but do not constitute compound words (*green house*). For plurals of compound words, generally add *s* to the last letter (*bookcases* and *Web sites*). However, when the first word of the compound is more important to its meaning than the last, the first word takes the *s* (*editors in chief*). Possessives are formed by adding *'s* to the end of the compound word (the *editor in chief's* desk, the *pipeline's* diameter, the *post office's* hours). See also possessive case.

conciseness

Conciseness means that extraneous words, phrases, clauses, and sentences have been removed from writing without sacrificing clarity or appropriate detail. Conciseness is not a synonym for brevity; a long report may be concise, while its abstract may be brief and concise. Conciseness is always desirable, but brevity may or may not be desirable in a given passage, depending on the writer's purpose. Although concise sentences are not guaranteed to be effective, wordy sentences always sacrifice some of their readability and coherence.

C

Causes of Wordiness

Modifiers that repeat an idea implicit or present in the word being modified contribute to wordiness by being redundant. See also reason is [because].

basic essentials	*completely* finished
final outcome	*present* status

Coordinated synonyms that merely repeat each other contribute to wordiness.

each and every	*basic and fundamental*
finally and for good	*first and foremost*

Excess qualification also contributes to wordiness.

perfectly clear	*completely* accurate

Expletives, relative pronouns, and relative adjectives, although they have legitimate purposes, often result in wordiness.

WORDY *There are* [expletive] many Web designers *who* [relative pronoun] are planning to attend the conference, *which* [relative adjective] is scheduled for May 13–15.

CONCISE Many Web designers plan to attend the conference scheduled for May 13–15.

Circumlocution (a long, indirect way of expressing things) is a leading cause of wordiness. See also gobbledygook.

WORDY The payment to which a subcontractor is entitled should be made promptly so that in the event of a subsequent contractual dispute we, as general contractors, may not be held in default of our contract by virtue of nonpayment.

CONCISE Pay subcontractors promptly. Then, if a contractual dispute occurs, we cannot be held in default of our contract because of nonpayment.

When conciseness is overdone, writing can become choppy and ambiguous. (See also telegraphic style.) Too much conciseness can produce a style that is not only too brief but also too blunt, especially in correspondence.

C

Writer's Checklist: Achieving Conciseness

Wordiness is understandable when you are **writing a draft**, but it should not survive **revision**.

☑ Use **subordination** to achieve conciseness.

- The *five-page* laboratory report was carefully ~~documented, and it covered five pages.~~ *documented.*

☑ Avoid **affectation** by using simple words and phrases.

WORDY It is the policy of the company to provide Web access to enable employees to conduct the online communication necessary to discharge their responsibilities; such should not be utilized for personal communications or nonbusiness activities.

CONCISE Employee Web access should be used only for appropriate company business.

☑ Eliminate redundancy.

WORDY Postinstallation testing, which is offered to all our customers at no further cost to them whatsoever, is available with each Line Scan System One purchased from this company.

CONCISE Free postinstallation testing is offered with each Line Scan System One.

☑ Change the passive **voice** to the active voice and the indicative **mood** to the imperative mood whenever possible.

WORDY Bar codes normally are used when an order is intended to be displayed on a computer, and inventory numbers normally are used when an order is to be placed with the manufacturer.

CONCISE Use bar codes to display the order on a computer, and use inventory numbers to place the order with the manufacturer.

☑ Eliminate or replace wordy introductory phrases or pretentious words and phrases (*in the case of, it may be said that, it appears that, needless to say*).

REPLACE	**WITH**
in order to; with a view to	to
due to the fact that; for the reason that; owing to the fact that; the reason for	because
by means of; by using; in connection with; through the use of	by; with
at this time; at this point in time; at present; at the present	now

Writer's Checklist: Achieving Conciseness (continued)

☑ Do not overuse intensifiers, such as *very, more, most, best, quite, great, really,* and *especially.* Instead provide specific and useful details.

☑ Use the search-and-replace command to find and revise wordy expressions, including *to be* and unnecessary helping verbs such as *will.*

conclusions

The conclusion of a document ties the main ideas together and can clinch a final significant point. This final point may, for example, make a prediction or judgment, summarize the key findings of a study, or recommend a course of action. Figure C–7 is a conclusion from a proposal to reduce health-care costs by increasing employee fitness through health-club subsidies. Notice that it summarizes key points, makes a recommendation, and points out several benefits of implementing the recommendation.

The way you conclude depends on your purpose, your readers' needs, and the context. For example, a committee report about possible locations for a new production facility might end with a recommendation that is realistic given the circumstances; a lengthy sales proposal might conclude persuasively with a summary of the proposal's salient points and the company's main strong points. The following examples are typical concluding strategies.

RECOMMENDATION

These results indicate that you need to alter your testing procedure to eliminate the impurities we found in specimens A through E.

SUMMARY

As this report describes, we would attract more recent graduates with the following strategies:

1. Establish a Web site where students can register and submit online résumés.
2. Increase our advertising in local student newspapers and our attendance at college career fairs.
3. Expand our local co-op program.

JUDGMENT

Based on the scope and degree of the tornado's damage, the current construction code for roofing on light industrial facilities is inadequate.

C

Conclusion and Recommendation
I recommend that ABO, Inc., participate in the corporate membership program at AeroFitness Clubs, Inc., by subsidizing employee memberships. By subsidizing memberships, ABO shows its commitment to the importance of a fit workforce. Club membership allows employees at all five ABO warehouses to participate in the program. The more employees who participate, the greater the long-term savings in ABO's health-care costs. Building and equipping fitness centers at all five warehouse sites would require an initial investment of nearly $2.5 million. These facilities would also occupy valuable floor space—on average, 4,000 square feet at each warehouse. Therefore, this option would be costly.

Enrolling employees in the corporate program at AeroFitness would allow them to receive a one-month free trial membership. Those interested in continuing could then join the club and pay half of the one-time membership fee of $900 and receive a 30 percent discount on the $600 yearly fee. The other half of the membership fee ($450) would be paid for by ABO. If employees leave the company, they would have the option of purchasing ABO's share of the membership to continue at AeroFitness or selling their half of the membership to another ABO employee wishing to join AeroFitness.

Implementing this program will help ABO, Inc., reduce its health-care costs while building stronger employee relations by offering employees a desirable benefit. If this proposal is adopted, I have some additional thoughts about publicizing the program to encourage employee participation. I look forward to discussing the details of this proposal with you and answering any questions you may have.

FIGURE C–7. Conclusion

IMPLICATION

Although our estimate calls for a substantially higher budget than in the three previous years, we believe that it is reasonable given our planned expansion.

PREDICTION

Although I have exceeded my original estimate for equipment, I have reduced my original labor estimate; therefore, I will easily stay within the original bid.

The concluding statement may merely present ideas for consideration, call for action, or deliberately provoke thought.

IDEAS FOR CONSIDERATION

The new security procedures at our facility become effective the first of the year. These procedures are reasonable given our sensitive projects, but some visitors and new employees may consider them intrusive. Please bear in mind the potential concerns of your new hires as well as professional associates who visit our facility.

CALL FOR ACTION

Please submit your revised safety procedures recommendations if you wish them to be considered for the agency's new safety manual. If you have not responded to my previous messages because of special circumstances, I will be glad to work with you personally.

THOUGHT-PROVOKING STATEMENT

Can we continue to accept the losses incurred by inefficiency? Or should we consider steps to control it now?

Be especially careful not to introduce a new topic when you conclude. A conclusion should always relate to and reinforce the ideas presented earlier in your writing. Moreover, the conclusions must be consistent with what the introduction promised the report would examine (its purpose) and how it would do so (its method).

For guidance about the location of the conclusion section in a report, see formal reports. For letter and other short closings, see correspondence and entries on specific types of documents throughout this book.

conjunctions

A conjunction connects words, phrases, or clauses and can also indicate the relationship between the elements it connects.

A *coordinating conjunction* joins two sentence elements that have identical functions. The coordinating conjunctions are *and*, *but*, *or*, *for*, *nor*, *yet*, and *so*.

- Nature *and* technology affect petroleum prices. [joins two nouns]
- To hear *and* to listen are two different things. [joins two phrases]
- I would like to include the survey, *but* that would make the report too long. [joins two clauses]

Coordinating conjunctions in the titles of books, articles, plays, and movies should not be capitalized unless they are the first or last word in the title.

- Our library contains *Consulting and Financial Independence* as well as *So You Want to Be a Consultant?*

Occasionally, a conjunction may begin a sentence; in fact, conjunctions can be strong transitional words and at times can provide emphasis. See also transition.

- I realize that the project is more difficult than expected and that you have encountered staffing problems. *But* we must meet our deadline.

Correlative conjunctions are used in pairs. The correlative conjunctions are *either . . . or*, *neither . . . nor*, *not only . . . but also*, *both . . . and*, and *whether . . . or.*

- The inspector will arrive *either* on Wednesday *or* on Thursday.

A *subordinating conjunction* connects sentence elements of different relative importance, normally independent and dependent clauses. The most frequently used subordinating conjunctions are *so*, *although*, *after*, *because*, *if*, *where*, *than*, *since*, *as*, *unless*, *before*, *that*, *though*, and *when.*

- I left the office *after* finishing the report.

A *conjunctive adverb* has the force of a conjunction because it joins two independent clauses. The most common conjunctive adverbs are *however*, *moreover*, *therefore*, *further*, *then*, *consequently*, *besides*, *accordingly*, *also*, and *thus.*

- The engine performed well in the laboratory; *however,* it failed under road conditions.

connotation / denotation

The *denotations* of a word are its literal meanings, as defined in a dictionary. The *connotations* of a word are its meanings and associations beyond its literal definition. For example, the denotations of *Hollywood* are "a district of Los Angeles" and "the U.S. movie industry as a whole"; its connotations are for many "romance, glittering success, and superficiality."

Often words have particular connotations for readers within professional groups and organizations. Choose words with both the most accurate denotations and the most appropriate connotations for the context. See also defining terms and word choice.

consensus

C

Because *consensus* means "harmony of opinion" among most of those in a group, the phrases *consensus of opinion* and *general consensus* defeat conciseness. The word *consensus* can be used to refer only to a group, never to one or two people.

- The ~~general~~ consensus ~~of opinion~~ among scientists is that the drug is ineffective.

context

Simply put, context is the environment or circumstances in which writers produce documents and within which readers interpret their meanings. Everything is written in a context, as illustrated in many entries and examples throughout this book. This entry considers specifically the effect of context on documents and suggests how you can be aware of it as you write.

The context for any document, such as a proposal or report, is determined by interrelated events or circumstances both inside and outside an organization.* For example, when you write a proposal to fund a project within your company, the economic condition of that company is part of the context that will determine how your proposal is received. If the company has recently laid off a dozen employees, the management may not be inclined to approve a proposal to expand into a new market—regardless of how well the proposal is written.

When you correspond with someone, the events that prompted you to write shape the context of the message and will affect what you say and how you say it. If you write to a customer in response to a complaint, for example, the tone and approach of your adjustment letter will be determined by the context—what you find when you investigate the issue. Is your company fully or partly at fault? Has the customer incorrectly used a product? contributed to a problem? The answers to such questions reveal the context. Or, if you write a manual for auto service technicians, other questions will reveal the context. What are the lighting and other physical conditions in garages? Will these physical conditions affect the layout and design of the manual? What potentially dangerous situations might the technicians encounter?

*For a detailed illustration of specific documents that have been affected by factors within particular contexts, see Linda Driskill, "Understanding the Writing Context in Organizations," in *Central Works in Technical Communication*, ed. Johndan Johnson-Eilola and Stuart Selber (New York: Oxford UP, 2004), 55–69.

C

Each time you write, you should keep in mind and even visualize the context so that you are sure your document will achieve its purpose. The following questions are starting points to help you become aware of the context, how it will influence your approach and your readers' interpretation of what you have written, and how it will affect the decisions you need to make during the writing process. See also "Five Steps to Successful Writing."

- What is your professional relationship with your readers and how might that affect the tone, style, and scope of your writing?
- What is "the story" behind the immediate reason you are writing; that is, what series of events or perhaps previous documents led to your need to write?
- What is the preferred medium of your readers? (See selecting the medium.)
- What specific factors, such as competition, finance, and regulation, are recognized within an organization or a department as important?
- What is the corporate culture in which your readers work, and what are its key values that you might find in its mission statement?
- What are the professional relationships among the specific readers who will receive a document?
- What current events within or outside an organization or a department may influence how readers interpret your writing?
- What national cultural differences might affect your readers' expectations or interpretations of the document? See global communication.

As these questions suggest, context is very specific each time you write and often involves, for example, the history of a specific organization or your past dealings with individual readers.

Context and Openings

Because context is so important, it is often helpful to remind your reader in some way of the context for your writing, as in the following opening for a cover letter to a proposal.

- During our meeting on improving quality last week, you mentioned that we have in the past required usability testing only for documents going to high-profile clients because of the costs involved. As I considered that practice, the idea occurred to me that

> we might try less extensive usability testing for many of our other clients. Because you suggested that you would welcome new ideas, I have proposed in the attached document a method of limited usability testing for a broad range of clients in order to improve overall quality.

Of course, as described in introductions, providing context for a reader may require only a brief background statement or short reminder.

- Several weeks ago a manager noticed a recurring problem in the software developed by Datacom Systems. Specifically, error messages repeatedly appeared when, in fact, no specific trouble . . .
- Jane, as I promised in my e-mail yesterday, I've attached the personnel budget estimates for the next fiscal year.

As the last example suggests, always provide some context when you send an e-mail with attachments.

continual / continuous

Continual implies "happening over and over" or "frequently repeated." (Writing well requires *continual* practice.) *Continuous* implies "occurring without interruption" or "unbroken." (The *continuous* roar of the machinery was deafening.)

contractions

A contraction is a shortened spelling of a word or phrase with an apostrophe substituting for the missing letter or letters (*cannot/can't; have not/haven't; will not/won't; it is/it's*). Contractions are often used in speech and informal writing; they are generally not appropriate in reports, proposals, and formal correspondence. See also technical writing style.

copyright

Copyright establishes legal protection for literary, dramatic, musical, artistic, and other intellectual works in printed or electronic form; it gives the copyright owner exclusive right to reproduce, distribute, perform, or display a work. Copyright protects all original works from the moment of their creation, regardless of whether they are published or contain a notice of copyright (©).

ETHICS NOTE If you plan to reproduce copyrighted material in your own publication or on your Web site, you must obtain permission from the copyright holder. To do otherwise is a violation of U.S. law. ✦

Permissions

To seek permission to reproduce copyrighted material, you must write to the copyright holder. In some cases it is the author; in other cases it is the editor or publisher of the work. For Web sites, read the site's "terms-of-use" information (if available) and e-mail your request to the appropriate party. State specifically which portion of the work you wish to reproduce and how you plan to use it. The copyright holder has the right to charge a fee and specify conditions and limits of use.

Exceptions

Some copyrighted material may be reproduced without permission. For example, a small amount of material from a copyrighted source may be used, especially for educational purposes, without permission or payment as long as the use satisfies the "fair-use" criteria, as described at the U.S. Copyright Office Web site, <www.copyright.gov/>.

In the workplace, employees often borrow from in-house manuals, reports, and other company documents. Because the company is considered the author of works prepared by its employees on the job, using such boilerplate material is not a violation of copyright.

Works created by or for U.S. government agencies are in the public domain—that is, they are not copyrighted. Such works, including text and visuals, can be reproduced without permission unless the information is classified or otherwise protected. As with other material you use in your own work, however, you should give appropriate credit to the source from which material is taken, as described in plagiarism and documenting sources.

correspondence

DIRECTORY

The process of writing letters or memos (or e-mail that functions as either) involves many of the same steps that go into writing most other documents, as described in "Five Steps to Successful Writing." One important consideration in correspondence is the impression you convey to readers. To convey a professional image—of yourself and your company or organization—take particular care with the tone and style of your writing.

Writing Style and Accuracy

Correspondence may vary from an informal (or casual) style that you might use with a colleague you know well to a formal (or restrained) style that would be appropriate to a client you do not know.

INFORMAL It worked! The new process is better than we had dreamed.

RESTRAINED You will be pleased to know that the new process is more effective than we had expected.

You will probably find yourself using the restrained style more frequently than the casual one. Remember that an overdone attempt to sound casual or friendly can sound insincere. However, do not adopt so formal a style that your writing reads like a legal contract. See affectation.

AFFECTED Per our dialogue yesterday, we no longer possess an original copy of the brochure requested. Please be advised that a PDF copy is attached to this e-mail.

IMPROVED We are out of original copies of the brochure we discussed yesterday, so I am attaching a PDF copy to this e-mail.

The improved version is both less stuffy and more concise. Do not be so concise, however, that you become blunt. Responding to a vague written request with "Your request was unclear" or "I don't understand" could offend your reader. Instead, ask for more information and establish goodwill to encourage your reader to provide the information.

- I will need more information before I can answer your request. Specifically, can you give me the title and the date of the report you are looking for?

Although this version is a bit longer, it is more tactful and will elicit a faster response. See also telegraphic style.

Check for accuracy: Facts, figures, and dates that are incorrect or misleading may cost time, money, and goodwill. Incorrect punctuation or grammar and unconventional usage can undermine your credibility. Remember that when you sign a letter, initial a memo, or press the "send" key for an e-mail, you are accepting responsibility for it. Therefore, allow yourself time to review correspondence carefully before sending it.

Audience: Tone and Goodwill

Correspondence is always more personal than reports or other forms of workplace writing because it is written directly to another person. To achieve a conversational style, imagine your reader sitting across the desk from you and write to the reader as if you were talking face-to-face.

Take into account your reader's needs and feelings. Ask yourself, "How might I feel if I received this letter?" and then tailor your message accordingly. Remember, an impersonal and unfriendly letter to a customer or client can tarnish the image of you and your organization, but a thoughtful and sincere letter can enhance it. Suppose, for example, you received a request from a client who forgot to enclose the special registration page with a damaged software program CD. In a response to that client, you might write the following:

- We must receive a copy of your registration page before we can issue a new NTX Program CD.
  [The writer's needs are emphasized: "*We* must."]

If you consider how you might keep the client's goodwill, you might word the request this way:

- Please mail or fax the registration page so that we can issue a new NTX Program CD.
  [This is polite, but the writer's needs are emphasized: "so that *we* can issue."]

You can put the reader's needs and interests foremost by writing from the reader's perspective. Often, doing so means using the words *you* and *your* rather than *we, our, I,* and *mine*—a technique called the "you" viewpoint. Consider the following revision:

- So you can receive your new NTX Program CD promptly, please mail or fax the registration page.
  [The reader's needs are emphasized with *you* and *your.*]

This revision stresses the reader's benefit and interest. By emphasizing the reader's needs, the writer will be more likely to accomplish the purpose: to get the reader to act.

Keep in mind that, if overdone, goodwill and the "you" viewpoint can produce writing that is fawning and insincere. Messages that are full of excessive praise and inflated language may be ignored—or even resented—by the reader.

EXCESSIVE You are just the kind of client that deserves the finest service that anyone can offer—and you deserve our best deal. Knowing how careful you are at making decisions, I know you'll think about the advantages of using our consulting service.

REASONABLE From our earlier correspondence, I can understand your need for reliable service—we strive to give all our priority clients our full attention. After you have reviewed our proposal, I am confident you will appreciate our "five-star" consulting option.

Writer's Checklist: Using Tone to Build Goodwill

Use the following guidelines to achieve a tone that builds goodwill with your recipients.

☑ Be respectful, not demanding.

DEMANDING Submit your answer in one week.

RESPECTFUL I would appreciate your answer within one week.

☑ Be modest, not arrogant.

ARROGANT My attached report is thorough, and I'm sure that you won't be able to continue without it.

MODEST The attached report contains details of the refinancing options, and I hope you find it useful.

☑ Be polite, not sarcastic.

C

Writer's Checklist: Using Tone to Build Goodwill (continued)

SARCASTIC I just now received the shipment we ordered six months ago. I'm sending it back—we can't use it now. Thanks a lot!

POLITE I am returning the shipment we ordered on March 12. Unfortunately, it arrived too late for us to be able to use it.

☑ Be positive and tactful, not negative and condescending.

NEGATIVE Your complaint about our prices is way off target. Our prices are definitely not any higher than those of our competitors.

TACTFUL Thank you for your suggestion concerning our prices. We believe, however, that our prices are competitive with, and in some cases are below, those of our competitors.

Good-News and Bad-News Patterns

Although the relative directness of correspondence may vary, it is generally more effective to present good news directly and bad news indirectly, especially if the stakes are high.* This principle is based on the fact that readers form their impressions and attitudes very early and that you as a writer may want to subordinate the bad news to reasons that make the bad news understandable. Further, if you are writing international correspondence, be aware that far more cultures are generally indirect in business messages than are direct.

Consider the thoughtlessness and direct rejection in Figure C–8. Although the message is concise and uses the pronouns *you* and *your*,

Dear Ms. Mauer:

Your application for the position of Cardiovascular Technician at Southtown Cardiac Associates has been rejected. We have found someone more qualified than you.

Sincerely,

FIGURE C–8. A Poor Bad-News Message

*Gerald J. Alred, "'We Regret to Inform You': Toward a New Theory of Negative Messages," in *Studies in Technical Communication,* ed. Brenda R. Sims (Denton: University of North Texas and NCTE, 1993), 17–36.

the writer does not consider how the recipient is likely to feel as she reads the rejection. Its pattern is (1) the bad news, (2) an explanation, and (3) the closing.

A better general pattern for bad news is (1) an opening that provides context (often called a "buffer"), (2) an explanation, (3) the bad news, and (4) a goodwill closing. (See also refusal letters.) The opening introduces the subject and establishes a professional tone. The body provides an explanation by reviewing the facts that make the bad news understandable. Although bad news is never pleasant, information that either puts the bad news in perspective or makes it seem reasonable maintains goodwill between the writer and the reader. The closing should reinforce a positive relationship through goodwill or helpful information. Consider, for example, the courteous rejection shown in Figure C–9. It carries the same disappointing news as does the letter in Figure C–8, but the writer is careful to thank the reader for her time and effort, explain why she was not accepted for the job, and offer her encouragement.

Presenting good news is, of course, easier. Present good news in your opening. By presenting the good news first, you increase the likelihood that the reader will pay careful attention to details, and you achieve goodwill from the start. The pattern for good-news messages should be (1) a good-news opening, (2) an explanation of facts, and (3) a goodwill closing. Figure C–10 is an example of an effective good-news message.

Openings and Closings

To focus the relevance of any correspondence for the reader, identify your subject in the opening.

- Yesterday, I received your letter and the pager, number AJ 50172. I sent the pager to our quality-control department for tests.

Your closing should let the reader know what he or she should do next and reinforce goodwill.

- Thanks again for the report, and let me know if you want me to send you a copy of the test results.

Because a closing is in a position of emphasis, consider it carefully. Before you simply use a routine closing (If you have further questions, please let me know), consider how you might make your closing work for you. It may be helpful to provide prompts to which the reader can respond.

- If you would like further information, such as a copy of the questionnaire we used, please e-mail me at delgado@prn.com.

See also conclusions and introductions.

C

Southtown Cardiac Associates
3221 Ryan Road San Diego, CA 92217
Phone: (714) 321-1579 Fax: (714) 321-1222

November 6, 2006

Ms. Barbara L. Mauer
157 Beach Drive
San Diego, CA 92113

Dear Ms. Mauer:

Context (or "buffer")

Thank you for your time and effort in applying for the position of Cardiovascular Technician at Southtown Cardiac Associates.

Explanation leading to bad news

Because we need someone who can assume the duties here with a minimum of training, we have selected an applicant with several years of experience.

Goodwill

Your college record is excellent, and I encourage you to apply for another position with us in the future.

Sincerely,

Mary Hernandez

Mary Hernandez
Office Manager

FIGURE C–9. A Courteous Bad-News Message

Format and Design

Although word-processing software provides templates for correspondence, it may not provide specific dimensions and spacing. To achieve a professional appearance, center the letter on the page vertically and horizontally. Although one-inch margins are the default standard in many word-processing programs, it is more important to establish a

C

Southtown Cardiac Associates
3221 Ryan Road San Diego, CA 92217
Phone: (714) 321-1579 Fax: (714) 321-1222

November 6, 2006

Ms. Barbara L. Mauer
157 Beach Drive
San Diego, CA 92113

Dear Ms. Mauer:

Good news

Please accept our offer of the position of Cardiovascular Technician at Southtown Cardiac Associates at the salary of $51,350.

Explanation

If the terms we discussed in the interview are acceptable to you, please come in at 9:30 a.m. on November 15. At that time, we will ask you to complete our benefits form, in addition to . . .

Goodwill

I, as well as the others in the office, look forward to working with you. Everyone was favorably impressed with you during your interview.

Sincerely,

Mary Hernandez

Mary Hernandez
Office Manager

FIGURE C–10. A Good-News Message

picture frame of blank space surrounding the page of text. When you use organizational letterhead stationery, consider the bottom of the letterhead as the top edge of the paper. The right margin should be approximately as wide as the left margin. To give a fuller appearance to very short letters, increase both margins to about an inch and a half.

C

Use your computer's full-page or print-preview feature to check for proportion.

The two most common formats for business letters are the full-block style shown in Figure C–11 and the modified-block style shown in Figure C–12. In the *full-block style*, which should be used only with letterhead, the entire letter is aligned at the left margin. In the *modified-block style*, the return address, date, and complimentary closing begin at the center of the page and the other elements are aligned at the left margin. All other letter styles are variations of the full-block and modified-block styles.

If your employer requires a particular format, use it. Otherwise, follow the guidelines provided here, and review the examples shown in Figures C–11 and C–12.

Heading. Place your full return address and the date in the heading. Because your name appears at the end of the letter, it need not be included in the heading. Spell out words like *street*, *avenue*, *first*, and *west* rather than abbreviating them. You may either spell out the name of the state in full or use the standard Postal Service abbreviation. The date usually goes directly beneath the last line of the return address. Do not abbreviate the name of the month. Begin the heading about two inches from the top of the page. If you are using company letterhead that gives the address, enter only the date, three lines below the last line of printed copy.

Inside Address. Include the recipient's full name, title, and address in the inside address, two to six lines below the date, depending on the length of the letter. The inside address should be aligned with the left margin, and the left margin should be at least one inch wide.

Salutation. Place the salutation, or greeting, two lines below the inside address and align it with the left margin. In most business letters, the salutation contains the recipient's personal title (such as *Mr.*, *Ms.*, *Dr.*) and last name, followed by a colon. If you are on a first-name basis with the recipient, use only the first name in the salutation.

Address women as *Ms.*, unless they have expressed a preference for *Miss* or *Mrs.* However, professional titles (such as *Professor*, *Senator*, *Major*) take precedence over *Ms.* If you do not know whether the recipient is a man or a woman, use a title appropriate to the context of the letter. The following are examples of the kinds of salutations you may find suitable: *Dear Customer*, *Dear Colleague*, *Dear IT Professional.*

If you are writing to a large company and do not know the name or title of the recipient, you may address the letter to an appropriate department or identify the subject in a subject line and use no salutation.

Letterhead	520 Niagara Street Braintree, MA 02184 Phone: (781) 787-1175 Fax: (781) 787-1213 E-mail: mail@evans.com 
Date	May 16, 2006
Inside address	Mr. George W. Nagel Director of Operations Boston Transit Authority 57 West City Avenue Boston, MA 02210
Salutation	Dear Mr. Nagel:
Body	Enclosed is our final report evaluating the safety measures for the Boston Intercity Transit System. We believe that the report covers the issues you raised in our last meeting and that you will be pleased with the results. However, if you have any further questions, we would be happy to meet with you again at your convenience. We would also like to express our appreciation to Mr. L. K. Sullivan of your committee for his generous help during our trips to Boston.
Complimentary close	Sincerely,
Signature	*Carolyn Brown*
Typed name Title	Carolyn Brown, Ph.D. Director of Research
End notations	CB/ls Enclosure: Final Safety Report cc: ITS Safety Committee Members

FIGURE C–11. Full-Block-Style Letter (with Letterhead)

C

Center

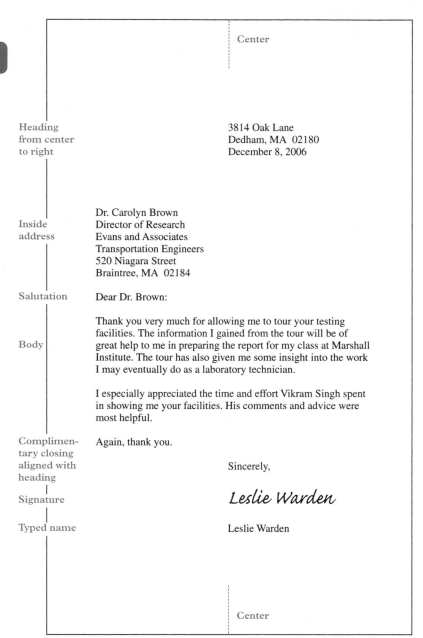

Heading from center to right

3814 Oak Lane
Dedham, MA 02180
December 8, 2006

Inside address

Dr. Carolyn Brown
Director of Research
Evans and Associates
Transportation Engineers
520 Niagara Street
Braintree, MA 02184

Salutation

Dear Dr. Brown:

Body

Thank you very much for allowing me to tour your testing facilities. The information I gained from the tour will be of great help to me in preparing the report for my class at Marshall Institute. The tour has also given me some insight into the work I may eventually do as a laboratory technician.

I especially appreciated the time and effort Vikram Singh spent in showing me your facilities. His comments and advice were most helpful.

Again, thank you.

Complimentary closing aligned with heading

Sincerely,

Signature

Leslie Warden

Typed name

Leslie Warden

Center

FIGURE C–12. Modified-Block-Style Letter (without Letterhead)

C

- National Business Systems
  501 West National Avenue
  Minneapolis, MN 55407

  Attention: Customer Relations Department

  I am returning three pagers that failed to operate. . . .

- National Business Systems
  501 West National Avenue
  Minneapolis, MN 55407

  Subject: Defective Parts for SL-100 Pagers

  I am returning three pagers that failed to operate. . . .

When a person's first name could be either feminine or masculine, one solution is to use both the first and last names in the salutation (*Dear Pat Smith:*). Avoid "To Whom It May Concern" because it is impersonal and dated.

For multiple recipients, the following salutations are appropriate:

- Dear Professor Allen and Dr. Rivera: [two recipients]
- Dear Ms. Becham, Ms. Moore, and Mr. Stein: [three recipients]
- Dear Colleagues: [*Members*, or other suitable collective term]

Body. The body of the letter should begin two lines below the salutation (or any element that precedes the body, such as a subject or an attention line). Single-space within paragraphs, and double-space between paragraphs. To provide a fuller appearance to a very short letter, you can increase the side margins or increase the font size. You can also insert extra space above the inside address, the typed (signature) name, and the initials of the person keying the letter—but do not exceed twice the recommended space for each of these elements.

Complimentary Closing. Type the complimentary closing two spaces below the body. Use a standard expression like *Sincerely*, *Sincerely yours*, or *Yours truly.* (If the recipient is a friend as well as a professional associate, you can use a less formal closing, such as *Best wishes* or *Best regards* or, simply, *Best.*) Capitalize only the initial letter of the first word, and follow the expression with a comma. Place your full name four lines below, aligned with the closing. On the next line include your title, if it is appropriate to do so. Sign the letter in the space between the complimentary closing and your name.

Second Page. If a letter requires a second page, always carry at least two lines of the body text over to that page. Use plain (nonletterhead) paper of quality equivalent to that of the letterhead stationery for the second page. It should have a header with the recipient's name, the page number, and the date. The header can go in the upper left-hand corner or across the page, as shown in Figure C–13.

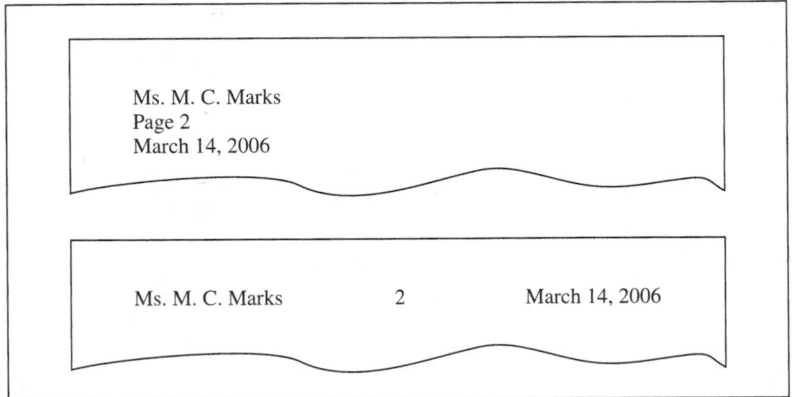

Ms. M. C. Marks
Page 2
March 14, 2006
Ms. M. C. Marks
2
March 14, 2006

FIGURE C–13. Headers for the Second Page of a Letter

End Notations. Business letters sometimes require additional information that is placed at the left margin, two spaces below the typed name and title of the writer in a long letter, four spaces below in a short letter.

Reference initials identify the person keying the letter if that person is not the writer. Show the letter writer's initials in capital letters, followed by a slash mark, and then the initials of the person keying the letter in lowercase letters, as shown in Figure C–11. When the writer is also the person keying the letter, no initials are needed.

Enclosure notations indicate that the writer is sending material along with the letter (an invoice, an article, and so on). Note that you must mention the enclosure in the body of the letter. Enclosure notations may take several forms:

- Enclosure: Final Safety Report
- Enclosures (2)
- Enc. or Encs.

Copy notation (cc:) tells the reader that a copy of the letter is being sent to the named recipients (see Figure C–11). Use a blind-copy notation (bcc:) when you do not want the addressee to know that a copy is being sent to someone else. A blind-copy notation appears only on the copy, not on the original (bcc: Dr. Brenda Shelton).

For additional details on letter format and design, you may wish to consult a guide such as *The Gregg Reference Manual* (Tenth Edition) by William A. Sabin.

Writer's Checklist: Writing Correspondence

- ☑ Establish your purpose, analyze your reader's needs, determine your scope, and consider the context.
- ☑ Prepare an outline, even if it is only a list of points to be covered in the order you want to cover them (see outlining).
- ☑ Write the first draft (see writing a draft).
- ☑ Allow for a cooling-off period prior to revision or seek a colleague's advice, especially for correspondence that addresses a problem.
- ☑ Revise the draft, checking for key problems in clarity and coherence.
- ☑ Use the appropriate or prescribed format.
- ☑ Check for accuracy: Make sure that all facts, figures, and dates are correct.
- ☑ Check for appropriate punctuation and use effective proofreading techniques.
- ☑ Remember that when you sign a letter, initial a memo, or send an e-mail, you are accepting responsibility for it.

cover letters

A cover letter (sometimes called a *transmittal*) accompanies a report, a proposal, an electronic file, or other material that you send to someone by mail, fax, or e-mail. It identifies an item that is being sent, the person to whom it is being sent, and the reason that it is being sent; it provides a permanent record for both the writer and the reader. For cover letters to résumés, see application letters.

Your opening should explain what is being sent and, if necessary, and why. Then you might highlight or briefly summarize the information enclosed or attached. A cover letter for a proposal, for example, might point out sections in the proposal of particular interest to the reader and go on to present a key point or two explaining why the writer's firm is the best one to do the job. This paragraph could also mention the conditions under which the proposal was prepared, such as limitations of time or budget. The closing paragraph might contain acknowledgments, offer additional assistance, or express the hope that the material will fulfill its purpose.

C

The example in Figure C–14 is brief, to the point, and slightly informal because the writer understands that this reader needs no further explanation of the attached document.

From: Susan Wong <swong@omni.com>
To: Mario Espinoza <mespinoza@lcp.com>
Sent: Tuesday, May 23, 2006 10:47 AM
Subject: Final Report for Installation of Pollution-Control Equipment
Attachment: Eastern Installation Report.doc

Dear Mario:

Attached is the final report on our installation of pollution-control equipment at Eastern Chemical Company, which we send with Eastern's permission. Please call me at the number below (x1206) or reply to this message if I can answer any questions.

Sincerely,

Susan

Susan Wong, Ph.D.
Technical Services Manager
Ecology Systems and Services
Technical Services = 800-351-2121

FIGURE C–14. Brief Cover Letter (Sent as E-mail)

The example in Figure C–15 is a bit more detailed and formal; it touches on the manner in which the information was gathered. In this message, the writer wishes to provide context to explain a delay as well as to acknowledge those who helped prepare the report. See also correspondence.

credible / creditable

Something is *credible* if it is believable. (The statistics in this report are *credible*.) Something is *creditable* if it is worthy of praise or credit. (The lead engineer did a *creditable* job.)

Ecology Systems
P.O. Box 413
Lansing, MI 48824
www.ecology.com
800-351-2121

May 23, 2006

Roger Hammersmith, Vice President
Waterford Paper Company
P.O. Box 413
Waterford, WI 53474

Dear Mr. Hammersmith:

Enclosed is the final report on our installation of your Ecology Contaminant Control System. As the report indicates, this system meets all state and federal guidelines and is 60 percent more cost-effective than the system it replaces.

The report is the result of several meetings with Diana Biel, Manager of ESS Pollution Control Division, and her staff as well as an extensive study of all your related systems. The report was delayed by the transfer of key staff in Building A. We believe, however, that you will find the report useful not only in understanding your particular installation but also . . .

We would like to thank you for the opportunity to work with you and your excellent staff. If you have any questions or need more information, please call me or e-mail me at the address below.

Sincerely,

Philip Snow

Philip Snow, Manager
Pollution Control Division
psnow@ecology.com

Enclosure

FIGURE C–15. Long Cover Letter (for a Report)

criteria / criterion

C

Criterion is a singular noun meaning "an established standard for judging or testing." *Criteria* and *criterions* are both acceptable plural forms of *criterion*, but *criteria* is generally preferred.

critique

A *critique* is a written or an oral evaluation of something. Avoid using *critique* as a verb meaning "criticize."

- Please ^*prepare a*^ critique ^*of*^ his job description.

D

dangling modifiers

Phrases that do not clearly and logically refer to the correct noun or pronoun are called *dangling modifiers*. Dangling modifiers usually appear at the beginning of a sentence as an introductory phrase.

DANGLING *While eating lunch*, my computer crashed. [*Who* was eating lunch?]

CORRECT While *I* was eating lunch, my computer crashed.

Dangling modifiers can appear at the end of the sentence as well.

DANGLING The program gains efficiency by *eliminating the superfluous instructions*. [*Who* eliminates the superfluous instructions?]

CORRECT The program gains efficiency *when you* eliminate the superfluous instructions.

To correct a dangling modifier, add the appropriate subject to either the dangling modifier or the main clause.

DANGLING After finishing the research, the proposal was easy to write. [The appropriate subject is *I*, but it is not stated in either the dangling phrase or the main clause.]

CORRECT After *I* finished the research, the proposal was easy to write. [The pronoun *I* is now the subject of an introductory clause.]

CORRECT After finishing the research, *I* found the proposal easy to write. [The pronoun *I* is now the subject of the main clause.]

For a discussion of misplaced modifiers, see modifiers.

dashes

D

The dash (—) can perform all the punctuation duties of linking, separating, and enclosing. The dash, sometimes indicated by two consecutive hyphens, can also indicate the omission of letters. (Mr. A—admitted his error.)

Use the dash cautiously to indicate more emphasis, informality, or abruptness than the other punctuation marks would show. A dash can emphasize a sharp turn in thought.

- The project will end May 15—unless we receive additional funding.

A dash can indicate an emphatic pause.

- The project will begin—after we are under contract.

Sometimes, to emphasize contrast, a dash is used with *but*.

- We completed the survey quickly—*but* the results were not accurate.

A dash can be used before a final summarizing statement or before repetition that has the effect of an afterthought.

- It was hot near the ovens—steaming hot.

Such a statement may also complete the meaning of the clause preceding the dash.

- We try to speak as we write—or so we believe.

Dashes set off parenthetical elements more sharply and emphatically than commas. Unlike dashes, parentheses tend to reduce the importance of what they enclose. Compare the following sentences:

- Only one person—the president—can authorize such activity.
- Only one person, the president, can authorize such activity.
- Only one person (the president) can authorize such activity.

Dashes can be used to set off parenthetical elements that contain commas.

- Three of the applicants—John Evans, Rosalita Fontiana, and Kyong-Shik Choi—seem well qualified for the job.

The first word after a dash is capitalized only if it is a proper noun.

data / datum

In much informal writing, *data* is considered a collective singular noun. In formal and scholarly writing, however, *data* is generally used as a plural, with *datum* as the singular form. Base your decision on whether your readers should consider the data as a single collection or as a group of individual facts. Whatever you decide, be sure that your pronouns and verbs agree in number with the selected usage.

- The *data are* voluminous. *They indicate* a link between cigarette smoking and lung cancer. [formal]
- The *data is* now ready for evaluation. *It is* in the mail. [less formal]

See also agreement and English, varieties of.

dates

In the United States, full dates are generally written in the month-day-year format, with a comma preceding and following the year.

- November 30, 2015, is the payoff date.

Do not use commas in the day-month-year format, which is used in many parts of the world and by the U.S. military.

- Note that 30 November 2015 is the payoff date.

No commas are used if only the month and year or day are included.

- The target date of May 2007 is optimistic, so I would like to meet on March 4 to discuss our options.

When writing days of the month without the year, use the cardinal number (March 4) rather than the ordinal number (March 4th). Of course, in speech or presentations, use the ordinal number ("March fourth").

Avoid the strictly numerical form for dates (11/30/07) because the date is not always immediately clear, especially in international correspondence. Writing out the name of the month makes the entire date immediately clear to all readers.

Centuries often cause confusion with numbers because their spelled-out forms, which are not capitalized, do not correspond with their numeral designations. The twentieth century, for example, is the 1900s: 1900–1999.

When the century is written as a noun, do not use a hyphen.

- During the twentieth century, technology transformed business practices.

D

When the centuries are written as adjectives, however, use hyphens.

- Twenty-first-century technology relies on dependable power sources.

defective / deficient

If something is *defective,* it is faulty. (The wiring was *defective.*) If something is *deficient,* it is lacking or is incomplete in an essential component. (The patient's diet was *deficient* in calcium.)

defining terms

Defining key terms and concepts is often essential for clarity. Terms can be defined either formally or informally, depending on your purpose, your readers, and the context.

A *formal definition* is a form of classification. You define a term by placing it in a category and then identifying the features that distinguish it from other members of the same category.

TERM	CATEGORY	DISTINGUISHING FEATURES
An *annual* is	a plant	that completes its life cycle, from seed to natural death, in one growing season.

An *informal definition* explains a term by giving a more familiar word or phrase as a synonym.

- Plants have a *symbiotic,* or *mutually beneficial,* relationship with certain kinds of bacteria.

State definitions positively; focus on what the term *is* rather than on what it is not.

NEGATIVE	In a legal transaction, *real property* is not personal property.
POSITIVE	*Real property* is legal terminology for the right or interest a person has in land and the permanent structures on that land.

For a discussion of when negative definitions are appropriate, see definition method of development.

Avoid circular definitions, which merely restate the term to be defined and therefore fail to clarify it.

CIRCULAR *Spontaneous combustion* is fire that begins spontaneously.

REVISED *Spontaneous combustion* is the self-ignition of a flammable material through a chemical reaction.

In addition, avoid "is when" and "is where" definitions. Such definitions fail to include the category and are too indirect.

a medical procedure in which
- A *biopsy* is ~~when~~ ^ a tissue sample is removed for testing.

In technical writing, as illustrated in this example, definitions often contain the purpose ("testing") of what is defined (*biopsy*).

definite / definitive

Definite and *definitive* both apply to what is precisely defined, but *definitive* more often refers to what is complete and authoritative. (Once we receive a *definite* proposal, our attorney can provide a *definitive* legal opinion.)

definition method of development

DIRECTORY

Definition is often essential to clarity and accuracy. Although defining terms may be sufficient, at other times definitions need to be expanded through (1) extended definition, (2) definition by analogy, (3) definition by cause, (4) definition by components, (5) definition by exploration of origin, and (6) negative definition. See also methods of development.

D

Extended Definition

When more than a phrase or a sentence or two is needed to explain an idea, use an extended definition, which explores a number of qualities of the item being defined. How an extended definition is developed depends on your readers' needs and on the complexity of the subject. Readers familiar with a topic might be able to handle a long, fairly complex definition, whereas readers less familiar with a topic might require simpler language and more basic information.

The easiest way to give an extended definition is with specific examples. Examples give readers easy-to-picture details that help them see and thus understand the term being defined.

- Form, which is the shape of landscape features, can best be represented by both small-scale features, such as *trees* and *shrubs*, and by large-scale elements, such as *mountains* and *mountain ranges*.

Definition by Analogy

Another useful way to define a difficult concept, especially when you are writing for nonspecialists, is to use an analogy to link the unfamiliar concept with a simpler or more familiar one. An analogy can help the reader understand an unfamiliar term by showing its similarities with a more familiar term. In the following description of radio waves in terms of their length (long) and frequency (low), notice how the writer develops an analogy to show why a low frequency is advantageous.

- The low frequency makes it relatively easy to produce a wave having virtually all its power concentrated at one frequency. Think, for example, of a group of people lost in a forest. If they hear sounds of a search party in the distance, they all will begin to shout for help in different directions. Not a very efficient process, is it? But suppose all the energy that went into the production of this noise could be concentrated into a single shout or whistle. Clearly the chances that the group will be found would be much greater.

Definition by Cause

Some terms are best defined by an explanation of their causes. In the following example from a professional journal, a nurse describes an apparatus used to monitor blood pressure in severely ill patients. Called an *indwelling catheter*, the device displays blood-pressure readings on an oscilloscope and on a numbered scale. Users of the device, the writer explains, must understand what a *dampened wave form* is.

- The dampened wave form, the smoothing out or flattening of the pressure wave form on the oscilloscope, is usually caused by an obstruction that prevents blood pressure from being freely transmitted to the monitor. The obstruction may be a small clot or bit of fibrin at the catheter tip. More likely, the catheter tip has become positioned against the artery wall and is preventing the blood from flowing freely.

Definition by Components

Sometimes a formal definition of a concept can be made simpler by breaking the concept into its component parts. In the following example, the formal definition of *fire* is given in the first paragraph, and the component parts are given in the second.

FORMAL DEFINITION	Fire is the visible heat energy released from the rapid oxidation of a fuel. A substance is "on fire" when the release of heat energy from the oxidation process reaches visible light levels.
COMPONENT PARTS	The classic fire triangle illustrates the elements necessary to create fire: *oxygen*, *heat*, and *burnable material* (*fuel*). Air provides sufficient oxygen for combustion; the intensity of the heat needed to start a fire depends on the characteristics of the burnable material. A burnable substance is one that will sustain combustion after an initial application of heat to start the combustion.

Definition by Exploration of Origin

Under certain circumstances, the meaning of a term can be clarified and made easier to remember by an exploration of its origin. Medical terms, because of their sometimes unfamiliar Greek and Latin roots, benefit especially from an explanation of this type. Tracing the derivation of a word also can be useful when you want to explain why a word has favorable or unfavorable associations, particularly if your goal is to influence your reader's attitude toward an idea or an activity. See also persuasion.

- Efforts to influence legislation generally fall under the head of *lobbying*, a term that once referred to people who prowled the lobbies of houses of government, buttonholing lawmakers and trying to get them to take certain positions. Lobbying today is all of this, and much more, too. It is a respected—and necessary—activity. It tells the legislator which way the winds of public opinion are blowing, and it helps inform [legislators] of the implications of certain bills, debates, and resolutions [that they must face].

  —Bill Vogt, *How to Build a Better Outdoors*

Negative Definition

In some cases, it is useful to point out what something is not to clarify what it is. A negative definition is effective only when the reader is familiar with the item with which the defined item is contrasted. If you say "*x* is not *y*," your readers must understand the meaning of *y* for the explanation to make sense. In a crane operator's manual, for instance, a negative definition is used to show that, for safety reasons, a hydraulic crane cannot be operated in the same manner as a lattice boom crane.

- A hydraulic crane is *not* like a lattice boom crane [a friction machine] in one very important way. In most cases, the safe lifting capacity of a lattice boom crane is based on the *weight needed to tip the machine*. Therefore, operators of friction machines sometimes depend on signs that the machine might tip to warn them of impending danger. This practice is very dangerous with a hydraulic crane. . . .

  —*Operator's Manual* (Model W-180), Harnishfeger Corporation

description

The key to effective description is the accurate presentation of details, whether for simple or complex descriptions. In Figure D–1, notice that the simple description contained in the purchase order includes five specific details in addition to the part number.

Complex descriptions, of course, involve more details. In describing a mechanical device, for example, describe the whole device and its function before giving a detailed description of how each part works. The description should conclude with an explanation of how each part contributes to the functioning of the whole.

PURCHASE ORDER

PART NO.	DESCRIPTION	QUANTITY
IW 8421	Infectious-waste bags, 12″ × 14″, heavy-gauge polyethylene, red double closures with self-sealing adhesive strips	5 boxes containing 200 bags per box

FIGURE D–1. Simple Description

In descriptions intended for readers who are unfamiliar with the topic, details are crucial. For these readers, show or demonstrate (as opposed to "tell") primarily through the use of images and details. Notice the use of color, shapes, and images in the following description of a company's headquarters. The writer assumes that the reader knows such terms as *colonial design* and *haiku fountain.*

- Their corporate headquarters, which reminded me of a rural college campus, are located north of the city in a 90-acre lush green wooded area. The complex consists of five three-story buildings of colonial design. The buildings are spaced about 50 feet apart and are built in a U shape surrounding a reflection pool that frames a striking haiku fountain. . . .

You can also use analogy, as described in figures of speech, to explain unfamiliar concepts in terms of familiar ones, such as "U shape" in the previous example or "armlike block" in Figure D–2.

Visuals can be powerful aids in descriptive writing, especially when they show details too intricate to explain completely in words. The example in Figure D–2 uses an illustration to help describe a mechanical assembly for an assembler/repairperson. Note that the description concentrates on the number of pieces—their sizes, shapes, and dimensions—and on their relationship to one another to perform their function. It also specifies the materials of which the hardware is made. Because the description is illustrated (with identifying labels) and is intended for technicians who have been trained on the equipment, it does not require the use of "bridging devices" to explain the unfamiliar in relation to the familiar. Even so, an important term is defined (*chad*), and crucial alignment dimensions are specified (± .003″). The illustration is labeled appropriately and integrated with text of the description.

design (*see* **layout and design**)

despite / in spite of

Although there is no literal difference between *despite* and *in spite of, despite* suggests an effort to avoid blame.

- *Despite* our best efforts, the plan failed.
  [We are not to blame for the failure.]
- *In spite of* our best efforts, the plan failed.
  [We did everything possible, but failure overcame us.]

D

The *die block assembly* shown in Figure 1 consists of two machined block sections, eight code punch pins, and a feed punch pin. The larger section, called the die block, is fashioned of a hard, non-corrosive beryllium-copper alloy. It houses the eight code punch pins and the smaller feed punch pin in nine finely machined guide holes.

Number of features and relative size differences noted

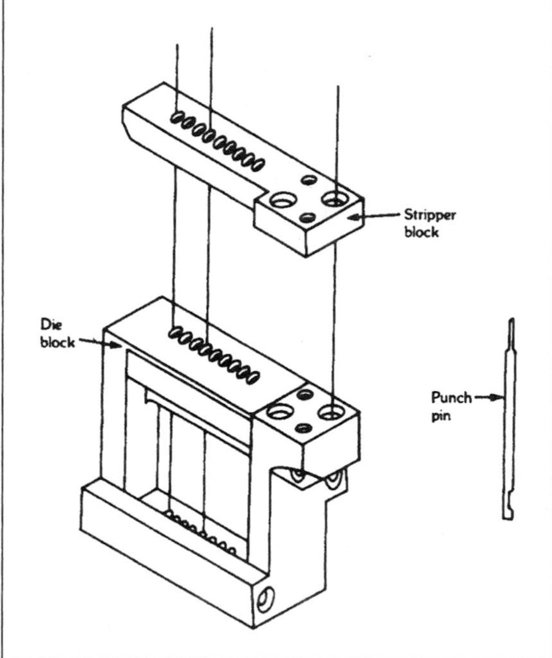

Figure 1. Die Block Assembly

The guide holes in the upper part of the die block are made smaller to conform to the thinner tips of the feed punch pins. Extending over the top of the die block and secured to it at one end is a smaller, armlike block called the *stripper block*. The stripper block is made from hardened tool steel, and it also has been drilled through with nine finely machined guide holes. It is carefully fitted to the die block at the factory so that its holes will be precisely above those in the die block and so that the space left between the blocks will measure .015" (± .003"). The residue from the punching operation, called chad, is pushed out through the top of the stripper block and guided out of the assembly by means of a plastic *residue chad collector* and *chad collector extender*.

Figure of speech

Definition

FIGURE D–2. Complex Description

Despite and *in spite of* (both meaning "notwithstanding") should not be blended into *despite of.*

In spite of
- ~~Despite of~~ our best efforts, the plan failed.

diagnosis / prognosis

Because they sound somewhat alike, these words are often confused with each other. *Diagnosis* means "an analysis of the nature of something" or "the conclusions reached by such analysis." (The meteorological *diagnosis* was that the pollution resulted from natural contaminants.) *Prognosis* means "a forecast or prediction." (The meteorologist's *prognosis* is that contaminant levels will increase over the next ten years.)

dictionaries

Dictionaries give more than just information about words' meanings. As illustrated in Figure D–3, they often provide words' etymologies (origins and history), forms, pronunciations, spellings, uses as idioms, and functions as parts of speech. For certain words, a dictionary lists synonyms and may also provide illustrations, if appropriate, such as tables, maps, photographs, and drawings. Dictionaries available on CD-ROM and the Web often include a human-voice pronunciation of each word. Excellent sites are also available that translate words from one language to another; one such site is "iTools" at <www.itools.com>.

Desk Dictionaries

Desk dictionaries are often abridged versions of larger dictionaries. There is no single "best" dictionary, but you should choose the most recent edition with upward of 125,000 entries. The following are considered good desk dictionaries.

The American Heritage College Dictionary, 4th ed. with CD-ROM, 2002

Microsoft Encarta World English Dictionary, free online edition at <www.encarta.com>, 2004

Merriam-Webster's Collegiate Dictionary, 11th ed. with CD-ROM, 2003

Random House Webster's College Dictionary, 2nd rev. and updated ed., 2000

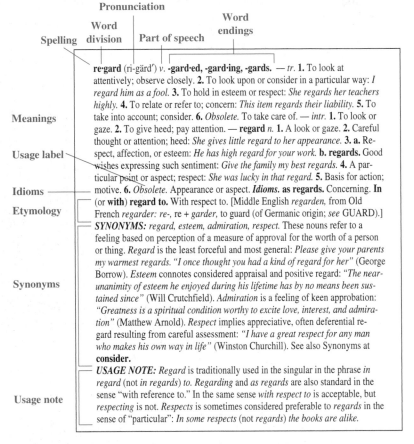

Pronunciation
Word endings
Spelling | Word division | Part of speech

re·gard (ri-gärd′) *v.* **-gard·ed, -gard·ing, -gards.** — *tr.* **1.** To look at attentively; observe closely. **2.** To look upon or consider in a particular way: *I regard him as a fool.* **3.** To hold in esteem or respect: *She regards her teachers highly.* **4.** To relate or refer to; concern: *This item regards their liability.* **5.** To take into account; consider. **6.** *Obsolete.* To take care of. — *intr.* **1.** To look or gaze. **2.** To give heed; pay attention. — **regard** *n.* **1.** A look or gaze. **2.** Careful thought or attention; heed: *She gives little regard to her appearance.* **3. a.** Respect, affection, or esteem: *He has high regard for your work.* **b. regards.** Good wishes expressing such sentiment: *Give the family my best regards.* **4.** A particular point or aspect; respect: *She was lucky in that regard.* **5.** Basis for action; motive. **6.** *Obsolete.* Appearance or aspect. ***Idioms.*** **as regards.** Concerning. **In** (or **with**) **regard to.** With respect to. [Middle English *regarden,* from Old French *regarder: re-,* re + *garder,* to guard (of Germanic origin; *see* GUARD).]

SYNONYMS: *regard, esteem, admiration, respect.* These nouns refer to a feeling based on perception of a measure of approval for the worth of a person or thing. *Regard* is the least forceful and most general: *Please give your parents my warmest regards. "I once thought you had a kind of regard for her"* (George Borrow). *Esteem* connotes considered appraisal and positive regard: *"The near-unanimity of esteem he enjoyed during his lifetime has by no means been sustained since"* (Will Crutchfield). *Admiration* is a feeling of keen approbation: *"Greatness is a spiritual condition worthy to excite love, interest, and admiration"* (Matthew Arnold). *Respect* implies appreciative, often deferential regard resulting from careful assessment: *"I have a great respect for any man who makes his own way in life"* (Winston Churchill). See also Synonyms at **consider.**

USAGE NOTE: *Regard* is traditionally used in the singular in the phrase *in regard* (not *in regards*) *to. Regarding* and *as regards* are also standard in the sense "with reference to." In the same sense *with respect to* is acceptable, but *respecting* is not. *Respects* is sometimes considered preferable to *regards* in the sense of "particular": *In some respects* (not *regards*) *the books are alike.*

Meanings
Usage label
Idioms
Etymology
Synonyms
Usage note

FIGURE D–3. Dictionary Entry

Unabridged Dictionaries

Unabridged dictionaries provide complete and authoritative linguistic information. The printed versions are impractical for desk use because of their size, and the CD and online subscription versions are expensive; libraries make available both versions of these important reference sources.

- *The Oxford English Dictionary*, 2nd ed., is the standard historical dictionary of the English language. Its 20 volumes contain over 500,000 words and give the chronological developments of over 240,000 words, providing numerous examples of uses and sources. It is also available on CD-ROM and online.

- *Webster's Third New International Dictionary, Unabridged*, CD-ROM edition, contains over 450,000 entries. Word meanings are listed in historical order, with the current meaning given last. This dictionary does not list biographical and geographical names, nor does it include usage information.

ESL Dictionaries

English-as-a-second-language (ESL) dictionaries are more helpful to the nonnative speaker than are regular English dictionaries or bilingual dictionaries. The pronunciation symbols in ESL dictionaries are based on the international phonetic alphabet rather than on English phonetic systems, and useful grammatical information is included in both the entries and special grammar sections. In addition, the definitions usually are easier to understand than those in regular English dictionaries; for example, a regular English dictionary defines *opaque* as "impervious to the passage of light," while an ESL dictionary defines the word as "not allowing light to pass through." The definitions in ESL dictionaries also are usually more thorough than those in bilingual dictionaries. For example, a bilingual dictionary might indicate that *obstacle* and *blockade* are synonymous—but not indicate that only *obstacle* can be used for abstract meanings. (Lack of money can be an *obstacle* [not a *blockade*] to a college education.)

The following dictionaries and references provide helpful information for nonnative speakers of English.

Longman Advanced American Dictionary with CD-ROM, 2002

Longman Dictionary of American English with CD-ROM, 2004

Longman American Idioms Dictionary with CD-ROM, Karen Stern, 2000

Oxford American Wordpower Dictionary for Learners of English, Ruth Urbom, 1998

Subject Dictionaries

For the meanings of words too specialized for a general dictionary, a subject dictionary is useful. Subject dictionaries define terms used in a particular field, such as medicine, geography, architecture, or consumer affairs. Definitions in subject dictionaries are generally more detailed and comprehensive than those found in general dictionaries. One well-known example is *Stedman's Medical Dictionary* (Lippincott Williams & Wilkins).

Although subject dictionaries are specialized and offer detailed definitions of field-specific terms, they are written in language that is straightforward enough to be understood by nonspecialists.

differ from / differ with

D

Differ from suggests that two things are not alike. (Our earlier proposal *differs from* the current one.) *Differ with* indicates disagreement between persons. (The architect *differed with* the contractor on the proposed site.)

different from / different than

In formal writing, the preposition *from* is used with *different.* (The Quantum PC is *different from* the Macintosh computer.) *Different than* is used when it is followed by a clause. (The job cost was *different than* we had estimated it.)

direct address

Direct address refers to a sentence or phrase in which the person being spoken or written to is explicitly named. It is often used in speech and in e-mail messages. Notice that the person's name in a direct address is set off by commas.

- *John*, call me as soon as you arrive at the airport.
- Call me, *John*, as soon as you arrive at the airport.

discreet / discrete

Discreet means "having or showing prudent or careful behavior." (Because the matter was personal, he asked Bob to be *discreet.*) *Discrete* means something is "separate, distinct, or individual." (Several *discrete* strands of copper wire form the cable.)

disinterested / uninterested

Disinterested means "impartial, objective, unbiased."

- Like good judges, scientists should be passionately interested in the problems they tackle but completely *disinterested* when they seek to solve those problems.

Uninterested means simply "not interested."

- Despite Asha's enthusiasm, her manager remained *uninterested* in the project.

division-and-classification method of development

An effective method of development for a complex subject is either to divide it into manageable parts and then discuss each part separately (division) or to classify (or group) individual parts into appropriate categories and discuss each category separately (classification). See also instructions and process explanation.

Division

You might use division to describe a physical object, such as the parts of a fax machine; to examine an organization, such as a company; or to explain the components of a system, such as the Internet. The emphasis in division as a method of development is on breaking down a complex whole into a number of like units—it is easier to consider smaller units and to examine the relationship of each to the other than to attempt to discuss the whole. The basis for division depends, of course, on your subject and your purpose.

If you were a financial planner describing the types of mutual funds available to your investors, you could divide the variety available into three broad categories: money market funds, bond funds, and stock funds. Such division would be accurate, but it would be only a first-level grouping of a complex whole. The three broad categories could, in turn, be subdivided into additional groups based on investment strategy, as follows:

Money market funds
- Taxable money market
- Tax-exempt money market

Bond funds
- Taxable bonds
- Tax-exempt bonds
- Balanced (mix of stocks and bonds)

Stock funds
- Balanced
- Equity-income
- Domestic growth
- Growth and income
- International growth
- Small-capitalization
- Aggressive growth
- Specialized

Specialized stock funds could be further subdivided as follows:

Specialized funds

- Communications
- Energy
- Environmental services
- Financial services
- Gold
- Health services
- Technology
- Utilities
- Worldwide capital goods

D

Classification

The process of classification is the grouping of a number of units (such as people, objects, or ideas) into related categories. Consider the following list:

triangular file	steel tape ruler	needle-nose pliers
vise	pipe wrench	keyhole saw
mallet	tin snips	C-clamps
rasp	hacksaw	plane
glass cutter	ball-peen hammer	steel square
spring clamp	claw hammer	utility knife
crescent wrench	folding extension ruler	slip-joint pliers
crosscut saw	tack hammer	utility scissors

To group the items in the list, you would first determine what they have in common. The most obvious characteristic they share is that they all belong in a carpenter's tool chest. With that observation as a starting point, you can begin to group the tools into related categories. Pipe wrenches belong with slip-joint pliers because both tools grip objects. The rasp and the plane belong with the triangular file because all three tools smooth rough surfaces. By applying this kind of thinking to all the items in the list, you can group (classify) the tools according to function (Figure D–4).

To classify a subject, you must first sort the individual items into the largest number of comparable groups. For explaining the functions of carpentry tools, the classifications (or groups) in Figure D–4 (smoothing, hammering, measuring, gripping, and cutting) are excellent. For recommending which tools a new homeowner should buy first, however, those classifications are not helpful—each group contains tools that a new homeowner might want to purchase right away. To give homeowners advice on purchasing tools, you probably would classify the types of repairs they most likely will have to do (plumbing, painting, etc.). That classification could serve as a guide to tool purchase.

Once you have established the basis for the classification, apply it consistently, putting each item in only one category. For example, it might seem logical to classify needle-nose pliers as both a tool that cuts and a tool that grips because most needle-nose pliers have a small section for cutting wires. However, the primary function of needle-nose

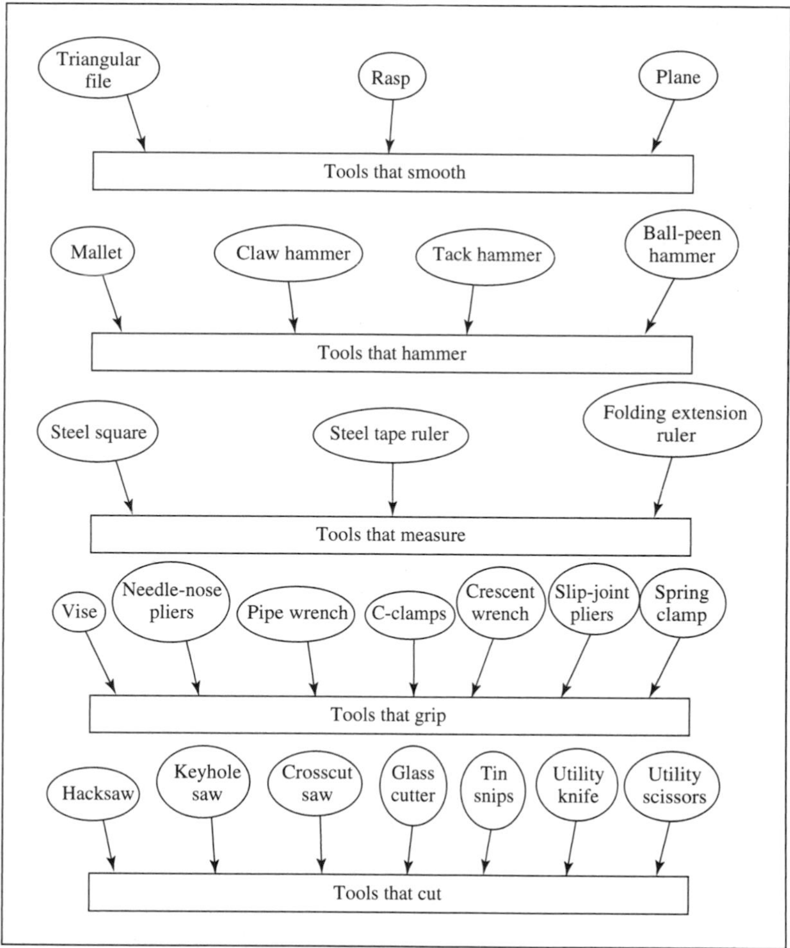

FIGURE D–4. Classification (Tools Placed into Categories)

pliers is to grip. So listing them only under "tools that grip" would be consistent with the basis used for listing the other tools.

documentation

The word *documentation* refers to the formal credit given to sources used or quoted in research papers, trade journal articles, and reports—that is, in documenting sources. (See also bibliographies, copyright, and plagiarism.) The term *documentation* may also refer to information that is recorded, or documented, on paper or in electronic files. The manuals and specifications that manufacturers provide to customers are examples of documentation. Even flowcharts and drawings are considered documentation.

D

documenting sources

DIRECTORY

Documenting sources achieves three important purposes:

- It allows readers to locate and consult the sources used and to find further information on the subject.
- It enables writers to support their assertions and arguments in such documents as proposals, reports, and trade journal articles.
- It helps writers to give proper credit to others and thus avoid plagiarism by identifying the sources of facts, ideas, quotations, and paraphrases. See also paraphrasing.

This entry shows citation models and sample pages for three principal documentation systems: APA, IEEE, and MLA. The following examples compare these three styles for citing a book by one author: *Project Management for the Technical Professional* by Michael S. Dobson, which was published in 2001 by Project Management Institute in Newtown Square, Pennsylvania.

- The American Psychological Association (APA) system of citation is often used in the social sciences. It is referred to as an author/date method of documentation because parenthetical in-text citations and a references list (at the end of the paper) in APA style emphasize the author(s) and date of publication so that the currency of the research is clear.

APA IN-TEXT CITATION

(Author's Last Name, Year)

(Dobson, 2001)

APA REFERENCES ENTRY

Author's Last Name, Initials. (Year). *Title in italics.* Place of Publication: Publisher.

Dobson, M. S. (2001). *Project management for the technical professional.* Newtown Square, PA: Project Management Institute.

- The system set forth in the *IEEE (Institute of Electrical and Electronics Engineers, Inc.) Standards Style Manual* is often used for the production of technical documents and standards in areas ranging from computer engineering, biomedical technology, and telecommunications to electric power, aerospace, and consumer electronics. The IEEE system is referred to as a number-style method of documentation because bibliographical reference numbers and a numbered bibliography identify the sources of information.

IEEE IN-TEXT CITATION

[Bibliographical reference number corresponding to a bibliography entry]

[B1]

IEEE BIBLIOGRAPHY ENTRY

[Bibliographical reference number]. Author's Last Name, First and Middle Initial (or Full First Name), *Title in Italics.* Place of Publication: Publisher, Date of Publication, Pages.

[B1]. Dobson, M. S., *Project Management for the Technical Professional.* Newtown Square, PA: Project Management Institute, 2001, 102–112.

- The Modern Language Association (MLA) system is used in the humanities. MLA style uses parenthetical in-text citations and a list of works cited and places greater importance on the pages on which cited information can be found than on the publication date.

MLA IN-TEXT CITATION

(Author's Last Name Page Number)

(Dobson 162)

MLA WORKS-CITED ENTRY

Author's Last Name, First Name. Title Underlined. Place of Publication: Publisher, Date of Publication.

Dobson, Michael S. Project Management for the Technical Professional. Newtown Square, PA: Project Management Institute, 2001.

These systems are described in full detail in the following style manuals:

American Psychological Association. *Publication Manual of the American Psychological Association*. 5th ed. Washington, D.C.: American Psychological Association, 2001. See also <www.apastyle.org>.

IEEE Standards Style Manual. 2005 ed. New York: IEEE, 2005. See also <http://standards.ieee.org/guides/style/>.

Gibaldi, Joseph. *MLA Handbook for Writers of Research Papers*. 6th ed. New York: Modern Language Association of America, 2003. See also <www.mla.org>.

For additional bibliographic advice and documentation models for types of sources not included in this entry, consult these style manuals or those listed at the end of this entry under "Other Style Manuals." See also bibliographies and research.

APA Documentation

APA In-Text Citations. APA parenthetical documentation within the text of a paper gives a brief citation—in parentheses—of the author, year of publication, and relevant page numbers if it helps locate a passage in a lengthy document.

- World War II was the occasion of radar's first application in warfare, and Great Britain led the way in radar research (Butrica, 2000).
- According to Butrica (2000), the use of radar as an offensive and a defensive warfare agent made World War II the "first electronic war" (p. 2).

When APA parenthetical citations are needed midsentence, place them after the closing quotation marks and continue with the rest of the sentence.

- The "first electronic war" (Butrica, 2000, p. 2) was fought as much in the research laboratory as on the battlefield.

If the APA parenthetical citation follows a block quotation, place it after the final punctuation mark.

- . . . a close collaboration with the physics and technology staff is essential. (Minsky, 2000)

When a work has two authors, cite both names joined by an ampersand: (Maddie & Khoshaba, 2005). For the first citation of a work with three, four, or five authors, include all names. For subsequent citations and for works with six or more authors, include only the last name of the first author followed by *et al.* (not italicized). When two or more works by different authors are cited in the same parentheses, list the citations alphabetically and use semicolons to separate them: (Maddie & Khoshaba, 2005; Prague, 2003).

D
APA

APA Documentation Models

BOOKS

Single Author

Van Grembergen, W. (2001). *Information technology evaluation methods and management*. New York: Idea.

Multiple Authors

Testerman, J. O., Kuegler, T. J., Jr., & Dowling, P. J., Jr. (2000). *Web advertising and marketing* (3rd ed.). Rocklin, CA: Prima.

Corporate Author

Ernst and Young. (2001). *Ernst and Young's retirement planning guide* (3rd ed.). New York: Wiley.

Edition Other Than First

Van Horne, J. C. (2000). *Financial market rates and flows* (6th ed.). New York: Prentice Hall.

Multivolume Work

Standard and Poor. (2004). *Standard and Poor's register of corporations, directors and executives* (Vols. 1–3). New York: McGraw-Hill.

Work in an Edited Collection

Griswold, C. L. (2002). Happiness and Cypher's choice: Is ignorance bliss? In W. Irwin (Ed.), *The Matrix and philosophy* (pp. 126–137). Chicago: Open Court.

Encyclopedia or Dictionary Entry

Gibbard, B. G. (2003). Particle detector. In *World Book encyclopedia* (Vol. 15, pp. 186–187). Chicago: World Book.

D
APA

ARTICLES IN PERIODICALS (*See also* Electronic Sources)

Magazine Article

Limp, F. (2001, June). Enterprise deployment. *Business Geographics*, *26*, 18–27.

Journal Article

Loza, J. (2004). Business-community partnerships: The case for community organization capacity building. *Journal of Business Ethics*, *53*, 297–311.

Newspaper Article

Loftus, P. (2004, August 31). Small business: Enterprise. *Wall Street Journal*, p. B4.

Article with an Unknown Author

CFOs remain optimistic on economy. (2004, September 27). *The Business Journal*, p. 33.

ELECTRONIC SOURCES

Entire Web Site

The APA recommends that at minimum, a reference to a Web source should provide a document title or description, a date (of the publication or retrieval of the document), an address (URL) that links directly to the document or section, and an author, whenever possible. On the rare occasion that you need to cite multiple pages of a Web site (or the entire site), provide a URL that links to the site's homepage.

Society for Technical Communication. (2002). Retrieved September 20, 2004, from http://www.stc.org [no period after URLs]

Document on a Web Site, with an Author

Locker, K. O. (2002). *The history of the ABC*. Retrieved September 20, 2004, from the Association for Business Communication Web site: http://www.businesscommunication.org/about/history/history.html

Document on a Web Site, with a Corporate Author

General Motors. (2003). *Company profile*. Retrieved September 20, 2004, from http://www.gm.com/company/corp_info/profiles/

Document on a Web Site, with an Unknown Author

Forgotten inventors. (2001). PBS Online. Retrieved September 20, 2004, from http://www.pbs.org/wgbh/amex/telephone/sfeature/index.html

Article or Other Work from a Database

Goldbort, R. C. (2001, March). Scientific writing as an art and as a science. *Journal of Environmental Health, 63*(7), pp. 22–25. Retrieved September 20, 2004, from Expanded Academic ASAP database.

Article in an Online Periodical

Tiernen, R. (2001, April 18). Waiting for wireless. *SmartMoney.com*. Retrieved September 20, 2004, from http://www.smartmoney.com/techmarket/index.cfm?story=20010418

E-mail

Personal communications (including e-mail and messages from discussion groups and electronic bulletin boards) are not cited in an APA reference list. They can be cited in the text as follows: "According to J. D. Kahl (personal communication, October 2, 2001), Web pages need to reflect. . . ."

Publication on CD-ROM

Money 2004. (2004). [CD-ROM]. Redmond, WA: Microsoft.

MULTIMEDIA SOURCES (Electronic and Print)

Map or Chart

Asia. (2001). [Map]. Maps.com. Retrieved April 20, 2001, from http://www.maps.com/explore/atlas/political/asia.html

Wisconsin. (2000). [Map]. Chicago: Rand.

Film or Video

Lawrence, D. (Director), & Christopher, J. (Editor). (2001). *Emergency film group video* [Video]. Retrieved April 20, 2001, from http://www.efilmgroup.com/video1.rm

Massingham, G. (Director), & Christopher, J. (Editor). (2003). *Introduction to hazardous chemicals* [Motion picture]. Edgartown, MA: Emergency Film Group.

Radio or Television Program

Norris, R. (Host). (2001, April 3). Energy supplies. *All things considered* [Radio broadcast]. Boston: WGBH. Retrieved April 20, 2001, from http://www.npr.org/programs/atc

Novak, R. (Host). (2001, April 16). Do Americans really want a tax cut? *CNN: Crossfire* [Television broadcast]. Washington, DC: CNN.

D

APA

OTHER SOURCES

Published Interview

Gates, B. (2000, April 17). The view from the very top [Interview]. *Newsweek, 135*, 36–39.

Personal Communications

Personal communications such as letters, interviews, e-mail, and messages from discussion groups and electronic bulletin boards are not cited in an APA reference list. They can be cited in the text as follows: "According to J. D. Kahl (personal communication, October 2, 2001), Web pages need to reflect. . . ."

Brochure or Pamphlet

Library of Congress. U.S. Copyright Office. (2004). *Copyright registration for online works* [Brochure]. Washington, DC: U.S. Government Printing Office.

Government Document

U.S. Department of Energy. (2004). *The August 14, 2003, blackout one year later* (Technical Publication No. 137-2004-12176). Washington, DC: U.S. Government Printing Office.

Report Published in a Collection

Jacobs, R. L. (2003). Structured on-the-job training: unleashing employee experience in the workplace (Report No. ED-474710) *ERIC*, 346082, 1–28.

Report Published Separately

Bertot, J. C., & McClure, C. R. (2000). *Public libraries and the Internet 2000: Summary findings and data tables*. Washington, DC: U.S. Government Printing Office.

Unpublished Data

Wisniewski, K., & Hussar, D. (2002). [Oregon small business statistics, by county]. Unpublished raw data.

APA Sample Pages

Shortened title and page number.

Ethics Cases 14

One-inch margins. Text double-spaced.

This report examines the nature and disposition of the 3,458 ethics cases handled companywide by CGF's ethics officers and managers during 2004. The purpose of such reports is to provide the Ethics and Business Conduct Committee with the information necessary for assessing the effectiveness of the first year of CGF's Ethics Program (Davis, Marks, & Tegge, 2001). According to Matthias Jonas (2004), recommendations are given for consideration "in planning for the second year of the Ethics Program" (p. 152).

The Office of Ethics and Business Conduct was created to administer the Ethics Program. The director of the Office of Ethics and Business Conduct, along with seven ethics officers throughout CGF, was given the responsibility for the following objectives, as described by Rossouw (1997):

Long quote indented five to seven spaces, double-spaced, without quotation marks.

> Communicate the values, standards, and goals of CGF's Program to employees. Provide companywide channels for employee education and guidance in resolving ethics concerns. Implement companywide programs in ethics awareness and recognition. Employee accessibility to ethics information and guidance is the immediate goal of the Office of Business Conduct in its first year. (p. 1543)

The purpose of the Ethics Program, established by the Committee, is to "promote ethical business conduct through open communication and compliance with company ethics standards" (Jonas, 2001, p. 89). To accomplish this purpose, any ethics policy must ensure confidentiality and anonymity for employees who raise genuine ethics concerns. The procedure developed at CGF guarantees

In-text citation gives name, date, and page number.

FIGURE D–5. APA Sample Page

D

APA

Heading centered.

List alphabetized by authors' last names and double-spaced.

Hanging-indent style used for entries.

Ethics Cases 21

References

Davis, W. C., Marks, R., & Tegge, D. (2001). *Working in the system: Five new management principles.* New York: St. Martin's Press.

Hassab, J. C. (1997). *Systems management: People, computers, machines, materials.* New York: CRC.

Jonas, M. (2001). Ethics in organizational communication: A review of the literature. *Journal of Ethics and Communication, 29,* 79–99.

Jonas, M. (2004). The Internet and ethical communication: Toward a new paradigm. *Journal of Ethics and Communication, 32,* 147–177.

Library of Congress. U.S. Copyright Office. (2004). *Copyright registration for online works* [Brochure]. Washington, DC: U.S. Government Printing Office.

The one-minute manager. (1998). [CD-ROM]. Boston: Bedford/St. Martin's.

Rossouw, G. J. (1997). Business ethics in South Africa. *Journal of Business Ethics, 16,* 1539–1547.

FIGURE D–6. APA Sample List of References

IEEE Documentation

IEEE In-Text Citations. IEEE bracketed bibliographical reference numbers within the text of a paper provide sequential numbers that correspond to the full citation in the bibliography: [A1], [A2], etc. The letter within the brackets specifies the annex, or appendix, where the bibliography is located (Annex A, Annex B, etc.). In IEEE style, the bibliography is always placed in an annex. Note that "Annex A" is usually reserved for standards (or *normative references*), whereas "Annex B" is reserved for typical in-text (or *informative*) citations.

- As Peterson writes, preparing a videotape of measurement methods is cost effective and can expedite training [B1].
- The results of these studies have led even the most conservative managers to adopt technologies that will "catapult the industry forward" [B2].

The first bibliographic reference cited in the document should be marked with a footnote that reads as follows.

- [1]The numbers in brackets correspond to those of the bibliography in Annex B.

IEEE Documentation Models

BOOKS

Single Author

[B1] Van Grembergen, W., *Information Technology Evaluation Methods and Management*. New York: Idea, 2001, pp. 230–292.

Multiple Authors

[B2] Testerman, J. O., Kuegler, T. J., and Dowling, P. J., *Web Advertising and Marketing*, 3rd ed. Rocklin, CA: Prima, 2000, pp. 105–112.

Corporate Author

[B3] Ernst and Young, *Ernst and Young's Retirement Planning Guide*, 3rd ed. New York: Wiley, 2001, pp. 16–22.

Edition Other Than First

[B4] Van Horne, J. C., *Financial Market Rates and Flows*, 6th ed. New York: Prentice Hall, 2000, pp. 33–91.

Multivolume Work

[B5] Standard and Poor, *Standard and Poor's Register of Corporations, Directors and Executives*, vol. 2. New York: McGraw-Hill, 2004, pp. 367–401.

Work in an Edited Collection

[B6] Griswold, C. L., "Happiness and Cypher's Choice: Is Ignorance Bliss?" In *The Matrix and Philosophy*, edited by William Irwin, 126–137. Chicago: Open Court, 2002, pp. 129–130.

Encyclopedia or Dictionary Entry

[B7] *World Book Encyclopedia*, 2003 ed., s.v. "particle detector."

ARTICLES IN PERIODICALS (*See also* Electronic Sources)

Magazine Article

[B8] Limp, F., "Enterprise deployment," *Business Geographics*, pp. 18–27, June 2001.

Journal Article

[B9] Loza, J., "Business-community partnerships: The case for community organization capacity building," *Journal of Business Ethics*, vol. 53, pp. 287–311, 2004.

Newspaper Article

[B10] Loftus, P., "Small business: Enterprise," *Wall Street Journal*, sec. B, 31 Aug. 2004.

Article with an Unknown Author

[B11] "CFOs remain optimistic on economy," *Business Journal*, p. 33, 27 Sept. 2004.

ELECTRONIC SOURCES

The *IEEE Standards Style Manual*, 2005 ed., does not include guidelines for documenting electronic or multimedia sources. Because the editors of the *Style Manual* suggest that readers consult the *Chicago Manual of Style* for more information on how to list various sources, the following models are based on the guidelines set forth in the *Chicago Manual of Style*, 15th ed. (See *Chicago Manual of Style* listing on page 155.)

Entire Web Site

[B12] Association for Business Communication. *Association for Business Communication*. 2004. http://www.businesscommunication.org.

Short Work from a Web Site, with an Author

[B13] Locker, K. O. "The History of the ABC." *Association for Business Communication*. 2002. http://www.businesscommunication.org/about/history/history.html.

Short Work from a Web Site, with a Corporate Author

[B14] General Motors. "Company Profile." *General Motors*. 2003. http://www.gm.com/company/corp_info/profiles/.

Short Work from a Web Site, with an Unknown Author

[B15] "Forgotten Inventors." *PBS Online*. 2001. http://www.pbs.org/wgbh/amex/telephone/sfeature/index.html.

Article or Other Work from a Database

[B16] Goldbort, R. C. "Scientific Writing as an Art and as a Science." *Journal of Environmental Health* 63, no. 7 (2001): 21–25. http://infotrac.galegroup.com/.

Article in an Online Periodical

[B17] Tiernen, R. "Waiting for Wireless." *SmartMoney.com.* http://www.smartmoney.com/techmarket/index.cfm?story=20010418.

E-mail

The *IEEE Standards Style Manual* does not provide advice for citing personal communications such as conversations, letters, and e-mail; the *Chicago Manual of Style*, 15th edition, suggests that personal communications are usually cited in the text rather than listed in a bibliography.

MULTIMEDIA SOURCES (Print and Electronic)

Map or Chart

[B18] *Wisconsin*. Map. Chicago: Rand, 2000.

Film or Video

[B19] *Introduction to Hazardous Chemicals*. Directed by Gordon Massingham, edited by Jane Christopher. Emergency Film Group, Edgartown, MA, 2003. Videocassette.

Television Interview

[B20] Brown, M. M., Interview by Charlie Rose. *The Charlie Rose Show*. Public Broadcasting System, 6 December 2001.

OTHER SOURCES

Published Interview

[B21] Gates, B., "The View from the Very Top," Interview. *Newsweek*, pp. 36–39, April 17, 2000.

Personal Interview

[B22] Sariolgholam, M., Interview by author. Tape recording. Berkeley, Calif., November 29, 2000.

Personal Letter

Personal communications such as letters, e-mail, and messages from discussion groups and electronic bulletin boards are not cited in an IEEE bibliography.

Brochure or Pamphlet

[B23] Library of Congress, U.S. Copyright Office, *Copyright Registration for Online Works*. Washington, DC: GPO, 2004, pp. 1–3.

Government Document

[B24] U.S. Department of Energy, *The August 14, 2003, Blackout One Year Later*. Washington, DC: GPO, 2004.

Report Published in a Collection

[B1] Jacobs, R. L., "Structured On-the-Job Training: Unleashing Employee Experience in the Workplace," Report no. ED-474710, *ERIC*, vol. 346082, pp. 1–28, Oct. 2003.

Report Published Separately

[B25] Bertot, J., and McClure, C. R. *Public Libraries and the Internet 2000: Summary Findings and Data Tables*. Washington, DC: GPO, 2000.

IEEE Sample Pages

One-inch margins, single-spaced.

Standards in Business, June 2004

This report examines the nature and disposition of the 3,458 ethics cases handled companywide by CGF's ethics officers and managers during 2004. The purpose of such reports is to provide the Ethics and Business Conduct Committee with the information necessary for assessing the effectiveness of the first year of CGF's Ethics Program. According to Matthias Jonas, recommendations are given for consideration "in planning for the second year of the Ethics Program" [B1].[1]

Shortened title and date.

The Office of Ethics and Business Conduct was created to administer the Ethics Program. The director of the Office of Ethics and Business Conduct was given the responsibility for the following objectives, as described by Rossouw:

Long quote indented, set in smaller font size, without quotation marks.

> Communicate the values, standards, and goals of CGF's Program to employees. Provide companywide channels for employee education and guidance in resolving ethics concerns. Implement companywide programs in ethics awareness and recognition. Employee accessibility to ethics information and guidance is the immediate goal of the Office of Business Conduct in its first year [B2].

The purpose of the Ethics Program, according to Jonas, is to "promote ethical business conduct through open communication and compliance with company ethics standards" [B3].

Block paragraphs.

Major ethics cases were defined as those situations potentially involving serious violations of company policies or illegal conduct. Examples of major ethics cases included cover-up of defective workmanship or use of defective parts in products; discrimination in hiring and promotion; involvement in monetary or other kickbacks; sexual harassment; disclosure of proprietary or company information; theft; and use of corporate Internet resources for inappropriate purposes, such as conducting personal business, gambling, or access to pornography.

The effectiveness of CGF's Ethics Program during the first year of implementation is most evidenced by (1) the active participation of employees in the program and the 3,458 contacts employees made regarding ethics concerns through the various channels available to them, and (2) the action taken in the cases reported by employees,

Footnote explains first in-text citation.

[1]The numbers in brackets correspond to those in the bibliography in Annex B.

14

Page number at bottom right of page, one space below last line.

FIGURE D–7. IEEE Sample Page

D

IEEE

Standards in Business, June 2004

Bibliography labeled "informative" in 14-point type within parentheses.

Annex B

(informative)

Bibliography

Title in bold; 16-point type.

List in order of appearance in document.

Single-spaced block entries.

[B1] Jonas, M., "The internet and ethical communication: Toward a new paradigm," *Journal of Ethics and Communication*, vol. 32, pp. 147–177, Fall 2004.

[B2] Rossouw, G. J., "Business ethics in South Africa," *Journal of Business Ethics*, vol. 16, pp. 1539–1547, 1997.

[B3] Jonas, M., "Ethics in organizational communication: A review of the literature," *Journal of Ethics and Communication*, vol. 29, pp. 77–99, Summer 2001.

[B4] Davis, W. C., Marks, R., and Tegge, D., *Working in the System: Five New Management Principles*. New York: St. Martin's, 2001, pp. 18–42.

[B5] Hassab, J. C., *Systems Management: People, Computers, Machines, Materials*. New York: CRC, 1997, pp. 312–333.

[B6] Sariolgholam, M., Interview by author. Tape recording. Berkeley, Calif., November 29, 2000.

[B7] Library of Congress, U.S. Copyright Office, *Copyright Registration for Online Works*. Washington, DC: GPO, 2004, pp. 1–3.

[B8] Association for Business Communication. *Association for Business Communication*. 2004. http://www.businesscommunication.org.

[B9] *The One-Minute Manager*. CD-ROM. Boston: Bedford/St. Martin's, 1998.

20

FIGURE D–8. IEEE Sample Bibliography

MLA Documentation

MLA In-Text Citations. MLA parenthetical citation within the text of a paper gives a brief citation—in parentheses—of the author and relevant page numbers, separated only by a space.

- The results of these studies have led even the most conservative managers to adopt technologies that will "catapult the industry forward" (Peterson 183–84).
- As Peterson writes, preparing a videotape of measurement methods is cost-effective and can expedite training (151). [If the author is cited in the text, include only the page numbers in parentheses.]

If the parenthetical citation refers to a long, indented quotation, place it outside the punctuation of the last sentence.

- . . . a close collaboration with the physics and technology staff is essential. (Minsky 42)

If no author is named or if you are using more than one work by the same author, give a shortened version of the title in the parenthetical citation, unless you name the title in the text (a "signal phrase"). A citation for Thomas J. Peterson's book *The Pursuit of Wow: Every Person's Guide to Topsy-Turvy Times* would appear as (Peterson, Pursuit 93).

MLA Documentation Models

BOOKS

Single Author

Van Grembergen, Wim. Information Technology Evaluation Methods and Management. New York: Idea, 2001.

Multiple Authors

Testerman, Joshua O., Thomas J. Kuegler, Jr., and Paul J. Dowling, Jr. Web Advertising and Marketing. 3rd ed. Rocklin, CA: Prima, 2000.

Corporate Author

Ernst and Young. Ernst and Young's Retirement Planning Guide. 3rd ed. New York: Wiley, 2001.

Edition Other Than First

Van Horne, James C. Financial Market Rates and Flows. 6th ed. New York: Prentice Hall, 2000.

Multivolume Work

Standard and Poor. Standard and Poor's Register of Corporations, Directors and Executives. 3 vols. New York: McGraw, 2004.

Work in an Edited Collection

Griswold, Charles L. "Happiness and Cypher's Choice: Is Ignorance Bliss?" The Matrix and Philosophy. Ed. William Irwin. Chicago: Open Court, 2002. 126–37.

Encyclopedia or Dictionary Entry

Gibbard, Bruce G. "Particle Detector." World Book Encyclopedia. 2003 ed.

ARTICLES IN PERIODICALS (*See also* Electronic Sources)

Magazine Article

Limp, Fred. "Enterprise Development." Business Geographics June 2001: 19.

Journal Article

Loza, Jehan. "Business-Community Partnerships: The Case for Community Organization Capacity Building." Journal of Business Ethics 53 (2004): 287–311.

Newspaper Article

Loftus, Patrick. "Small Business: Enterprise." Wall Street Journal 31 Aug. 2004: B4.

Article with an Unknown Author

"CFOs Remain Optimistic on Economy." Business Journal 27 Sept. 2004: 33.

ELECTRONIC SOURCES

Entire Web Site

The MLA recommends that at minimum, a reference to a Web source should provide a document title or description, the date of the publication (posted or updated), the date that the publication was retrieved, an address (URL) that links directly to the document or section, and an author, whenever possible.

Society for Technical Communication. Jan. 2002. Society for Technical Communication. 20 Sept. 2005 <http://www.stc.org/>.

Short Work from a Web Site, with an Author

Locker, Kitty O. "The History of the Association for Business Communication." Association for Business Communication. Oct. 2002. Assn. for Business Communication. 20 Sept. 2005 <http://www.businesscommunication.org/about/history/history.html>.

Short Work from a Web Site, with a Corporate Author

General Motors. "Company Profile." General Motors. 2003. 20 Sept. 2004 <http://www.gm.com/company/corp_info/profiles/>.

Short Work from a Web Site, with an Unknown Author

"Forgotten Inventors." Forgotten Inventors. PBS Online. 2001. 19 Apr. 2001 <http://www.pbs.org/wgbh/amex/telephone/sfeature/index.htm>.

Article from a Database

Goldbort, Robert C. "Scientific Writing as an Art and as a Science." Journal of Environmental Health 63.7 (2001): 22. Expanded Academic ASAP. InfoTrac. Salem State Coll. Lib., Salem, MA. 19 Apr. 2001. <http://infotrac.galegroup.com>.

Article in an Online Periodical

Tiernan, Robert. "Waiting for Wireless." SmartMoney.com 18 Apr. 2001. 19 Apr. 2001 <http://www.smartmoney.com/techmarket/index.cfm?story=20010418>.

Publication on CD-ROM

Money 94. CD-ROM. Redmond, WA: Microsoft, 2004.

E-mail Message

Kahl, Jonathan D. "Re: Web page." E-mail to the author. 2 Oct. 2001.

MULTIMEDIA SOURCES (Print and Electronic)

Map or Chart

Wisconsin. Map. Chicago: Rand, 2000.

Film or Video

Massingham, Gordon, dir., and Jane Christopher, ed. Introduction to Hazardous Chemicals. Videocassette. Edgartown, MA: Emergency Film Group, 2003.

Television Interview

Brown, Mark Malloch. Interview. The Charlie Rose Show. By Charlie Rose. Public Broadcasting System. 6 Dec. 2001.

OTHER SOURCES

Published Interview

Gates, Bill. "The View from the Very Top." Interview. Newsweek 17 Apr. 2000: 36–39.

D

MLA

Personal Interview

Sariolgholam, Mahmood. Personal interview. 29 Nov. 2000.

Personal Letter

Pascatore, Monica. Letter to the author. 10 April 2002.

Brochure or Pamphlet

Library of Congress. US Copyright Office. Copyright Registration for Online Works. Washington: GPO, 2004.

Government Document

United States. Dept. of Energy. The August 14, 2003, Blackout One Year Later. Technical Pub. 137-2004-12176. Washington: GPO, 2004.

Report Published in a Collection

Jacobs, Ronald L. "Structured On-the-Job Training: Unleashing Employee Experience in the Workplace." *ERIC* 346082 (2003): 1–28. ED-474710.

Report Published Separately

Bertot, John Carlo, and Charles R. McClure. Public Libraries and the Internet 2000: Summary Findings and Data Tables. Washington: GPO, 2000.

Lecture or Speech

McKinney, Scott. Lecture. Demarest Hall, Hobart and William Smith Colls., Geneva, NY. 3 May 2002.

Krug, Steve. "Don't Make Me Think: The Art of Designing User-Friendly Websites." New England School of Art & Design at Suffolk U. Boston Public Lib. 3 Apr. 2001.

D

MLA

MLA Sample Pages

Author's last name and page number.

Litzinger 14

One-inch margins. Text double-spaced.

This report examines the nature and disposition of the 3,458 ethics cases handled companywide by CGF's ethics officers and managers during 2004. The purpose of such reports is to provide the Ethics and Business Conduct Committee with the information necessary for assessing the effectiveness of the first year of CGF's Ethics Program (Davis, Marks, and Tegge 142). According to Matthias Jonas, recommendations are given for consideration "in planning for the second year of the Ethics Program" ("Internet" 152).

The Office of Ethics and Business Conduct was created to administer the Ethics Program. The director of the Office of Ethics and Business Conduct, along with seven ethics officers throughout CGF, was given the responsibility for the following objectives, as described by Rossouw:

Long quote indented one inch (or 10 spaces), double-spaced, without quotation marks.

> Communicate the values, standards, and goals of CGF's Program to employees. Provide companywide channels for employee education and guidance in resolving ethics concerns. Implement companywide programs in ethics awareness and recognition. Employee accessibility to ethics information and guidance is the immediate goal of the Office of Business Conduct in its first year. (1543)

In-text citations give author name and page number. Title used when multiple works by same author cited.

The purpose of the Ethics Program, according to Jonas, is to "promote ethical business conduct through open communication and compliance with company ethics standards" ("Ethics" 89). To accomplish this purpose, any ethics policy must ensure confidentiality

FIGURE D–9. MLA Sample Page

D MLA

Heading centered.

Litzinger 21

Works Cited

List alphabetized by authors' last names or title and double-spaced.

Hanging-indent style used for entries.

Association for Business Communication. Aug. 2004. Assn. for Business Communication. 16 Mar. 2005 <http://www.businesscommunication.org/>.

Davis, W. C., Roland Marks, and Diane Tegge. Working in the System: Five New Management Principles. New York: St. Martin's, 2001.

Hassab, Joseph C. Systems Management: People, Computers, Machines, Materials. New York: CRC, 1997.

Jonas, Matthias. "Ethics in Organizational Communication: A Review of the Literature." Journal of Ethics and Communication 29 (2001): 79–99.

---. "The Internet and Ethical Communication: Toward a New Paradigm." Journal of Ethics and Communication 32 (2004): 147–77.

Library of Congress. US Copyright Office. Copyright Registration for Online Works. Washington: GPO, 2004.

The One-Minute Manager. CD-ROM. Boston: Bedford, 1998.

Rossouw, George J. "Business Ethics in South Africa." Journal of Business Ethics 16 (1997): 1539–47.

Sariolgholam, Mahmood. Personal interview. 29 Nov. 2000.

FIGURE D–10. MLA Sample List of Works Cited

Other Style Manuals

Many professional societies, publishing companies, and other organizations publish manuals that prescribe bibliographic reference formats for their publications or for publications in their fields. Many of these style manuals are well known and widely used.

General

The Chicago Manual of Style. 15th ed. Chicago: University of Chicago Press, 2003. See also <www.press.uchicago.edu/Misc/Chicago/cmosfaq/about.html>.

Biology

Council of Science Editors. *Scientific Style and Format: The CBE Manual for Authors, Editors, and Publishers.* 6th ed. New York: Cambridge University Press, 1994. See also <www.councilscienceeditors.org>.

Chemistry

American Chemical Society. *ACS Style Guide: A Manual for Authors and Editors.* 2nd ed. Washington, D.C.: American Chemical Society, 1998. See also <www.acs.org>.

Geology

Bates, Robert L., Rex Buchanan, and Marla Adkins-Heljeson, eds. *Geowriting: A Guide to Writing, Editing, and Printing in Earth Science.* 5th ed. Alexandria, Va.: Amer. Geological Inst., 1995. See also <www.agiweb.org>.

Government Documents

United States Government Printing Office. *Style Manual.* Washington, D.C.: U.S. Government Printing Office, 2000. See also <www.gpo.gov>.

Medicine

American Medical Association. *American Medical Association Manual of Style.* 9th ed. Baltimore: Williams, 1998. See also <www.ama-assn.org>.

Physics

American Institute of Physics, Publication *Board. Style Manual for Guidance in the Preparation of Papers.* 4th ed. New York: American Institute of Physics, 1990. See also <www.aip.org/pubservs/style.html>.

Science and Technical Writing

National Information Standards Organization. *Scientific and Technical Reports—Elements, Organization, and Design.* Bethesda, Md.: National Information Standards Organization, 1995. See also <www.niso.org>.

Rubens, Phillip, ed. *Science and Technical Writing: A Manual of Style.* 2nd ed. New York: Routledge, 2001.

double negatives

A double negative is the use of an additional negative word to reinforce an expression that is already negative. In writing and speech, avoid such constructions.

UNCLEAR	We don't have none. [This sentence literally means that we have some.]
CLEAR	We have none.

Barely, *hardly*, and *scarcely* cause problems because writers sometimes do not recognize that those words are already negative.

- The corporate policy ~~doesn't~~ hardly covers the problem.

Not unfriendly, *not without*, and similar constructions are not double negatives because in such constructions two negatives are meant to suggest the gray area between negative and positive meanings. Be careful how you use such constructions; they can be confusing to the reader and should be used only if they serve a purpose.

- He is *not unfriendly*.
  [He is neither hostile nor friendly.]
- It is *not without* regret that I offer my resignation.
  [I have mixed feelings rather than only regret.]

The correlative <u>conjunctions</u> *neither* and *nor* may appear together in a clause without creating a double negative, so long as the writer does not attempt to use the word *not* in the same clause.

- It was ~~not~~ neither, as a matter of fact, ~~neither~~ his duty nor his desire to fire the employee.

- It was not, as a matter of fact, ~~neither~~ *either* his duty ~~nor~~ *or* his desire to fire the employee.

- She ~~did not~~ neither ~~care~~ *cared* about nor ~~notice~~ *noticed* the error.

Negative forms are full of traps that often entice writers into logic errors, as illustrated in the following example:

> **ILLOGICAL** The book reveals *nothing* that has *not* already been published in some form, but some of it is, I believe, very little known.

In this sentence, "some of it" logically can refer only to "*nothing* that has *not* already been published." The sentence can be corrected by stating the idea in more positive writing.

> **LOGICAL** Everything in the book has been published in some form, but some of it is, I believe, very little known.

drawings

A drawing can show an object's appearance or illustrate instructions. It can emphasize the significant parts or functions of a mechanism, omit what is not significant, and focus on details or relationships that a photograph cannot reveal. Think about your need for drawings during the preparation and research stages of writing. Include them in your outline, indicating approximately where each should appear throughout the outline with "drawing of . . ." enclosed in brackets. For integrating drawings into your text, see visuals.

The types of drawings discussed in this entry are conventional line drawings, exploded-view drawings, cutaway drawings, schematic diagrams, and clip-art images. Each type of drawing has unique advantages; the type of drawing you use should be determined by its purpose and what your readers might need or expect.

A conventional line drawing, like that in Figure D–11, is appropriate if your readers need a representation of an object's general appearance or an overview of a series of steps.

An exploded-view drawing, like that in Figure D–12, can be useful when you need to show the proper sequence in which parts fit together or to show the details of individual parts. Figure D–12 shows a page from a manual for purchasers of a Xerox machine, instructing them to safely unpack the machine and its key parts.

PULL
the pin: Some extinguishers require releasing a lock latch, pressing a puncture lever, or taking another first step.

FIGURE D–11. Conventional Line Drawing Illustrating Instructions

A cutaway drawing, like the one in Figure D–13, can be useful when you need to show the internal parts of a device or structure and illustrate their relationship to the whole.

Schematic diagrams (or schematics) are drawings that portray the operation of electrical and mechanical systems, using lines and symbols rather than physical likenesses. Schematics are useful in instructions or descriptions—for example, when you need to emphasize the relationships among parts without giving their precise proportions.

The schematic diagram in Figure D–14 depicts the process of trapping particulates from coal-fired power plants. Note that the particulate filter is depicted both symbolically and as an enlarged drawing to show relevant structural details. Schematic symbols generally represent equipment, gauges, and electrical or electronic components.

If you need only general-interest images to illustrate newsletters and brochures or to create presentation slides, use noncopyrighted clip-art drawings from your word-processing program or from thousands of noncopyrighted Web and library sources. Figure D-15 shows examples of clip-art images.

ETHICS NOTE Be careful not to use drawings from copyrighted sources without permission and proper documentation, especially from the Web—the same copyright laws that apply to printed material also apply to Web-based graphics. See also documenting sources and plagiarism. ✦

D

Installation

As you unpack the WorkCentre, familiarize yourself with its contents. After the WorkCentre is installed, and the Ready Indicator is lit, the WorkCentre is ready to make copies.

IMPORTANT: Save the carton and packing materials. They should be used to repack the WorkCentre if it has to be shipped for servicing or in case you move.

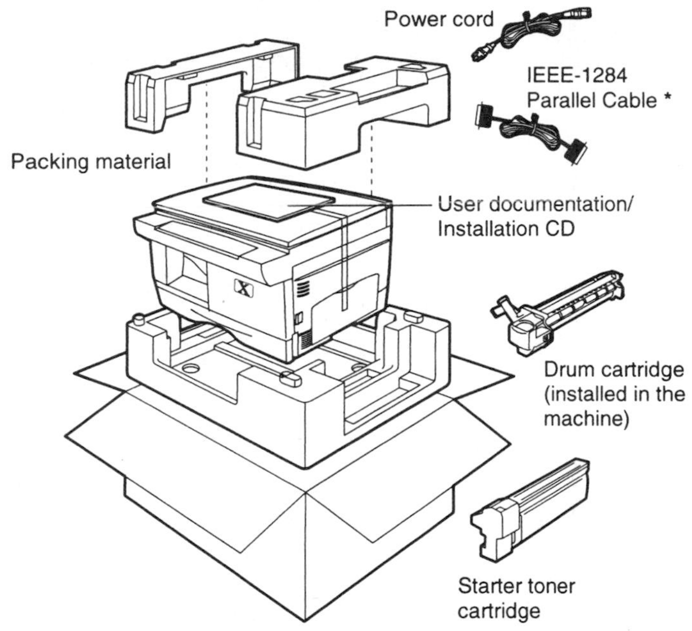

* **Note:** To ensure reliability of the WorkCentre use the IEEE-1284 compliant parallel cable that is supplied with the machine. Only cables labeled "IEEE-1284" can be used with your WorkCentre.

FIGURE D–12. Exploded-View Drawing

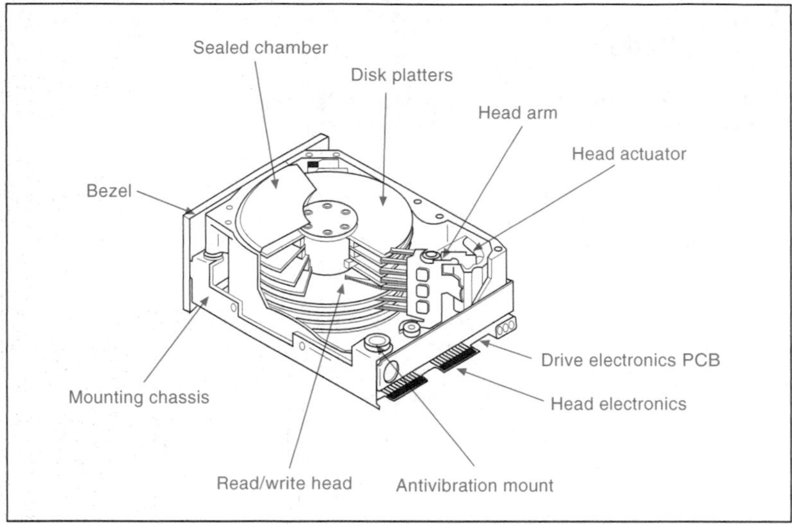

FIGURE D–13. Cutaway Drawing

Writer's Checklist: Creating and Using Drawings

- ☑ Seek the help of professional graphics specialists for drawings that require a high degree of accuracy and precision.
- ☑ Show equipment and other objects from the point of view of the person who will use them.
- ☑ When illustrating a subsystem, show its relationship to the larger system of which it is a part.
- ☑ Draw the different parts of an object in proportion to one another or indicate that certain parts are enlarged.
- ☑ When a sequence of drawings is used to illustrate a process, arrange them from left to right or from top to bottom on the page.
- ☑ Label parts in the drawing so that the text references to them are clear and consistent.
- ☑ Depending on the complexity of what is shown, label the parts themselves, as in Figure D–12, or use a letter/number key.
- ☑ Refer to the figure in the text and do not place the figure prior to that text reference.

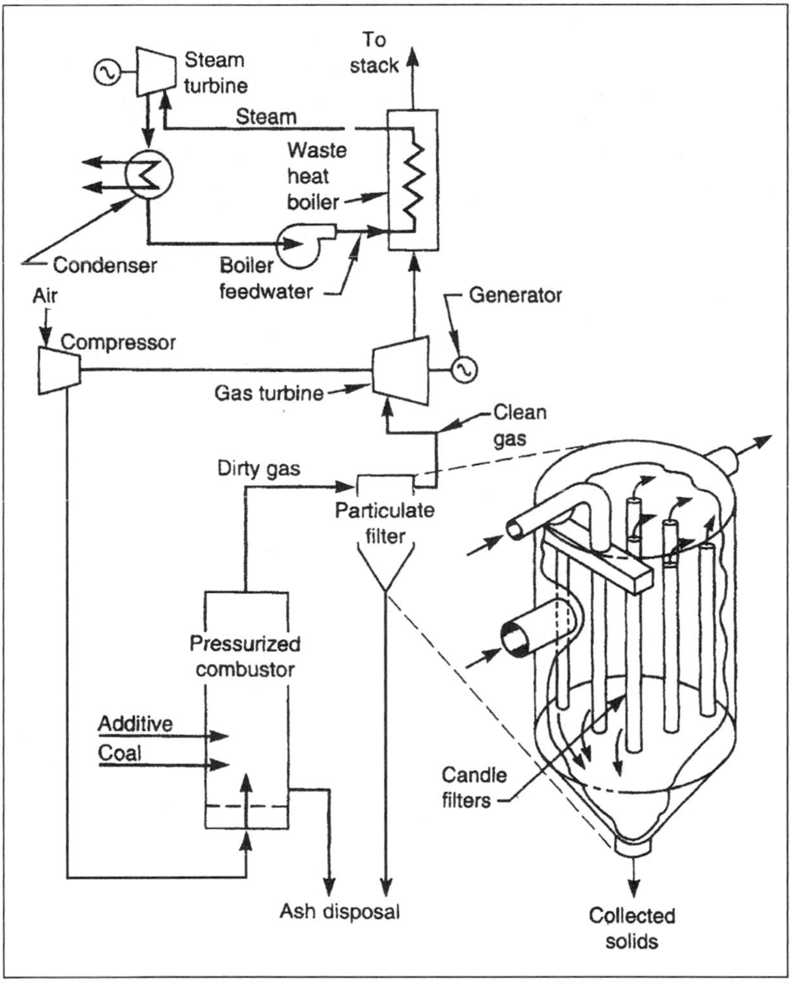

FIGURE D–14. Schematic Diagram

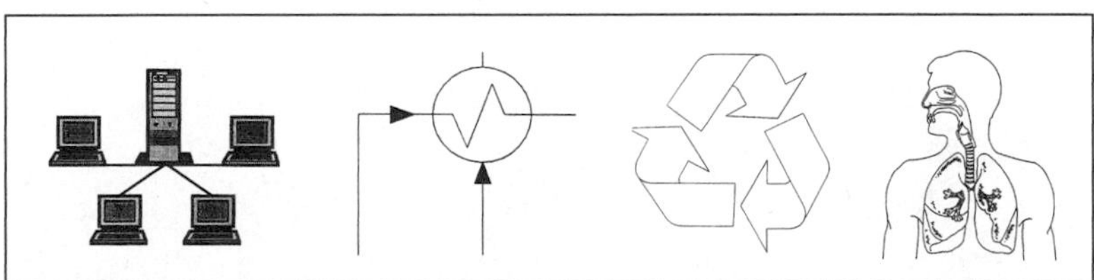

FIGURE D–15. Clip-Art Images

due to / because of

D

Due to (meaning "caused by") is acceptable following a linking verb.

- His short temper was *due to* stress from the project.

Due to is not acceptable, however, when it is used with a nonlinking verb to replace *because of*.

- He left work ~~due to~~ *because of* illness.

E

each

When *each* is used as a subject, it takes a singular verb or pronoun.

- *Each* of the reports *is* to be submitted ten weeks after *it* is assigned.

When *each* refers to a plural subject, it takes a plural verb or pronoun. See also agreement.

- The reports *each have* company logos on *their* covers.

The phrase *each and every* is redundant. See also conciseness.

economic / economical

Economic refers to the production, development, and management of material wealth. (The euro had an *economic* impact on several countries.) *Economical* simply means "not wasteful or extravagant." (Employees should be as *economical* as possible in their equipment purchases.)

editing (*see* revision *and* proofreading)

e.g. / i.e.

The abbreviation *e.g.* stands for the Latin *exempli gratia*, meaning "for example"; *i.e.* stands for the Latin *id est*, meaning "that is." Because the English expressions (*for example* and *that is*) are clear to all readers, avoid the Latin *e.g.* and *i.e.* abbreviations except to save space in notes and visuals.

If you must use *i.e.* or *e.g.*, punctuate them as follows. If *i.e.* or *e.g.* connects two independent clauses, a semicolon should precede it and a comma should follow it.

- The conference reflected international viewpoints*; i.e.,* Germans, Italians, Japanese, Pakistanis, and Americans gave presentations.

If *i.e.* or *e.g.* connects a noun and an appositive, a comma should precede it and follow it.

- The conference included speakers from five countries*, i.e.,* Germany, Italy, Japan, Pakistan, and the United States.

Do not italicize either abbreviation.

ellipses

An ellipsis is the omission of words from quoted material; it is indicated by three spaced periods called *ellipsis points* (. . .). When you use ellipsis points, omit marks of internal punctuation at the point of omission, unless they are necessary for clarity or the omitted material comes at the end of a quoted sentence.

ORIGINAL TEXT	"Promotional material is sometimes charged for, particularly in high-volume distribution to schools, although prices for these publications are much lower than the development costs when all factors are considered."
WITH OMISSION AND ELLIPSIS POINTS	"Promotional material is sometimes charged for . . . although prices for these publications are much lower than the development costs. . . ."

Notice in the preceding example that the final period is retained and what remains of the quotation is grammatically complete. When the omitted part of the quotation is preceded by a period, retain the period and add the three ellipsis points after it, as in the following example.

ORIGINAL TEXT	"Of the 172 major ethics cases reported, 57 percent were found to involve unsubstantiated concerns. Misinformation was the cause of unfounded concerns of misconduct in 72 cases. Forty-four cases, or 26 percent of the total cases reported, involved incidents partly substantiated by ethics officers as serious misconduct."

WITH OMISSION AND ELLIPSIS POINTS	"Of the 172 major ethics cases reported, 57 percent were found to involve unsubstantiated concerns. . . . Forty-four cases, or 26 percent of the total cases reported, involved incidents partly substantiated by ethics officers as serious misconduct."

Do not use ellipsis points when the beginning of a quoted sentence is omitted. Notice in the following example that the comma is dropped to prevent a grammatical error. See also quotations.

- The ethics report states that "26 percent of the total cases reported involved incidents partly substantiated by ethics officers as serious misconduct."

e-mail

DIRECTORY

E-mail (or email) functions in the workplace as a medium to send information, elicit discussions, collect opinions, and transmit documents and files of all types electronically. Although e-mail may take the form of informal notes, you should generally follow the same guidelines for writing style and strategy described in the entries correspondence and memos. See also selecting the medium.

Review and Confidentiality

E-mail is such a quick and easy way to communicate that you need to work to avoid the temptation of sending a first draft without revision. As with all correspondence, your message should include all crucial details and be free of grammatical or factual errors, ambiguities, or unintended implications. See proofreading.

Be especially careful when sending messages to superiors in your organization or to people outside the organization. Time you spend reviewing your e-mail before you send it can save a great deal of time and embarrassment correcting misunderstandings caused by a careless message. One helpful strategy is to write the draft and revise your e-mail before filling in the "To" line with the address of your recipient. Be careful as well to observe the rules of netiquette (Inter*net* + *etiquette*) in the *Writer's Checklist* that follows.

ETHICS NOTE Keep in mind the issue of confidentiality when sending e-mail. Remember, e-mail can be intercepted by someone other than the intended recipient, and e-mail messages are never truly deleted. Most companies back up and save all company e-mail and are legally entitled to monitor e-mail. Companies can also be compelled legally to provide e-mail in a court of law. Consider the content of all your messages in the light of these possibilities, and carefully review your text before you click "Send." ✦

E

Writer's Checklist: Observing Workplace Netiquette

To maintain a high level of professionalism in workplace e-mail, observe the rules of netiquette.

☑ Use company e-mail only for appropriate business.

- Do not send or forward jokes or humorous stories, use biased language, or discuss office gossip.
- Do not send *flames* (e-mails that contain abusive, obscene, or derogatory language) to attack someone.
- Do not send *spam* (mass-distributed e-mail that often promotes personal projects and interests).

☑ Respond to incoming e-mail promptly. If you receive an assignment by e-mail that will take a few days or longer to complete, send a courtesy response saying so.

☑ Be scrupulous about typing e-mail addresses and otherwise ensuring that the intended recipient gets the message.

☑ Send an attachment only after verifying that your recipient wants or needs the file and that your recipient's software will accept it.

☑ Include a message that functions as a cover letter for all e-mails with attachments.

☑ Consider posting a large file at an Internet server and supplying the file's address so that your recipient may download the file at his or her convenience.

☑ Do not write in ALL UPPERCASE LETTERS; such a message is difficult to read and is considered the equivalent of shouting. Likewise, do not write in all lowercase letters; it is considered lazy and too informal for professional work.

☑ Avoid e-mail abbreviations used in personal e-mail and chat rooms (BTW for *by the way*, for example).

Writer's Checklist: Observing Workplace Netiquette (continued)

☑ Do not use emoticons (keyboard characters used to create sideways faces conveying emotions) for business and professional e-mail. For advice on providing typographic emphasis, review the next section on design considerations.

DIGITAL TIPS **SENDING AN E-MAIL ATTACHMENT**

Large files, like graphics, slow the transmission speed of e-mail messages, and the recipient's software or Internet provider may not be able to accept large attachments. Consider using a compression software utility like WinZip at <http://winzip.com/>, which can reduce the file size by 80 percent or more. **Caution**: Viruses can be embedded in e-mail attachments, so regularly update your virus-scanning software and do not open attachments from anyone you do not know. For more on this topic, see *<bedfordstmartins.com/alredtech>* and select *Digital Tips*, "Sending an E-mail Attachment."

Design Considerations

Some e-mail systems allow you to use sophisticated typographical features, such as various fonts and bullets. These options increase your e-mail file size and may display unpredictably in other e-mail systems. Unless you are sure your recipient's software will display your formatted message correctly, set your e-mail software to send messages in "plain text" and use alternative highlighting devices. For example, capital letters or asterisks, used sparingly, can substitute for boldface, italics, and underlines as emphasis.

- Dr. Wilhoit's suggestions benefit doctors AND patients.
- Although the proposal is sound in *theory*, it will never work in *practice*.

Intermittent underlining can replace solid underlining or italics when referring to published works in an e-mail message.

- My report follows the format given in the _Handbook of Technical Writing_.

Keep in mind the following additional design considerations when sending e-mail.

- Break the text into short paragraphs to avoid dense blocks of text.
- Consider providing an overview at the top in a brief paragraph for messages that run longer than a screen of text.

E

- Send as attachments documents containing such formatted elements as tables and bulleted lists that do not transmit well in e-mails.
- Place your response to someone else's message at the beginning (or top) of the e-mail window so that recipients can see your response immediately.
- When replying to a message, quote only relevant parts. If your system does not distinguish the quoted text, note it with a greater-than symbol (>).
- Use the "cc:" (copy) and "bcc:" (blind copy) address lines sparingly and thoughtfully because some recipients use their placement in the address lists to filter their mail. See also correspondence.
- Provide a concise phrase in the subject line that describes the topic of your message; recipients use subject lines to prioritize and file their incoming messages. See also memos.

Salutations, Closings, and Signature Blocks

E-mail can function as a letter, memo, or personal note; therefore, you must adapt your salutation and complimentary closing to your readers and context. Unless your employer requires certain forms, use the following guidelines:

- When e-mail functions as a traditional letter, use the standard letter salutation (*Dear Ms. Tucker:* or *Dear Docuform Customer:*) and a closing (*Best wishes,* or *Sincerely,*).
- When e-mail functions as a memo to a large group within an organization or a formal announcement to a group, you do not need a salutation or closing.
- When e-mail functions as a memo to individuals or small groups inside an organization, you may wish to adopt a more personal greeting (*Dear Andy,* or *Dear Project Colleagues,*).
- When e-mail functions as a personal note to a friend or close colleague, you can use an informal greeting or only a first name (*Hi Mike,* or *Hello Jenny,* or *Bill,*) and a closing (*Take care,* or *Cheers,*).

Be aware that in some cultures, business correspondents do not use first names as quickly as they do in American correspondence. See international correspondence.

Because e-mail does not provide letterhead with standard addresses and contact information, many companies and individual writers include *signature blocks* (also called *signatures*) at the bottom of their messages. Signature blocks, which writers can preprogram to appear on every e-mail they send, supply information that company letterhead usually provides as well as links to Web sites. If your organization requires a certain format, adhere to that standard. Otherwise, use the pattern shown in Figure E–1.

==============================		
Daniel J. Vasquez, Publications Manager	←	Name and Title
Medical Information Systems	←	Department or Division
TechCom Corporation	←	Company Name
P.O. Box 5413 Salinas, CA 93962	←	Mailing Address
Office Phone 888-229-4511 (x 341)	←	Phone Number
General Office Fax 888-229-1132	←	Fax Number
www.tcc.com	←	Web Address (URL)
==============================		

FIGURE E–1. E-mail Signature Block

For signature blocks, consider the following guidelines:

- Include as few lines as possible. Most netiquette guides advocate using five lines or fewer, but more are acceptable in business correspondence because it typically requires more contact information.
- Keep line length to 60 characters or fewer to avoid unpredictable line wraps.
- Test your signature block in plain-text e-mail systems to verify your format; centered text and vertical lines may look fine in one system but chaotic in another.
- Use highlighting cues, such as hyphens, equal signs, and white space, to separate the signature from the message.
- Avoid using quotations, aphorisms, and other messages ("May the Force be with you") in professional signatures.

DIGITAL TIPS LEAVING AN AWAY-FROM-DESK MESSAGE

Many e-mail systems allow you to create an away-from-desk automatic response. Your message should inform senders when you are expected back and, if necessary, whom they can contact in your absence. To learn how to set up an automatic response, see <*bedfordstmartins.com/alredtech*> and select *Digital Tips*, "Leaving an Away-from-Desk Message."

Writer's Checklist: Managing Your E-mail

- ☑ Check your in-box several times each day; try to clear your in-box by the end of the day.
- ☑ Set priorities for reading e-mail by skimming sender names and subject lines.
- ☑ Review all messages on a subject before you respond, so you don't waste time responding to an issue that is no longer relevant.
- ☑ Learn the advanced features of your system so that you can use filters that organize messages as they arrive.
- ☑ Create electronic folders for e-mail, using personal names and key topics.
- ☑ Copy yourself or save sent copies of important e-mails in your electronic folders.
- ☑ Use the search command to find topics and individual names.
- ☑ Print and file hard copies of crucial e-mail messages that are complex or that you will need for meetings.
- ☑ Keep an up-to-date address book.

emphasis

Emphasis is the principle of stressing the most important ideas in writing. It can be achieved with the careful use of position, climactic order, sentence length, sentence type, active voice, repetition, intensifiers, direct statements, long dashes, and mechanical devices.

Achieving Emphasis

Position. Place the idea in a conspicuous position. The first and last words of a sentence, paragraph, or document stand out in readers' minds.

- Moon craters are important to understanding the earth's history because they reflect geological history.

This sentence emphasizes *moon craters* simply because the term appears at the beginning of the sentence and *geological history* because it is at the end of the sentence. See also subordination.

Climactic Order. List the ideas or facts within a sentence in sequence from least to most important. See also lists.

- Over subsequent weeks the Human Resources Department worked diligently, management showed tact and patience, and the employees demonstrated remarkable support for the policy changes.

Sentence Length. Vary sentence length strategically. A very short sentence that follows a very long sentence or a series of long sentences stands out in the reader's mind, as in the short sentence ("We must raise our standards") that ends the following paragraph. See sentence construction.

- We have already reviewed the problem the quality-control department has experienced during the past year. We could continue to examine the causes of our problems and point an accusing finger at all the culprits beyond our control, but in the end it all leads to one simple conclusion. We must raise our standards.

Sentence Type. Vary sentences by using a compound sentence, a complex sentence, or a simple sentence. See sentence variety.

- The report submitted by the committee was carefully illustrated, and it covered five pages of single-spaced copy.
  [This compound sentence carries no special emphasis because it contains two coordinate independent clauses.]
- The committee's report, which was carefully illustrated, covered five pages of single-spaced copy.
  [This complex sentence emphasizes the size of the report.]
- The carefully illustrated report submitted by the committee covered five pages of single-spaced copy.
  [This simple sentence emphasizes that the report was carefully illustrated.]

Active Voice. Use the active voice to emphasize the performer of an action: Make the performer the subject of the verb.

- Our department designed the new system.
  [The performer, *our department*, is the subject of the verb, *designed*.]

Repetition. Repeat keywords and key phrases, as in the use of the word *remains* and the phrase *come and go* in the following sentence.

- Similarly, atoms *come and go* in a molecule, but the molecule *remains*; molecules *come and go* in a cell, but the cell *remains*; cells *come and go* in a body, but the body *remains*; persons *come and go* in an organization, but the organization *remains*.
  —Kenneth Building, *Beyond Economics*

Intensifiers. Although you can use intensifiers (*most*, *much*, *very*) for emphasis, this technique is so easily abused that it should be used with caution.

- The final proposal is *much* more persuasive than the first one.

E

Direct Statements. Use direct statements, such as "most important," "foremost," or someone's name in a direct address.

- Most important, keep in mind that everything you do affects the company's bottom line.
- John, I believe we should rethink our plans.

Long Dashes. Use a dash to call attention to a particular word or statement.

- The job will be done—after we are under contract.

Mechanical Devices. Use *italics*, **bold type**, underlining, and CAPITAL LETTERS—but use them sparingly because overuse can create visual clutter. See also layout and design, capitalization, and italics.

English as a second language

DIRECTORY

Learning to write well in a second language takes a great deal of effort and practice. The most effective way to improve your command of written English is to read widely beyond the reports and professional articles your job requires, such as magazines, newspapers, articles, novels, biographies, and any other writing that interests you. In addition, listen carefully to native speakers on television, on radio, and in person. Do not hesitate to consult a native speaker of English, especially for important writing tasks, such as e-mails, memos, and reports. Focus on those particular areas of English that give you trouble. See also global communication.

Count and Mass Nouns

Count nouns refer to things that can be counted (tables, pencils, projects, reports). *Mass nouns* (also called *noncount nouns*) identify things that cannot be counted (electricity, water, air, loyalty, information). This distinction can be confusing with words like *electricity* and *water*. Although we can count kilowatt-hours of electricity and bottles of water, counting becomes inappropriate when we use the words *electricity* and *water* in a general sense, as in "*Water* is an essential resource." Following is a list of common mass nouns.

acid	coffee	information	precision
advice	education	knowledge	research
air	electricity	loyalty	technology
anger	equipment	machinery	transportation
biology	furniture	money	water
business	health	news	weather
clothing	honesty	oil	work

The distinction between whether something can or cannot be counted determines the form of the noun to use (singular or plural), the kind of article that precedes it *(a, an, the,* or no article), and the kind of limiting adjective it requires (such as *fewer* or *less* and *much* or *many*). See also fewer/less.

Articles

This discussion of articles applies only to common nouns (not to proper nouns, such as the names of people) because count and mass nouns are always common nouns.

The general rule is that every count noun must be preceded by an article (*a, an, the*), a demonstrative adjective (*this, that, these, those*), a possessive adjective (*my, your, her, his, its, their*), or some expression of quantity (such as *one, two, several, many, a few, a lot of, some, no*). The article, adjective, or expression of quantity appears either directly in front of the noun or in front of the whole noun phrase.

- Beth read *a* report last week. [article]
- *Those* reports Beth read were long. [demonstrative adjective]
- *Their* report was long. [possessive adjective]
- *Some* reports Beth read were long. [indefinite adjective]

The articles *a* and *an* are used with count nouns that refer to one item of the whole class of like items.

- Matthew has *a* pen.
  [Matthew could have *any* pen.]

The article *the* is used with nouns that refer to a specific item that both the reader and the writer can identify.

- Matthew has *the* pen.
  [Matthew has a *specific* pen that is known to both the reader and the writer.]

When making generalizations with count nouns, writers can either use *a* or *an* with a singular count noun or use no article with a plural count noun. Consider the following generalization using an article.

- An egg is a good source of protein.
  [any egg, all eggs, eggs in general]

However, the following generalization uses a plural count noun with no article.

- Eggs are good sources of protein.
  [any egg, all eggs, eggs in general]

When you are making a generalization with a mass noun, do not use an article in front of the mass noun.

- Sugar is bad for your teeth.

Gerunds and Infinitives

Nonnative writers of English are often puzzled by which form of a <u>verbal</u> (a verb used as another part of speech) to use when it functions as the direct object of a <u>verb</u>—or a <u>complement</u>. No structural rule exists for distinguishing between the use of an infinitive and a gerund as the object of a verb. Any specific verb may take an infinitive as its object, others may take a gerund, and yet others take either an infinitive or a gerund. At times, even the base form of the verb is used.

- He enjoys *working*. [gerund as a complement]
- She promised *to fulfill* her part of the contract.
  [infinitive as a complement]
- The president had the manager *assign* her staff to another project.
  [basic verb form as a complement]

To make such distinctions accurately, rely on what you hear native speakers use or what you read. You might also consult a reference book for ESL students.

Adjective Clauses

Because of the variety of ways adjective clauses are constructed in different languages, they can be particularly troublesome for nonnative writers of English. The following guidelines will help you form adjective clauses correctly.

Place an adjective clause directly after the noun it modifies.

- The tall woman ^*who is standing across the room* is a vice president of the company^. ~~who is standing across the room.~~

The adjective clause *who is standing across the room* modifies *woman*, not *company*, and thus comes directly after *woman*.

Avoid using a relative pronoun with another pronoun in an adjective clause.

- The man who ~~he~~ sits at that desk is my boss.

Present Perfect Verb Tense

In general, use the present perfect tense to refer to events completed in the past that have some implication for the present.

PRESENT PERFECT	She *has revised* that report three times. [She might revise it again.]

When a specific time is mentioned, however, use the simple past.

SIMPLE PAST	I *wrote* the letter yesterday morning. [The action, *wrote*, does not affect the present.]

Use the present perfect with a *since* or *for* phrase to describe actions that began in the past and continue in the present.

- This company *has been* in business *for* seventeen years.
- This company *has been* in business *since* 1985.

Present Progressive Verb Tense

The present progressive tense is especially difficult for those whose native language does not use this tense. The present progressive tense is used to describe some action or condition that is ongoing (or in progress) in the present and may continue into the future.

PRESENT PROGRESSIVE	I *am searching* for an error in the document. [The search is occurring now and may continue.]

E

In contrast, the simple present tense more often relates to habitual actions.

SIMPLE PRESENT	I *search* for errors in my documents. [I regularly search for errors, but I am not necessarily searching now.]

See "ESL Tips for Using the Progressive Form" in the entry tense.

ESL Entries

Most of the entries in this handbook may interest writers of English as a second language; however, the specific entries listed in the Contents by Topic under "ESL Trouble Spots" address issues that often cause problems.

WEB LINK **ENGLISH AS A SECOND LANGUAGE**

For Web sites and electronic grammar exercises intended for speakers of English as a second language, see <*bedfordstmartins.com/alredtech*> and select *Links for Handbook Entries* and *Exercise Central*.

English, varieties of

Written English includes two broad categories: standard and nonstandard. Standard English is used in business, industry, government, education, and all professions. It has rigorous and precise criteria for capitalization, punctuation, spelling, and usage.

Nonstandard English does not conform to such criteria; it is often regional in origin, or it reflects the special usages of a particular ethnic or social group. As a result, although nonstandard English may be vigorous and colorful, its usefulness as a means of communication is limited to certain contexts and to people already familiar and comfortable with it in those contexts. It rarely appears in printed material except for special effect. Nonstandard English is characterized by inexact or inconsistent punctuation, capitalization, spelling, diction, and usage choices; thus, you should avoid using it in workplace writing.

Colloquial English

Colloquial English is spoken English or writing that uses words and expressions common to casual conversation (We need to get him up to speed). Colloquial English is appropriate to some kinds of writing (personal letters, notes, some e-mail) but not to most workplace writing.

Dialectal English

Dialectal English is a social or regional variety of the language that is comprehensible to people of that social group or region but may be incomprehensible to outsiders. Dialect, which is usually nonstandard English, involves distinct word choice, grammatical forms, and pronunciations. For example, residents of southern Louisiana who descended from French colonists speak a dialect often referred to as Cajun.

Localisms

A localism is a regional wording or phrasing. For example, a large sandwich on a long split roll is variously known throughout the United States as a *hero*, *hoagie*, *grinder*, *poor boy*, *submarine*, and *torpedo.* Such words normally should be avoided in workplace writing because not all readers will be familiar with the local meanings.

Slang

Slang is an informal vocabulary composed of figures of speech and colorful words used in humorous or extravagant ways. There is no objective test for slang, and many standard words are given slang applications. For instance, slang may be a familiar word used in a new way (*chill* meaning "relax"), or it may be a new word (*wonk* meaning "someone who works or studies excessively").

Most slang is short-lived and has meaning only for a narrow audience. Sometimes, however, slang becomes standard because the word fills a legitimate need. *Skyscraper* and *date* (as in "go on a date"), for example, were once considered slang expressions. Nevertheless, although slang may be valid in informal and personal writing or fiction, it generally should be avoided in workplace writing. See also jargon and technical writing style.

equal / unique / perfect

Logically, *equal* (meaning "having the same quantity or value as another"), *unique* (meaning "one of a kind"), and *perfect* (meaning "a state of highest excellence") are words with absolute meanings and therefore should not be compared. However, colloquial usage of *more* and *most* as modifiers of *equal*, *unique*, and *perfect* is so common that an absolute prohibition on such use is impossible.

- Our system is *more unique* [or *more perfect*] than theirs.

E

Some writers try to overcome the problem by using *more nearly equal* (*more nearly unique*, *more nearly perfect*). When clarity and preciseness are critical, the use of comparative degrees with *equal*, *unique*, and *perfect* can be misleading. It is best to avoid using comparative degrees with absolute terms. See also comparison.

MISLEADING	Ours is a *more equal* percentage split than theirs.
PRECISE	Our percentage split is 51–49; theirs is 54–46.

etc.

Etc. is an abbreviation for the Latin *et cetera*, meaning "and others" or "and so on." Therefore, do not use the redundant phrase *and etc.* Likewise, do not use *etc.* at the end of a series introduced by the phrases *such as* and *for example*—those phrases already indicate unnamed items of the same category. Use *etc.* with a logical progression (1, 2, 3, etc.) and when at least two items are named.

- The sorting machine processes coins (~~for example~~ pennies, nickels, ~~and~~ etc.), and then packages them for redistribution.

Otherwise, avoid *etc.* because the reader may not be able to infer what other items a list might include.

VAGUE	He will bring legal pads, paper clips, etc., to the trade show.
CLEAR	He will bring legal pads, paper clips, and other office supplies to the trade show.

Do not italicize *etc.*

ethics in writing

Ethics refers to the choices we make that affect others for good or ill. Ethical issues are inherent in writing and speaking because what we write and say can influence others. Further, how we express ideas affects our readers' perceptions of us and our company's ethical stance.

ETHICS NOTE Obviously, no book can describe how to act ethically in every situation, but this entry describes some typical ethical lapses to watch for during revision.* In other entries throughout this book, ethical issues are highlighted using the symbols surrounding this paragraph. ✦

Avoid language that attempts to evade responsibility. Some writers use the passive voice because they hope to avoid responsibility or obscure an issue.

- It has been decided. [*Who* has decided?]
- Several mistakes were made. [*Who* made them?]

Avoid deceptive language. Do not use words with more than one meaning to circumvent the truth. Consider the company document that stated, "A nominal charge will be assessed for using our facilities." When clients objected that the charge was actually very high, the writer pointed out that the word *nominal* means "the named amount" as well as "very small." In that situation, clients had a strong case in charging that the company was attempting to be deceptive. Various abstract words, technical and legal jargon, and euphemisms are unethical when they are used to mislead readers or to hide a serious or dangerous situation, even though technical or legal experts could interpret them as accurate. See also word choice.

Do not deemphasize or suppress important information. Not including information that a reader would want to have, such as potential safety hazards or hidden costs for which a customer might be responsible, is unethical. Such omissions may also be illegal. Use layout and design features like type size, bullets, lists, and footnotes to highlight information that is important to readers. Do not hide information in dense paragraphs with small type and little white space, as is common in credit card contracts. Such a design will defeat all but the most persistent readers.

Do not emphasize misleading or incorrect information. Similarly, avoid the temptation to highlight a feature or service that readers would find attractive but that is available only with some product models or at extra cost. (See also logic errors and positive writing.) Readers could justifiably object that you have given them a false impression to sell a product or service, especially if you also deemphasize the extra cost or other special conditions.

In general, treat others—individuals, companies, groups—with fairness and with respect. Avoid language that is biased, racist, or sexist or that perpetuates stereotypes. See also biased language.

In technical writing, guard against reporting false, fabricated, or plagiarized research and test results. As an author, a technical reviewer, or an editor, your ethical obligation is to correct or point out any misrepresentations of fact before publication, whether the publication

*Adapted from Brenda R. Sims, "Linking Ethics and Language in the Technical Communication Classroom," *Technical Communication Quarterly* 2.3 (Summer 1993): 285–99.

is a trade journal article, test report, or product handbook. (See also laboratory reports.) The stakes of such ethical oversights are high because of the potential risk to the health and safety of others. Those at risk can include unwary consumers and workers injured because of faulty products or unprotected exposure to toxic materials.

Finally, be aware that both plagiarism and violations of copyright are not only unethical but also can have serious legal and professional consequences for you in the classroom and on the job. You should review those entries carefully to make sure you have a clear understanding of these issues.

E

Writer's Checklist: Writing Ethically

Ask yourself the following questions:

☑ *Am I willing to take responsibility, publicly and privately, for what the document says?* Will you stand behind what you have written to your employer? to your family and friends?

☑ *Is the document honest and truthful?* Scrutinize findings and conclusions carefully. Make sure that the data support them.

☑ *Am I acting in my employer's best interest? my client's or the public's best interest? my own best long-term interest?*

☑ *Does the document violate anyone's rights?* Have people from different backgrounds review your writing.

☑ *Am I ethically consistent in my writing?* Apply consistently the principles outlined here and those you have assimilated throughout your life to meet this standard.

☑ *What if everybody acted or communicated in this way?* If you were the intended audience, would the message be acceptable and respectful?

If the answers to these questions do not come easily, consider seeking the advice of a trusted colleague to review and comment on what you have written. If information is confidential and you have serious concerns, consider a review by the company's legal staff or attorney.

WEB LINK PROFESSIONAL CODES OF ETHICS

Professional groups often create ethical codes of practice for their members. One example is the National Society of Professional Engineers (NSPE), which represents individual engineering professionals and licensed engineers. For links to the NSPE's code as well as codes of ethics for other fields, see <*bedfordstmartins.com/alredtech*> and select *Links for Handbook Entries*.

euphemisms

A euphemism is an inoffensive substitute for a word or phrase that could be distasteful, offensive, or too blunt: *passed away* for *died*; *previously owned* or *preowned* for *used*; *lay off* or *downsize* for *terminate* or *fire*. Used judiciously, euphemisms can help you avoid embarrassing or offending someone.

ETHICS NOTE Euphemisms can also hide the facts of a situation (*incident* or *event* for *accident*) or be a form of affectation if used carelessly. Avoid them especially in international correspondence and other forms of global communication where their meanings could be not only confusing but also misleading. See also ethics in writing. ✦

everybody / everyone

Both *everybody* and *everyone* are usually considered singular and take singular verbs and pronouns.

- *Everyone* here *leaves* at 4:30 p.m.
- *Everybody* at the meeting made *his or her* proposal separately.

However, the meaning can be obviously plural.

- *Everyone* thought the report should be revised, and I really couldn't blame *them*.

Although normally written as one word, *everyone* is written as two words to emphasize each individual in a group. (*Every one* of the team members contributed to this discovery.) See also agreement.

exclamation marks

The exclamation mark (!) indicates strong feeling, urgency, elation, or surprise (Hurry! Great! Wow!). (See also interjections.) However, it cannot make an argument more convincing, lend force to a weak statement, or call attention to an intended irony.

An exclamation mark can be used after a whole sentence or an element of a sentence.

- This meeting—please note it well!—concerns our budget deficit.

When used with quotation marks, the exclamation mark goes outside, unless what is quoted is an exclamation.

- The manager yelled, "Get in here!" Then Ben, according to Ray, "jumped like a kangaroo"!

In instructions and manuals, the exclamation mark is often used in cautions and warnings (Danger! Stop!). See also emphasis.

executive summaries

An executive summary consolidates the principal points of a report or proposal. Executive summaries differ from abstracts in that readers scan abstracts to decide whether to read the work in full. However, an executive summary may be the only section of a longer work read by many readers, so it must accurately and concisely reflect the original document. It should restate the document's purpose, scope, methods, findings, conclusions, and recommendations as well as explain how results were obtained or the reasons for the recommendations. Executive summaries tend to be about 10 percent of the length of the documents they summarize and generally follow the same sequence.

Write the executive summary so that it can be read independently of the report or proposal. Executive summaries may occasionally include a figure, table, or footnote if that information is essential to the summary. Do not refer by number to figures, tables, or references contained elsewhere in the document.

The sample executive summary in Figure E–2 is from a 35-page report on the potential human health hazards of industrial-use cooling towers constructed of materials containing asbestos fibers. It was written by a research laboratory for specialists at the Nuclear Regulatory Commission who are responsible for assessing whether the tower construction materials pose a potential health hazard. The specialists will use the report to decide whether to authorize the continued use of certain construction materials. Note the appropriate use of technical terms and mathematical notations for the intended audience.

E

EXECUTIVE SUMMARY

Cooling towers are water-recycling devices that can reduce the quantity of waste heat discharged to surface waters by dissipating up to 95% of the heat directly to the atmosphere. Cooling towers are used for air-conditioning large buildings and for cooling in various industries—increasingly, at steam-electric power plants.

One essential component of wet cooling towers is the "fill," the material through which the hot water flows and by which the evaporative (cooling) process is enhanced. Fill can be composed of a variety of materials, including asbestos paper or asbestos cement. Release of asbestos fibers from these cooling towers into the ambient air or into surface waters is thus a possibility, and may create a potential human health hazard. (Occupational exposure to asbestos is known to result in a number of diseases, including lung and digestive-system cancers.)

Purpose and Methodology

To aid in assessing the environmental impact of industries that use cooling towers containing asbestos fill, the Nuclear Regulatory Commission sponsored a study by the Argonne National Laboratory to investigate the quantities of asbestos fibers that could be released into the environment from such cooling towers. The study included a review of the human-health-effects literature on asbestos; collection of information from cooling-tower vendors and users; collection of water, sediment, and air samples at operating cooling towers and analysis for asbestos-fiber content using electron microscope techniques; and estimation of the quantities of fibers that could be re-leased to the ambient air due to drift from cooling towers. Results of the study are discussed in this document and have relevance to any site using cooling towers.

Scope and Findings

At 10 of the 18 cooling-tower sites sampled, chrysotile-asbestos fibers were detected in cooling-tower basin and blowdown waters. Concentrations ranged up to 10^8 fibers/liter, with mass concentrations ranging to 37 μg/liter. At all but two sites, 53% to 100% of the chrysotile fibers were less than 5 μm long; aspect ratios ranged from 3.5 to 1700. The major source of these fibers, at least in nonurban areas, appears to be the components of the towers rather than the air drawn through the towers or the water taken into the towers to replace evaporative and other losses. Ash- or other material-settling ponds appear to reduce, by several orders of magnitude, asbestos-fiber concentrations in cooling-tower effluent.

Assuming a 100-fold dilution in a receiving stream, asbestos-fiber concentrations in public water supplies may range from "none

1

FIGURE E–2. Executive Summary. *Source*: U.S. Nuclear Regulatory Commission.

E

detected" to 10^6 fibers/liter as a result of effluent from cooling towers using asbestos fill. This is the same order of asbestos-fiber concentrations reported for some beverages and drinking water.

The bulk of the literature reviewed during the course of this study did not support a causal relationship between the presence of asbestos fibers in food, beverages, or drinking water and the occurrence of human cancers. However, the uncertainties involved in the epidemiological studies preclude any conclusions regarding "safe" concentrations of asbestos fibers. Also, at least one study indicated a statistical association between certain human cancers and asbestos in drinking water.

Conclusions and Recommendations

In view of the lack of consistent evidence linking adverse health effects to drinking water containing asbestos concentrations on the order of 10^6 fibers/liter or less, as well as ignorance of the health risks and resource costs related to the manufacture and use of other types of cooling-tower fill, it is not presently suggested that the use of asbestos fill be discontinued. However, it is strongly recommended that asbestos fibers in cooling-tower effluent to off-site receiving waters be kept as low as possible by adequate control of circulating water chemistry. Pilot studies should be initiated by industrial research groups on the feasibility of side-stream filtration as a means of reducing asbestos fibers in cooling-tower effluent to below detection limits. The effectiveness of asbestos-fiber removal by ash- or other material-settling ponds should be seriously investigated. It is also recommended that monitoring programs for asbestos in cooling-tower waters be carried out at sites where asbestos fill is used. This recommendation is based on the opinion that caution and surveillance are dictated by the present uncertainties in evaluating health risks from long-term consumption of barely detectable concentrations of asbestos in drinking water.

On the basis of estimates made in this study, monitoring for asbestos fibers in drift from cooling towers, under conditions similar to those at the study sites, does not appear to be warranted. Estimates made using drift-deposition models indicated that asbestos concentrations in air at ground level due to drift are one to several orders of magnitude less than a proposed ambient air-quality standard of 30 ng/m^3. Exceptions might be sites where older, mechanical-draft cooling towers with drift rates equal to or greater than 0.01% of the circulating water flow are in operation; in these cases, the ground-level concentrations of asbestos in the air within 250 m downwind of the towers may exceed Occupational Safety and Health Administration standards.

2

FIGURE E–2. Executive Summary (*continued*)

Writer's Checklist: Writing Executive Summaries

- ☑ Write the executive summary after you have completed the original document.
- ☑ Avoid using terminology that may not be familiar to your intended audience.
- ☑ Spell out all uncommon symbols and **abbreviations**.
- ☑ Make the summary concise, but do not omit transitional words and phrases (*however, moreover, therefore, for example, in summary*).
- ☑ Include only information discussed in the original document.
- ☑ Place the executive summary at the very beginning of the body of the report, as described in **formal reports**.

expletives

An expletive is a word that fills the position of another word, phrase, or clause. *It* and *there* are common expletives.

- *It* is certain that the O-ring will fail at these pressures.

In the example, the expletive *it* occupies the position of subject in place of the real subject, *that the O-ring will fail at these pressures.* Expletives are sometimes necessary to avoid **awkwardness**, but they are commonly overused, and most sentences can be better stated without them.

- ~~There were many~~ *Many* orders *were* lost because of a computer virus.

In addition to its use as a grammatical term, the word *expletive* means an exclamation or oath, especially one that is obscene.

explicit / implicit

An *explicit* statement is one expressed directly, with precision and clarity.

- He gave us *explicit* directions to the Wausau facility.

An *implicit* meaning is one that is not directly expressed.

- Although the CEO did not mention the lawsuit directly, the company's commitment to ethical practices was *implicit* in her speech.

exposition

Exposition, or *expository writing*, informs readers by presenting facts and ideas in direct and concise language; it usually relies less on colorful or figurative language than does writing meant to be expressive or persuasive. Expository writing attempts to explain to readers what the subject is, how it works, and how it relates to something else. Exposition is aimed at the readers' understanding rather than at their imagination or emotions; it is a sharing of the writer's knowledge. Exposition aims to provide accurate, complete information and to analyze it for the readers.

Because it is the most effective form of discourse for explaining difficult subjects, exposition is widely used in reports, memos, and other types of technical writing. To use exposition effectively, you must have a thorough knowledge of your subject. As with all writing, how much of that knowledge you pass on to your readers depends on the readers' needs and your purpose.

F

fact

Expressions containing the word *fact* ("due to the *fact* that," "except for the *fact* that," "as a matter of *fact*," or "because of the *fact* that") are often wordy substitutes for more accurate terms.

- *Because* ~~Due to the fact that~~ the technical staff has a high turnover rate, our training program has suffered.

Do not use the word *fact* to refer to matters of judgment or opinion.

- *In my opinion,* ~~It is a fact that~~ our research has improved because we now have a capable technical staff.

The word *fact* is, of course, valid when facts are what is meant.

- Our tests uncovered numerous *facts* to support your conclusion.

See also conciseness and logic errors.

fax (*see* selecting the medium)

feasibility reports

When organizations consider a new project—developing a new product or service, expanding a customer base, purchasing equipment, or moving operations—they first try to determine the project's chances for success. A feasibility report presents evidence about the practicality of a proposed project based on specific criteria. It answers such questions as the following: Is new construction or development necessary? Is sufficient staff available? What are the costs? What are the legal ramifications? Based on the findings of this analysis, the writer offers logical conclusions and recommends whether the project should be carried out. When feasibility reports stress specific steps that should be taken as a result of a study of a problem or an issue, they are often referred to as *recommendation reports*. In the condensed feasibility report shown in Figure F–1, a technology consulting firm conducts a feasibil-

Introduction

The purpose of this report is to determine which of two proposed options would best enable ACM Technology Consulting to upgrade its file servers and its Internet capacity to meet its increasing data and communication requirements.

Background. In October 2004, the Information Development and Technical Support Group at ACM put the MISSION System into operation. Since then, the volume of processing transactions has increased fivefold (from 1,000 to 5,000 updates per day). This increase has severely impaired system response time; in fact, average response time has increased from less than 10 seconds to 120 seconds. Further, our new Web-based client services system has increased exponentially the demand for processing speed and access capacity.

Scope. We have investigated two alternative solutions to provide increased processing capacity: (1) purchase of a second ARC 98 processor to supplement the first, and (2) purchase of an HRS 60/EP with PRS enterprise software and expandable peripherals to replace the current ARC 98. The two alternatives are evaluated here, according to cost and, to a lesser extent, according to expanded capacity for future operations.

Additional ARC 98 Processor

Purchasing a second ARC 98 processor would require additional annual maintenance costs, salary for an additional computer specialist, increased energy costs, and a one-time construction cost for necessary remodeling as well as installing Internet and other connections.

Annual maintenance costs	$35,000
Annual costs for computer specialist	75,000
Annual increased energy costs	7,500
Total annual operating costs	$117,500
Construction cost (one-time)	50,000
Total first-year costs	$167,500

The installation and operation of another ARC 98 processor are expected to produce savings in system reliability and readiness.

FIGURE F–1. Feasibility Report

System Reliability. A second ARC 98 would reduce current downtime periods from four to two per week. Downtime recovery averages 30 minutes and affects 40 users. Assuming that 50 percent of users require system access at a given time, we determined that the following reliability savings would result:

2 downtimes × 0.5 hours × 40 users × 50% × $50.00/hour overtime × 52 weeks = $52,000 annual savings

[*The feasibility report would also discuss the second option—purchase of the HRS 60/EP and its long-term savings.*]

Conclusion

A comparison of costs for both systems indicates that the HRS 60/EP would cost $2,200 more in first-year costs.

	ARC 98	*HRS 60/EP*
Net additional operating costs	$56,300	$84,000
One-time construction costs	50,000	24,500
First-year total	$106,300	$108,500

Installation of a second ARC 98 processor would permit the present information-processing systems to operate relatively smoothly and efficiently. It would not, however, provide the expanded processing capacity that the HRS 60/EP processor would for implementing new subsystems required to increase processing speed and Internet access.

Recommendation

The HRS 60/EP processor should be purchased because of the long-term savings and because its additional capacity and flexibility will allow for greater expansion in the future.

3

FIGURE F–1. Feasibility Report (*continued*)

ity study to determine how to upgrade its computer system and Internet capability.

Before beginning to write a feasibility report, analyze your readers' needs and the purpose of your study. Then write a purpose statement, such as "The purpose of this study is to determine the feasibility of expanding our research and development operations." See also brainstorming and collaborative writing.

Report Sections

Feasibility reports vary in length from several pages for small projects to dozens for very large projects. Regardless of length, every feasibility report should contain an introduction, a body, a conclusion, and a recommendation. See also proposals and formal reports for further advice on creating these sections.

F

Introduction. The introduction states the purpose of the report, describes the circumstances that led to the report, and includes any pertinent background information. It may also discuss the scope of the report, any procedures or methods used in the analysis of alternatives, and any limitations of the study.

Body. The body of the report presents a detailed review of all the alternatives under consideration. Examine each alternative according to specific criteria, such as cost and financing, availability of staff, and other relevant requirements, identifying the subsections with headings to guide readers.

Conclusion. The conclusion interprets the available options and points to one option as the best or most feasible.

Recommendation. The recommendation section clearly presents the writer's opinion on which alternative best meets the criteria as summarized in the conclusion.

WEB LINK **SAMPLE FEASIBILITY REPORTS**

For links to full feasibility reports, see <*bedfordstmartins.com/alredtech*> and select *Links for Handbook Entries*.

female / male

The terms *female* and *male* are usually restricted to scientific, legal, and medical contexts (a *female* suspect, a *male* patient). Keep in mind that the terms sound cold and impersonal. *Girl*, *woman*, and *lady* or *boy*, *man*, *gentleman* may be acceptable substitutes in other contexts, but be sensitive to their connotations involving age, dignity, and social position. See also biased language and connotation/denotation.

few / a few

In certain contexts, *few* carries more negative overtones than does the phrase *a few*.

NEGATIVE	There are *few* helpful ideas in the report.
POSITIVE	There are *a few* helpful ideas in the report.

F

fewer / less

Fewer refers to items that can be counted (count nouns).

- *Fewer* employees took the offer than we expected.

Less refers to mass quantities or amounts (mass nouns).

- Because we had *less* rain this year, the crop yield decreased.

See also English as a second language and nouns.

figuratively / literally

Literally means "actually" and is often confused with *figuratively*, which means "metaphorically." To say that someone "literally turned green with envy" would mean that the person actually changed color.

- In the winner's circle the jockey was, *figuratively* speaking, ten feet tall.
- When he said, "Let's bury our competitors," he did not mean it *literally*.

Avoid the use of *literally* to reinforce the importance of something.

- She was ~~literally~~ the best of the applicants.

figures of speech

A figure of speech is an imaginative expression that often compares two things that are basically not alike but have at least one thing in common. For example, if a device is cone-shaped and has an opening at the narrow end, you might say that it looks like a volcano.

Figures of speech can clarify the unfamiliar by relating a new concept to one with which readers are familiar. In that respect, they help establish understanding between the specialist and the nonspecialist. Figures of speech can also help translate the abstract into the concrete; in the process of doing so, they can also make writing more colorful and graphic. (See also abstract/concrete words.) A figure of speech must make sense, however, to achieve the desired effect.

ILLOGICAL Without the fuel of tax incentives, our economic engine would operate less efficiently.
[It would not operate at all without fuel.]

Figures of speech also must be consistent to be effective.

- We must get our research program *back on course*, and we are counting on you to ~~carry the ball.~~ *steer the effort.*

A figure of speech should not overshadow the point the writer is trying to make. In addition, it is better to use no figure of speech at all than to use a trite one. A surprise that comes "like a bolt out of the blue" seems stale and not much of a surprise at all. See also clichés.

Types of Figures of Speech

Analogies are comparisons that show the ways in which two objects or concepts are similar, often used to make one of them easier to understand. The following example explains a computer search technique by comparing it to the use of keywords in a dictionary.

- The search technique used in *indexed sequential processing* is similar to a search technique you might use to find the page on which a particular word is located in a dictionary. You might scan the keywords located at the top of each dictionary page that identify the first and last words on each page until you find the keywords that encompass the word you seek. Indexed sequential processing works the same way with computer files.

Hyperboles are gross exaggerations used to achieve an effect or emphasis.

- We were *dead* after working all night on the report.

Litotes are understatements, for emphasis or effect, achieved by denying the opposite of the point you are making.

- Two hundred dollars is not a small price for a book.

Metaphors are figures of speech that point out similarities between two things by treating them as though they were the same thing.

- He is the tech-support department's *utility infielder.*

Metonyms are figures of speech that use one aspect of a thing to represent it, such as *the blue* for the sky and *wheels* for a car.

- The *hard-hat* sector of the labor force was especially hurt by the terms of the contract.

Personification is a figure of speech that attributes human characteristics to nonhuman things or abstract ideas. We might refer, for example, to the *birth* of a planet or apply emotions to machines.

- She said that she was frustrated with the *stubborn* computer system.

Similes are direct comparisons of two essentially unlike things, linking them with the word *like* or *as.*

- His feelings about his competitor are so bitter that in recent conversations with his staff he returned to the subject compulsively, *like someone scratching an itch.*

Avoid figures of speech in global communication and international correspondence because those in other cultures may translate figures of speech literally and be confused by their meanings.

fine

When used in expressions such as "I feel *fine*" or "a *fine* day," *fine* is colloquial and, like the word *nice*, is too vague for technical writing. Use the word *fine* to mean "refined," "delicate," or "pure."

- A *fine* film of oil covered the surface of the water.
- *Fine* crystal is made in Austria.
- She made a *fine* distinction between the two possible meanings of the disputed passage.

first / firstly

First and *firstly* are both adverbs. Avoid *firstly* in favor of *first*, which sounds less stiff than *firstly*. The same is true of other ordinal numbers, such as *second*, *third*, and so on.

flammable / inflammable / nonflammable

Both *flammable* and *inflammable* mean "capable of being set on fire." Because the *in-* prefix usually causes the base word to take its opposite meaning (*incapable*, *incompetent*), use *flammable* instead of *inflammable* to avoid possible misunderstanding. (The cargo of gasoline is *flammable.*) *Nonflammable* is the opposite, meaning "not capable of being set on fire." (The asbestos suit was *nonflammable.*)

flowcharts

A flowchart is a diagram using symbols, words, or pictures to show the stages of a process in sequence from beginning to end. A flowchart provides an overview of a process and allows the reader to identify its essential steps quickly and easily. Flowcharts can take several forms. The steps might be represented by labeled blocks, as shown in Figure F–2; pictorial symbols, as shown in Figure F–3; or ISO (International Organization for Standardization) symbols, as shown in Figure F–4.

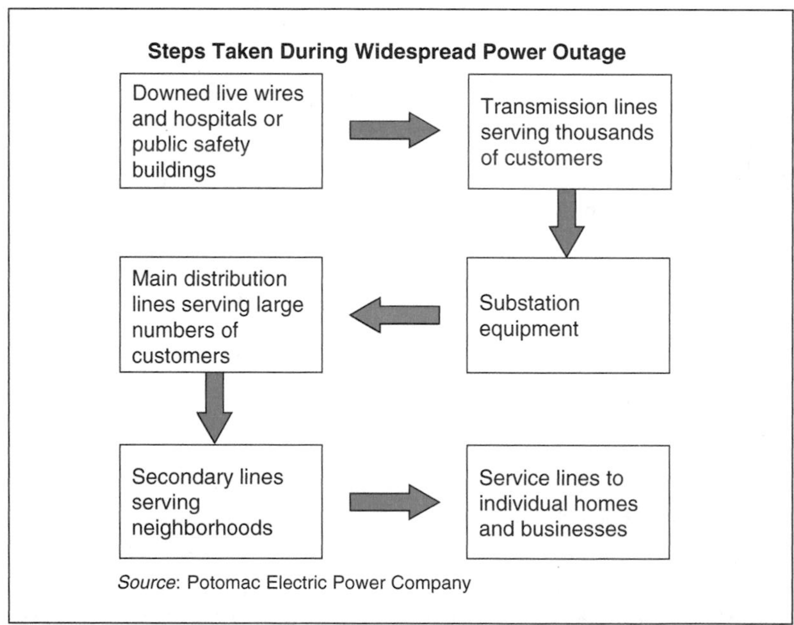

FIGURE F–2. Flowchart Using Labeled Blocks (Depicting Electric Utility Power Restoration Process)

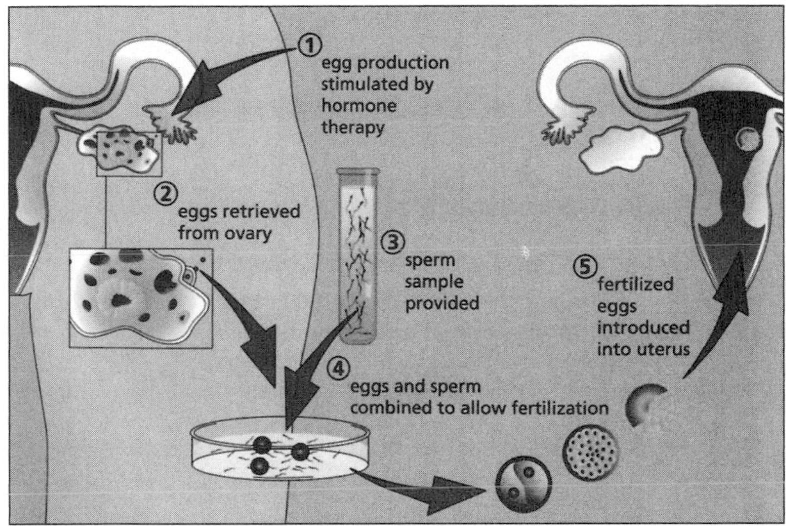

FIGURE F–3. Flowchart Using Pictorial Symbols. *Source*: FDA Consumer, Nov.—Dec. 2004. U.S. Food and Drug Administration.

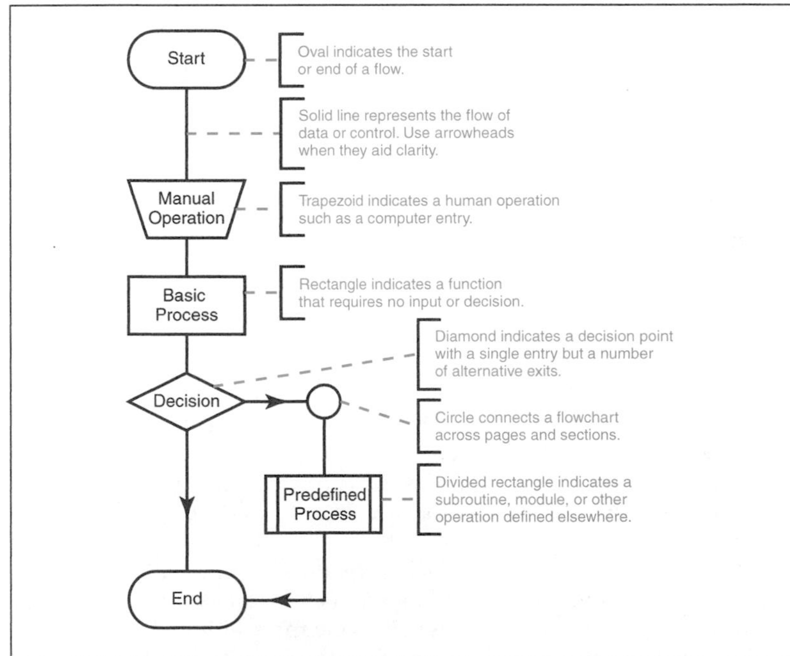

FIGURE F–4. Common ISO Flowchart Symbols (with Annotations).

F

Writer's Checklist: Creating Flowcharts

- ☑ Label each step in the process or identify each step with labeled blocks, pictorial representations, or standardized symbols.
- ☑ Follow the standard flow directions: left to right and top to bottom. When the flow is otherwise, indicate that with arrows.
- ☑ Include a key (or callouts) if the flowchart contains symbols your readers may not understand.
- ☑ Use standardized symbols for flowcharts that document computer programs and other information-processing procedures, as set forth in *Information Processing—Documentation Symbols and Conventions for Data, Program and System Flowcharts, Program Network Charts, and System Resources Charts*, ISO publication 5807-1985 (E).

For advice on integrating flowcharts into your text, see visuals. See also global graphics.

F

footnotes (*see* documenting sources)

forceful / forcible

Although *forceful* and *forcible* are both adjectives meaning "characterized by or full of force," *forceful* is usually limited to persuasive ability and *forcible* to physical force.

- John made a *forceful* presentation at the committee meeting.
- The police report listed three *forcible* entries into the building.

foreign words in English

The English language has a long history of borrowing words from other languages. Most borrowing occurred so long ago that we seldom recognize the borrowed terms (also called *loan words*) as being of foreign origin (*kindergarten* from German, *animal* from Latin, *church* from Greek).

Words not likely to be familiar to readers or not fully assimilated into the English language are set in italics (*sine qua non, coup de grâce, in res, in camera*). Most dictionaries offer guidance, although you should also be guided by the context. Even when foreign words have been fully assimilated, they often retain their diacritical marks (cliché, résumé, vis-à-vis). As foreign words are absorbed into English, their plural forms give way to English plurals (*agenda* becomes *agendas* and *formulae* becomes *formulas*).

Generally, foreign expressions should be used only if they serve a real need. (See also e.g./i.e. and etc.) The overuse of foreign words in an attempt to impress your reader or achieve elegance is affectation. Effective communication can be accomplished only if your readers understand what you write; therefore, choose foreign expressions only when they make an idea clearer.

F

foreword / forward

Although the pronunciation is the same, the spellings and meanings of these two words are quite different. The word *foreword* is a noun meaning "introductory statement at the beginning of a book or other work." (The director wrote a *foreword* for the proposal.) The word *forward* is an adjective or adverb meaning "at or toward the front."

- Move the lever to the *forward* position on the panel. [adjective]
- Turn the dial until the needle begins to move *forward.* [adverb]

formal reports

DIRECTORY

Formal reports are usually written accounts of major projects that require substantial research, and they often involve more than one writer. See also collaborative writing.

Most formal reports are divided into three primary parts—front matter, body, and back matter—each of which contains a number of elements. The number and arrangement of the elements may vary, depending on the subject, the length of the report, and the kinds of material covered. Further, many organizations have a preferred style for formal reports and furnish guidelines for report writers to follow. If you are not required to follow a specific style, use the format recommended in this entry. The following list includes most of the elements a formal report might contain, in the order they typically appear. (The items shown with page numbers appear in the sample formal report on pages 203–19.) Often, a cover letter or memo (only for internal documents) precedes the front matter and identifies the report by title, the person or persons to whom it is being sent, and the reason it was written, as shown on page 203.

Front Matter

The front matter serves several functions: It gives readers a general idea of the writer's purpose; it gives an overview of the type of information in the report; and it lists where specific information is covered in the report. Not all formal reports include every element of front matter described here. A title page and table of contents are usually mandatory, but the scope of the report and its intended audience determine whether the other elements are included.

Title Page. Although the formats of title pages may vary, they often include the following items:

- *The full title of the report.* The title describes the topic, scope, and purpose of the report, as described in titles.
- *The name of the writer(s), principal investigator(s), or compiler(s).* Sometimes contributors identify themselves by their job title in the organization or by their tasks in contributing to the report (Olivia Jones, Principal Investigator).
- *The date or dates of the report.* For one-time reports, the date shown is the date the report is distributed. For periodic reports (monthly, quarterly, or yearly), the subtitle shows the period that the report covers. Elsewhere on the title page, the date shown is the date the report is distributed.
- *The name of the organization for which the writer(s) works.*
- *The name of the organization to which the report is being submitted.* This information is included if the report is written for a customer or client.

The title page, as in the example on page 204, although unnumbered, is considered page i. The back of the title page, which is blank and unnumbered, is considered page ii, and the abstract falls on page iii. The body of the report begins with Arabic number 1, and a new chapter or large section typically begins on a new right-hand (odd-numbered) page. Reports with printing on only one side of each sheet can be numbered consecutively regardless of where new sections begin. Center page numbers at the bottom of each page throughout the report.

Abstract. An abstract, which normally follows the title page, highlights the major points of the report, as shown on page 205, enabling readers to decide whether to read the report.

Table of Contents. A table of contents lists all the major sections or headings of the report in their order of appearance, as shown on page 206, along with their page numbers.

List of Figures. When a report contains more than five figures, list them, along with their page numbers, in a separate section, beginning on a new page immediately following the table of contents. Number figures consecutively with Arabic numbers. Figures include all visuals contained in the report—drawings, photographs, maps, charts, and graphs.

List of Tables. When a report contains more than five tables, list them, along with their titles and page numbers, in a separate section immediately following the list of figures (if there is one). Number tables consecutively with Arabic numbers.

Foreword. A foreword is an optional introductory statement about a formal report or book that is written by someone other than the author(s). The foreword author is usually an authority in the field or an executive of the organization sponsoring the report. That author's name and affiliation appear at the end of the foreword, along with the date it was written. The foreword generally provides background information about the publication's significance and places it in the context of other works in the field. The foreword precedes the preface when a work has both.

Preface. The preface, another type of optional introductory statement, is written by the author(s) of the formal report. It may announce the work's purpose, scope, and context (including any special circumstances leading to the work). A preface may also specify the audience for a work, contain acknowledgments of those who helped in its preparation, and cite permission obtained for the use of copyrighted works. See also copyright.

List of Abbreviations and Symbols. When the report uses numerous abbreviations and symbols that readers may not be able to interpret, the front matter may include a section that lists symbols and abbreviations with their meanings.

Body

The body is the section of the report that describes in detail the methods and procedures used to generate the report, demonstrates how results were obtained, describes the results, draws conclusions, and, if appropriate, makes recommendations.

Executive Summary. The body of the report begins with the executive summary (if included), which provides a more complete overview of the report than an abstract does. As shown on pages 207–08, the executive summary states the report's purpose, provides background information, and describes the scope of the report; it also summarizes conclusions and recommendations.

Introduction. The introduction gives readers any general information, such as the report's purpose and scope, necessary to understand the detailed information in the rest of the report (see pages 209–11).

Text. The text of the body presents, as appropriate, the details of how the topic was investigated, how a problem was solved, what alternatives were explored, and how the best choice among them was selected. This information is enhanced by the use of visuals, tables, and references that both clarify the text and persuade the reader.

Conclusions. The conclusions section pulls together the results of the research and interprets the findings of the report, as shown on pages 217–18.

Recommendations. Recommendations, which are sometimes combined with the conclusions, as is true on pages 217–18, state what course of action should be taken based on the earlier arguments and conclusions of the study. The recommendations section may state, for example, "We should pursue new markets in . . ." or "I recommend we expand our marketing efforts on the Internet by. . . ."

F

Explanatory Notes. Occasionally, reports contain notes that amplify terms or points for some readers that would be a distraction for others. If such notes are not included as footnotes on the page where the term or point appears, they may appear in a final "Notes" section.

References (or Works Cited). A list of references or works cited appears in a separate section if the report refers to material in, or quotes directly from, a published work or other research sources, including online sources. If your employer has a preferred reference style, follow it; otherwise, use the guidelines provided in the entry documenting sources. For a relatively short report, place the references at the end of the body of the report, as shown on page 219. For a report with a number of sections or chapters, place the reference section at the end of each major section or chapter. In either case, title the reference or works-cited section as such and begin it on a new page. If a particular reference appears in more than one section or chapter, repeat it in full in each appropriate reference section. See also plagiarism and quotations.

DIGITAL TIPS CREATING STYLES AND TEMPLATES

In most word-processing programs, you can create styles—sets of formatting characteristics for text elements such as headings, paragraphs, and lists—that allow you to automate much of the work of formatting your report. Once you have specified your styles, save them as a template and use it each time you create a formal report. For step-by-step instructions, see <*bedfordstmartins.com/alredtech*> and select *Digital Tips*, "Creating Styles and Templates."

Back Matter

The back matter of a formal report contains supplementary material, such as where to find additional information about the topic (bibliography), and expands on certain subjects (appendixes). Other back-matter elements define the terms used (glossary) and provide information on how to easily locate information in the report (index). For very large formal reports, back-matter sections may be individually numbered.

Appendixes. An appendix clarifies or supplements the body with information that is too detailed or lengthy for the primary audience but that is relevant to secondary audiences.

Bibliography. A bibliography is an alphabetical list of all sources that were consulted (not just those cited) in researching the report. A bibliography is not necessary if the reference listing contains a complete list of sources.

Glossary. A glossary is an alphabetical list of specialized terms used in the report and their definitions.

Index. An index is an alphabetical list of all the major topics and subtopics discussed in the report. It cites the page numbers where discussion of each topic can be found and allows readers to find information on topics quickly and easily. The index is always the final section of a report. See also indexing.

Sample Formal Report

Figure F–5 shows the typical sections of a formal report. Keep in mind that the number and arrangement of the elements vary, depending on the specific subject and requirements of an organization or a client.

CGF Aircraft Corporation
Memo

To: Members of the Ethics and Business Conduct Committee
From: Susan Litzinger, Director of Ethics and Business Conduct *SL*
Date: March 4, 2005
Subject: Reported Ethics Cases, 2004

F

Enclosed is "Reported Ethics Cases: Annual Report, 2004." This report, required by CGF Policy CGF-EP-01, contains a review of the ethics cases handled by CGF ethics officers and managers during 2004, the first year of our Ethics Program.

Identifies topic

The ethics cases reported are analyzed according to two categories: (1) major ethics cases, or those potentially involving serious violations of company policy or illegal conduct, and (2) minor ethics cases, or those that do not involve serious policy violations or illegal conduct. The report also examines the mode of contact in all of the reported cases and the disposition of the substantiated major ethics cases.

Briefly summarizes content

It is my hope that this report will provide the Committee with the information needed to assess the effectiveness of the first year of CGF's Ethics Program and to plan for the coming year. Please let me know if you have any questions about this report or if you need any further information. I may be reached at (555) 211-2121 and by e-mail at sl@cgf.com.

Offers contact information

Enc.

FIGURE F–5. Formal Report (Cover Memo). Reprinted and adapted by permission of Susan Litzinger, a student at Pennsylvania State University, Altoona.

F

Full title

REPORTED ETHICS CASES
2004 Annual Report

Author's name and job title

Prepared by Susan Litzinger
Director of Ethics and Business Conduct

Report Distributed March 4, 2005

Company name

Prepared for
The Ethics and Business Conduct Committee
CGF Aircraft Corporation

No page number

FIGURE F–5. Formal Report (*continued*) (Title Page)

Reported Ethics Cases—2004

ABSTRACT

Summarizes purpose

Methods and scope

Conclusions and recommendations

This report examines the nature and disposition of 3,458 ethics cases handled companywide by CGF Aircraft Corporation's ethics officers and managers during 2004. The purpose of this annual report is to provide the Ethics and Business Conduct Committee with the information necessary for assessing the effectiveness of the Ethics Program's first year of operation. Records maintained by ethics officers and managers of all contacts were compiled and categorized into two main types: (1) major ethics cases, or cases involving serious violations of company policies or illegal conduct, and (2) minor ethics cases, or cases not involving serious policy violations or illegal conduct. This report provides examples of the types of cases handled in each category and analyzes the disposition of 30 substantiated major ethics cases. Recommendations for planning for the second year of the Ethics Program are (1) continuing the channels of communication now available in the Ethics Program, (2) increasing financial and technical support for the Ethics Hotline, (3) disseminating the annual ethics report in some form to employees to ensure employee awareness of the company's commitment to uphold its Ethics Policies and Procedures, and (4) implementing some measure of recognition for ethical behavior to promote and reward ethical conduct.

Lowercase Roman numerals used on front matter pages

iii

FIGURE F–5. Formal Report (*continued*) (Abstract)

F

Reported Ethics Cases — 2004

Uniform heading styles

TABLE OF CONTENTS

Indented subheads

Page numbers for each entry

iv

FIGURE F–5. Formal Report (*continued*) (Table of Contents)

Reported Ethics Cases—2004

EXECUTIVE SUMMARY

States purpose

This report examines the nature and disposition of the 3,458 ethics cases handled by the CGF Aircraft Corporation's ethics officers and managers during 2004. The purpose of this report is to provide CGF's Ethics and Business Conduct Committee with the information necessary for assessing the effectiveness of the first year of the company's Ethics Program.

Provides background information

Effective January 1, 2004, the Ethics and Business Conduct Committee (the Committee) implemented a policy and procedures for the administration of CGF's new Ethics Program. The purpose of the Ethics Program, established by the Committee, is to "promote ethical business conduct through open communication and compliance with company ethics standards." The Office of Ethics and Business Conduct was created to administer the Ethics Program. The director of the Office of Ethics and Business Conduct, along with seven ethics officers throughout the corporation, was given the responsibility for the following objectives:

- Communicate the values and standards for CGF's Ethics Program to employees.
- Inform employees about company policies regarding ethical business conduct.
- Establish companywide channels for employees to obtain information and guidance in resolving ethics concerns.
- Implement companywide ethics-awareness and education programs.

Employee accessibility to ethics information and guidance was available through managers, ethics officers, and an ethics hotline.

Describes scope

Major ethics cases were defined as those situations potentially involving serious violations of company policies or illegal conduct. Examples of major ethics cases included cover-up of defective workmanship or use of defective parts in products; discrimination in hiring and promotion; involvement in monetary or other kickbacks; sexual harassment; disclosure of proprietary or company information; theft; and use of corporate Internet resources for inappropriate purposes, such as conducting personal business, gambling, or access to pornography.

1

FIGURE F–5. Formal Report (*continued*) (Executive Summary)

Reported Ethics Cases—2004

Minor ethics cases were defined as including all reported concerns not classified as major ethics cases. Minor ethics cases were classified as informational queries from employees, situations involving coworkers, and situations involving management.

F

Summarizes conclusions

The effectiveness of CGF's Ethics Program during the first year of implementation is most evidenced by (1) the active participation of employees in the program and the 3,458 contacts employees made regarding ethics concerns through the various channels available to them, and (2) the action taken in the cases reported by employees, particularly the disposition of the 30 substantiated major ethics cases. Disseminating information about the disposition of ethics cases, particularly information about the severe disciplinary actions taken in major ethics violations, sends a message to employees that unethical or illegal conduct will not be tolerated.

Includes recommendations

Based on these conclusions, recommendations for planning the second year of the Ethics Program are (1) continuing the channels of communication now available in the Ethics Program, (2) increasing financial and technical support for the Ethics Hotline, the most highly utilized mode of contact in the ethics cases reported in 2004, (3) disseminating this report in some form to employees to ensure their awareness of CGF's commitment to uphold its Ethics Policies and Procedures, and (4) implementing some measure of recognition for ethical behavior, such as an "Ethics Employee of the Month" award to promote and reward ethical conduct.

Executive summary is about 10 percent of report length

2

FIGURE F–5. Formal Report (*continued*) (Executive Summary)

Reported Ethics Cases—2004

INTRODUCTION

Opening states purpose

This report examines the nature and disposition of the 3,458 ethics cases handled companywide by CGF's ethics officers and managers during 2004. The purpose of this report is to provide the Ethics and Business Conduct Committee with the information necessary for assessing the effectiveness of the first year of CGF's Ethics Program. Recommendations are given for the Committee's consideration in planning for the second year of the Ethics Program.

Subheads signal shifts in topic

Ethics and Business Conduct Policies and Procedures

Effective January 1, 2004, the Ethics and Business Conduct Committee (the Committee) implemented Policy CGF-EP-01 and Procedure CGF-EP-02 for the administration of CGF's new Ethics Program. The purpose of the Ethics Program, established by the Committee, is to "promote ethical business conduct through open communication and compliance with company ethics standards" (CGF, "Ethics and Conduct").

The Office of Ethics and Business Conduct was created to administer the Ethics Program. The director of the Office of Ethics and Business Conduct, along with seven ethics officers throughout CGF, was given the responsibility for the following objectives:

List identifies key points

- Communicate the values, standards, and goals of CGF's Ethics Program to employees.
- Inform employees about company ethics policies.
- Provide companywide channels for employee education and guidance in resolving ethics concerns.
- Implement companywide programs in ethics awareness, education, and recognition.
- Ensure confidentiality in all ethics matters.

Employee accessibility to ethics information and guidance became the immediate and key goal of the Office of Ethics and Business Conduct in its first year of operation. The following channels for contact were set in motion during 2004:

3

FIGURE F–5. Formal Report (*continued*) (Introduction)

F

Reported Ethics Cases — 2004

- Managers throughout CGF received intensive ethics training; in all ethics situations, employees were encouraged to go to their managers as the first point of contact.
- Ethics officers were available directly to employees through face-to-face or telephone contact, to managers, to callers using the ethics hotline, and by e-mail.
- The Ethics Hotline was available to all employees, 24 hours a day, 7 days a week, to anonymously report ethics concerns.

Confidentiality Issues

CGF's Ethics Policies ensures confidentiality and anonymity for employees who raise genuine ethics concerns. Procedure CGF-EP-02 guarantees appropriate discipline, up to and including dismissal, for retaliation or retribution against any employee who properly reports any genuine ethics concern.

Documentation of Ethics Cases

The following requirements were established by the director of the Office of Ethics and Business Conduct as uniform guidelines for the documentation by managers and ethics officers of all reported ethics cases:

Includes detailed methods

- Name, position, and department of individual initiating contact, if available
- Date and time of contact
- Name, position, and department of contact person
- Category of ethics case
- Mode of contact
- Resolution

Managers and ethics officers entered the required information in each reported ethics case into an ACCESS database file, enabling efficient retrieval and analysis of the data.

4

FIGURE F–5. Formal Report (*continued*) (Introduction)

Reported Ethics Cases—2004

Major/Minor Category Definition and Examples

Major ethics cases were defined as those situations potentially involving serious violations of company policies or illegal conduct. Procedure CGF-EP-02 requires notification of the Internal Audit and the Law Departments in serious ethics cases. The staffs of the Internal Audit and the Law Departments assume primary responsibility for managing major ethics cases and for working with the employees, ethics officers, and managers involved in each case.

Examples of situations categorized as major ethics cases:

- Cover-up of defective workmanship or use of defective parts in products
- Discrimination in hiring and promotion
- Involvement in monetary or other kickbacks from customers for preferred orders
- Sexual harassment
- Disclosure of proprietary customer or company information
- Theft
- Use of corporate Internet resources for inappropriate purposes, such as conducting private business, gambling, or gaining access to pornography

Organized by decreasing order of importance

Minor ethics cases were defined as including all reported concerns not classified as major ethics cases. Minor ethics cases were classified as follows:

- Informational queries from employees
- Situations involving coworkers
- Situations involving management

5

FIGURE F–5. Formal Report (*continued*) (Introduction)

F

Reported Ethics Cases—2004

ANALYSIS OF REPORTED ETHICS CASES

Reported Ethics Cases, by Major/Minor Category

Text introduces figure

CGF ethics officers and managers companywide handled a total of 3,458 ethics situations during 2004. Of these cases, only 172, or 5 percent, involved reported concerns of a serious enough nature to be classified as major ethics cases (see Figure 1). Major ethics cases were defined as those situations potentially involving serious violations of company policy or illegal conduct.

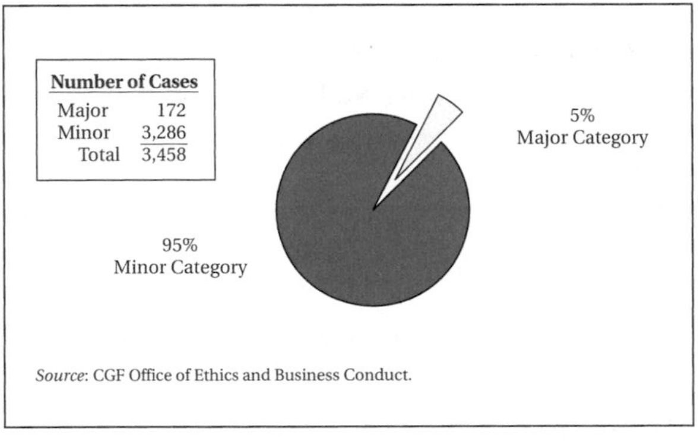

Number and title identify figure

Figure 1. Reported ethics cases, by major/minor category in 2004.

Major Ethics Cases

Of the 172 major ethics cases reported during 2004, 57 percent, upon investigation, were found to involve unsubstantiated concerns. Incomplete information or misinformation most frequently was discovered to be the cause of the unfounded concerns of misconduct in 98 cases. Forty-four cases, or 26 percent of the total cases reported, involved incidents partly substantiated by ethics officers as serious misconduct; however, these cases were discovered to also involve inaccurate information or unfounded issues of misconduct.

6

FIGURE F–5. Formal Report (*continued*) (Body)

Reported Ethics Cases — 2004

Only 17 percent of the total number of major ethics cases, or 30 cases, were substantiated as major ethics situations involving serious ethical misconduct or illegal conduct (CGF, "2004 Ethics Hotline Results") (see Figure 2).

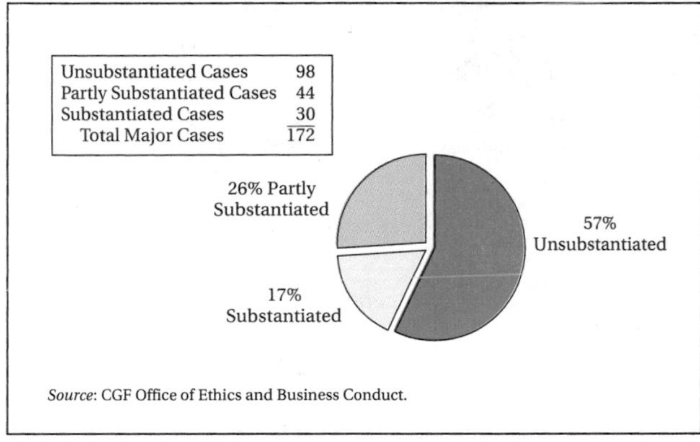

Figure 2. Major ethics cases in 2004.

Identifies source of infor-mation

Of the 30 substantiated major ethics cases, seven remain under investigation at this time, and two cases are currently in litigation. Disposition of the remainder of the 30 substantiated reported ethics cases included severe disciplinary action in five cases: the dismissal of two employees and the demotion of three employees. Seven employees were given written warnings, and nine employees received verbal warnings (see Figure 3).

7

FIGURE F–5. Formal Report (*continued*) (Body)

F

Reported Ethics Cases—2004

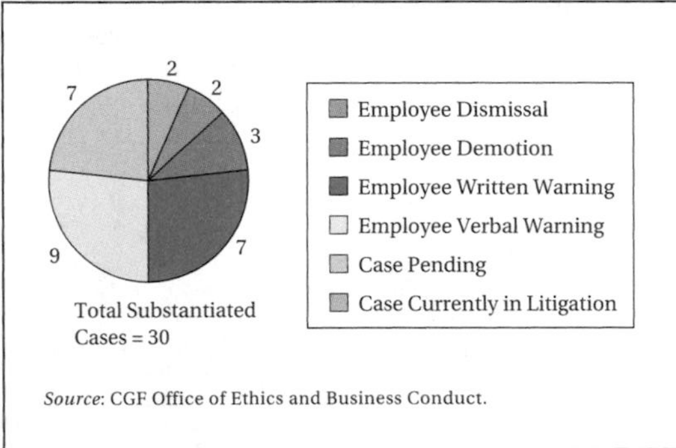

Figure 3. Disposition of substantiated major ethics cases in 2004.

Minor Ethics Cases

Minor ethics cases included those that did not involve serious violations of company policy or illegal conduct. During 2004, ethics officers and company managers handled 3,286 such cases. Minor ethics cases were further classified as follows:

Reports findings in detail

- Informational queries from employees
- Situations involving coworkers
- Situations involving management

As might be expected during the initial year of the Ethics Program implementation, the majority of contacts made by employees were informational, involving questions about the new policies and procedures. These informational contacts comprised 65 percent of all contacts of a minor nature and numbered 2,148. Employees made 989 contacts regarding ethics concerns involving coworkers and 149 contacts regarding ethics concerns involving management (see Figure 4).

8

FIGURE F–5. Formal Report (*continued*) (Body)

Reported Ethics Cases—2004

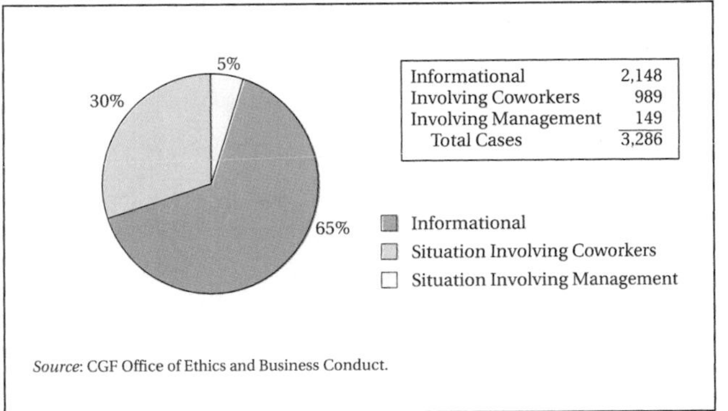

Figure 4. Minor ethics cases in 2004.

Mode of Contact

The effectiveness of the Ethics Program rested on the dissemination of information to employees and the provision of accessible channels through which employees could gain information, report concerns, and obtain guidance. Employees were encouraged to first go to their managers with any ethical concerns, because those managers would have the most direct knowledge of the immediate circumstances and individuals involved.

Assesses findings

Other channels were put into operation, however, for any instance in which an employee did not feel able to go to his or her manager. The ethics officers companywide were available to employees through telephone conversations, face-to-face meetings, and e-mail contact. Ethics officers also served as contact points for managers in need of support and assistance in handling the ethics concerns reported to them by their subordinates.

The Ethics Hotline became operational in mid-January 2004 and offered employees assurance of anonymity and confidentiality. The Ethics Hotline was accessible to all employees on a 24-hour, 7-day basis. Ethics officers companywide took responsibility on a rotational basis for handling calls reported through the hotline.

9

FIGURE F–5. Formal Report (*continued*) (Body)

F

Reported Ethics Cases—2004

In summary, ethics information and guidance was available to all employees during 2004 through the following channels:

- Employee to manager
- Employee telephone, face-to-face, and e-mail contact with ethics officer
- Manager to ethics officer
- Employee Hotline

Bulleted lists help organize and summarize information

The mode of contact in the 3,458 reported ethics cases was as follows (see Figure 5):

- In 19 percent of the reported cases, or 657, employees went to managers with concerns.
- In 9 percent of the reported cases, or 311, employees contacted an ethics officer.
- In 5 percent of the reported cases, or 173, managers sought assistance from ethics officers.
- In 67 percent of the reported cases, or 2,317, contacts were made through the Ethics Hotline.

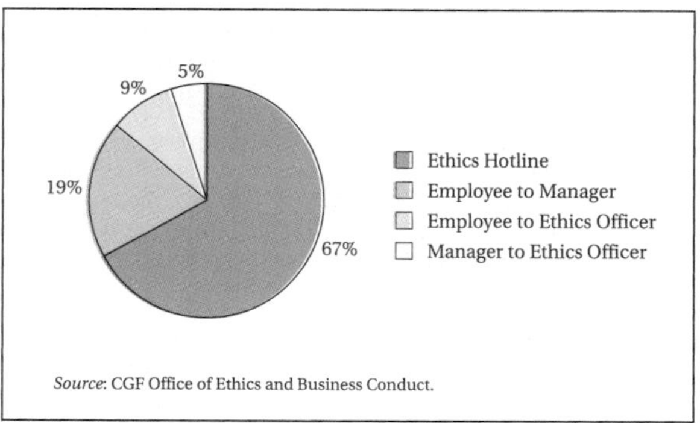

Figure 5. Mode of contact in reported ethics cases in 2004.

10

FIGURE F–5. Formal Report (*continued*) (Body)

Pulls together findings

CONCLUSIONS AND RECOMMENDATIONS

The effectiveness of CGF's Ethics Program during the first year of implementation is most evidenced by (1) the active participation of employees in the program and the 3,458 contacts employees made regarding ethics concerns through the various channels available to them, and (2) the action taken in the cases reported by employees, particularly the disposition of the 30 substantiated major ethics cases.

Uses sources for support

One of the 12 steps to building a successful Ethics Program identified by Frank Navran in *Workforce* magazine is an ethics communication strategy. Navran explains that such a strategy is crucial in ensuring

> that employees have the information they need in a timely and usable fashion and that the organization is encouraging employee communication regarding the values, standards and the conduct of the organization and its members. (Navran 119)

The 3,458 contacts by employees during 2004 attest to the accessibility and effectiveness of the communication channels that exist in CGF's Ethics Program.

An equally important step in building a successful ethics program is listed by Navran as "Measurements and Rewards," which he explains as follows:

> In most organizations, employees know what's important by virtue of what the organization measures and rewards. If ethical conduct is assessed and rewarded, and if unethical conduct is identified and dissuaded, employees will believe that the organization's principals mean it when they say the values and code of ethics are important. (Navran 121)

Interprets findings

Disseminating information about the disposition of ethics cases, particularly information about the severe disciplinary actions taken in major ethics violations, sends a message to employees that unethical or illegal conduct will not be tolerated. Making public the tough-minded actions taken in cases of ethical misconduct provides "a golden opportunity to make other employees aware that the behavior is unacceptable and why" (Ferrell and Gardiner 129).

FIGURE F–5. Formal Report (*continued*) (Conclusions and Recommendations)

F

Reported Ethics Cases — 2004

With these two points in mind, I offer the following recommendations for consideration for plans for the Ethics Program's second year:

Recommends specific steps

- Continuation of the channels of communication now available in the Ethics Program
- Increased financial and technical support for the Ethics Hotline, the most highly utilized mode of contact in the reported ethics cases in 2004
- Dissemination of this report in some form to employees to ensure employees' awareness of CGF's commitment to uphold its Ethics Policies and Procedures
- Implementation of some measure of recognition for ethical behavior, such as an "Ethics Employee of the Month," to promote and reward ethical conduct

To ensure that employees see the value of their continued participation in the Ethics Program, feedback is essential. The information in this annual review, in some form, should be provided to employees. Knowing that the concerns they reported were taken seriously and resulted in appropriate action by Ethics Program administrators would reinforce employee involvement in the program. While the negative consequences of ethical misconduct contained in this report send a powerful message, a means of communicating the *positive* rewards of ethical conduct at CGF should be implemented. Various options for recognition of employees exemplifying ethical conduct should be considered and approved.

Links recommendations to company goal

Continuation of the Ethics Program's successful 2004 operations, with the implementation of the above recommendations, should ensure the continued pursuit of the Ethics Program's purpose: "to promote a positive work environment that encourages open communication regarding ethics and compliance issues and concerns."

12

FIGURE F–5. Formal Report (*continued*) (Conclusions and Recommendations)

References appear on a separate page

Reported Ethics Cases—2004

WORKS CITED

CGF. "Ethics and Conduct at CGF Aircraft Corporation." 1 Jan. 2004. 11 Feb. 2005. <http://www.cgfac.com/aboutus/ethics.html>.

---. "2004 Ethics Hotline Investigation Results." 15 Jan. 2005. 11 Feb. 2005. <www.cgfac.com/html/ethics.html>.

Ferrell, O. C., and Gareth Gardiner. In Pursuit of Ethics: Tough Choices in the World of Work. Springfield, IL: Smith Collins, 1991.

Kelley, Tina. "Corporate Prophets, Charting a Course to Ethical Profits." New York Times 8 Feb. 1998: BU12.

Navran, Frank. "12 Steps to Building a Best-Practices Ethics Program." Workforce Sept. 1997: 117–22.

This report uses MLA style

13

F

FIGURE F–5. Formal Report (*continued*) (Works Cited)

format

Format refers to both the organization of information in a document and the physical arrangement of information on the page.

In one sense, format refers to the fact that some types of job-related writing, such as formal reports and correspondence, are characterized by conventions that govern the scope and placement of information. For example, in formal reports, the table of contents precedes the preface but follows the title page and the abstract. Likewise, although variations exist, parts of letters—such as inside address, salutation, and complimentary closing—typically are arranged in standard patterns.

Format also refers to the general physical appearance (as discussed in layout and design) of a finished document, whether printed or electronic. See also e-mail and Web design.

former / latter

Former and *latter* should be used to refer to only two items in a sentence or paragraph.

- The president and his aide emerged from the conference, the *former* looking nervous and the *latter* looking glum.

Because these terms make the reader look to previous material to identify the reference, they complicate reading and are best avoided.

forms design

Forms provide an economical and uniform way to gather, record, and evaluate data. They help respondents provide necessary information in the form of standardized answers that are easy for you to evaluate and tabulate. Figures F–6 and F–7 are two examples of forms.

Preparing a Form

An effective form makes it easy for one person to supply information and for another person to retrieve, record, and interpret that information. Ideally, a form should be self-explanatory to someone seeing it for the first time. When preparing a form, determine the kind of information you are seeking and arrange the questions in a logical order. See also questionnaires.

Annual Reappointment Form Effective July 1, 20___ to June 30, 20___

CHILDREN'S MEDICAL CENTER
1735 Chapel Street
Toledo, Ohio 43692

NAME:

List appointments or offices held, teaching positions, independent studies in medical or dental societies or other medical organizations, and any other professional recognitions you would like to have included in your file:

____
____
____
____
____
____
____

Open-ended questions

Do you wish a change in your privileges? Yes ☐ No ☐ If so, specify:

____

Labeled boxes

Have there been any changes in your board specialties? Yes ☐ No ☐
If yes: Date of Change ______________
Specialty Board ______________

Signature ______________
Date ______________

Signature and date

WHITE: Committee Chair BLUE: Human Resources
PINK: Employee

Routing instructions

FIGURE F–6. Form (for Staff Reappointment)

Choosing Online or Paper. Many organizations are moving their forms online. Forms especially well suited for online use include job applications, conference or seminar registrations, medical records, and order forms. Web-based and other online forms both standardize respondents' interfaces and link to databases that tabulate and interpret data. Online forms ensure that the form will be completed correctly

F

104-M S A

Section 125 Flexible Spending Account (FSA) Claim Form

Form title

Employee Name: ______________________

Social Security Number: _____-____-______

Name of Employer: ______________________

Employee Signature: ______________________

Writing lines

Instructions

Complete section below for medical, dental, or vision reimbursement

CLAIM TYPE I: MEDICAL CARE ACCOUNT

Amount of Expense Incurred: $______________________

Dates of Services: From: ______________ To: ______________

Instructions

Complete section below for reimbursement of care for your dependent provided by a child-care facility, adult dependent-care center, or caretaker

CLAIM TYPE II: DEPENDENT-CARE ACCOUNT

Amount of Expense Incurred: $______________________

Name of Dependent-Care Provider: ______________________

Provider Social Security or Federal ID Number: _____-____-______

Mail or fax form with documentation to:
Specialized Benefit Services, Inc.
P. O. Box 498
Framingham, MA 01702
Fax: (508) 877-1182
For additional claim forms: www.sbsclaims.com

Mailing and contact information

FIGURE F–7. Form (for a Medical Claim)

because their programs do not accept a form until all the necessary fields are completed. In addition, using online forms can eliminate the problems associated with distributing and collecting forms. However, using online forms can be difficult for people with limited computer literacy or access, so consider your readers carefully before opting to collect your responses online. See also Web design.

Writing Instructions. Place instructions at the beginning of the form or at the beginning of each section of the form and use headings or other design elements, such as the boldface type and shading in Figure F–7, to attract the readers' attention. When necessary, place instructions for distributing the various copies of multiple-copy forms at the bottom of the form and repeat them on every copy of the form, as in Figure F–6. Instructions for mailing or faxing the form should be clearly indicated, as in Figure F–7.

Choosing Response Types. Forms should ask questions in ways that are best suited to the types of data you hope to collect. The two main types of questions are open-ended questions and closed-ended questions.

- *Open-ended questions* allow respondents the freedom to choose their own words. Such questions are most appropriate if you wish to elicit answers you may not have anticipated (as in a complaint form) or if there are too many possible answers to use a multiple-choice format. However, the responses to open-ended questions are difficult to tabulate and analyze.
- *Closed-ended questions* provide a list of answers from which the respondent can select, limiting the range of possible responses. When you want to make sure you receive a standardized, easy-to-tabulate response, use any of the types of closed-ended questions that follow:
  - *Multiple choice*: Choose one (or sometimes more) from a preset list of options.
  - *Ranked choice*: Rank items according to preference, such as selecting vacation days or choosing job assignments.
  - *Forced choice*: Choose between two preset options, such as yes/no or male/female.

Wording Captions. Questions are normally presented as captions. Keep them brief, specific, and to the point; avoid unnecessary repetition by combining related information under an explanatory heading.

WORDY What make of car (or vehicle) do you drive? ________
What year was it manufactured? ________
What model is it? ________
What is the body style? ________

CONCISE Vehicle Information
Make ________ Year ________
Model ________ Body Style ________

F

If a requested date is other than the date on which the form is being filled out, the caption should read, for example, "Effective date" or "Date issued," rather than simply "Date." (See Figure F–6.) As in all writing, put yourself in your reader's place and imagine what sort of requests would be clear.

Sequencing Data. At the top of the form, clearly indicate preliminary information, such as the name of your organization, the title of your form, and any reference number. In the main portion of the form, include the entries you need to obtain the necessary data. At the end of the form, include space for a signature and a date.

Arrange requests for information in an order that will be the most logical to the person filling out the form.

- Sequence entries to fit the subject matter. For instance, a form requesting reimbursement for travel expenses would logically begin with the first day of the week (or month) and end with the last day of the appropriate period.
- If the response to one item is based on the response to another item, be sure the items appear in the correct order.
- Group requests for related information together whenever possible.
- For ease of reading, arrange entries from left to right and from top to bottom.

ETHICS NOTE Information gathered on forms can be of a sensitive or personal nature, so make sure to present questions in a way that is not invasive or illegal. Unless otherwise indicated on the form, the person filling out the form should have the expectation of confidentiality. If you are concerned about issues of confidentiality or legality, check your organization's policy or seek your instructor's advice. ✦

Designing a Form

You can design computer-generated forms specific to your needs with form-design-and-management software or with word-processing software. However you prepare the final version of a form, pay particular attention to design details, especially to the placement of entry lines and the amount of space allowed for responses.

Entry Lines. A form can be designed so that the person filling it out provides information on a writing line, in a writing block, or in square boxes. A *writing line* is simply a rule with a caption, as shown in Figure F–7.

A *writing block* is essentially the same as a writing line, except that each entry is enclosed in a ruled block, making it impossible for the respondent to associate a caption with the wrong line, as shown in Figure F–8. Some writing blocks arrange the captions horizontally, as in Figure F–8; other writing blocks list captions vertically, as shown in Figure F–9.

When it is possible to anticipate all likely responses, you can make the form easy to fill out by writing the question on the form, supplying a labeled box for each possible answer, and asking the respondent to

NAME		TELEPHONE
STREET ADDRESS		
CITY	STATE	ZIP CODE

FIGURE F–8. Writing Block for a Form

DESTINATION	
SOURCE	
SUPPLIER	
METHOD OF SHIPMENT	

FIGURE F–9. Vertical Captions for a Form

check the appropriate boxes. Such a design also makes it easy to tabulate the data. Be sure your questions are both simple and specific.

- Would your department order another MAX-PC? Yes ☐ No ☐

Spacing. Provide enough space to enable the person filling out the form to enter the data. Insufficient writing space makes it difficult for people to respond, resulting in responses that are hard to read. Reading responses that are too tightly spaced or that snake around the side of the form can cause eyestrain and errors. To ensure the usability of the form, have a coworker fill one out before printing the final version.

fortuitous / fortunate

When an event is *fortuitous*, it happens by chance or accident and without plan. Such an event may be lucky, unlucky, or neutral.

- My encounter with the general manager in Denver was entirely *fortuitous*; I had no idea he was there.

When an event is *fortunate*, it happens by good fortune or happens favorably.

- Our chance meeting had a *fortunate* outcome.

fragments (*see* **sentence fragments**)

functional shift

Many words shift easily from one part of speech to another, depending on how they are used. When they do, the process is called a *functional shift*, or a shift in function.

- It takes ten minutes to *walk* from the laboratory to the emergency room. However, the long *walk* from the laboratory to the emergency room reduces efficiency.
  [*Walk* shifts from verb to noun.]
- I talk to the Chicago office on the *phone* every day. He was concerned about the office *phone* expenses. He will *phone* the home office from London.
  [*Phone* shifts from noun to adjective to verb.]

- *After* we discuss the project, we will begin work. *After* lengthy discussions, we began work. The partners worked well together forever *after*.
  [*After* shifts from conjunction to preposition to adverb.]

Jargon is often the result of functional shifts. In hospitals, for example, an *attending physician* is referred to simply as the "attending" (a shift from an adjective to a noun). Likewise, in nuclear plant construction, a *reactor containment building* is called a "containment" (a shift from an adjective to a noun). Do not shift the function of a word indiscriminately merely to shorten a phrase or an expression. See also affectation and conciseness.

G

garbled sentences

A garbled sentence is one that is so tangled with structural and grammatical problems that it cannot be repaired. Garbled sentences often result from an attempt to squeeze too many ideas into one sentence.

- My job objectives are accomplished by my having a diversified background which enables me to operate effectively and efficiently, consisting of a degree in computer science, along with twelve years of experience, including three years in Staff Engineering-Packaging sets a foundation for a strong background in areas of analyzing problems and assessing economical and reasonable solutions.

Do not try to patch such a sentence; rather, analyze the ideas it contains, list them in a logical sequence, and then construct one or more entirely new sentences.

An analysis of the preceding example yields the following five ideas:

- My job requires that I analyze problems to find economical and workable solutions.
- My diversified background helps me accomplish my job.
- I have a computer-science degree.
- I have twelve years of job experience.
- Three of these years have been in Staff Engineering-Packaging.

Using those five ideas—together with parallel structure, sentence variety, subordination, and transition—the writer might have described the job as follows:

- My job requires that I analyze problems to find economical and workable solutions. Both my education and experience help me achieve this goal. Specifically, I have a computer-science degree and twelve years of job experience, three of which have been in the Staff Engineering-Packaging Department.

See also clarity, mixed constructions, and sentence construction.

gender

In English grammar, *gender* refers to the classification of nouns and pronouns as masculine, feminine, and neuter. The gender of most words can be identified only by the choice of the appropriate pronoun (*he*, *she*, *it*). Only these pronouns and a select few nouns (*buck/doe*) or noun forms (*heir/heiress*) reflect gender. Many such nouns have been replaced by single terms for both sexes. See also biased language and he/she.

Gender is important to writers because they must be sure that nouns and pronouns within a grammatical construction agree in gender.

TIPS FOR ASSIGNING GENDER

The English language has an almost complete lack of gender distinctions. That can be confusing for a nonnative speaker of English whose native language may assign gender. In the few cases in which English does make a gender distinction, there is a close connection between the assigning of gender and the sex of the subject. The few instances in which gender distinctions are made in English are summarized as follows:

Subject pronouns	he/she
Object pronouns	him/her
Possessive adjectives	his/her(s)
Some nouns	king/queen, boy/girl, bull/cow, etc.

When a noun, such as *doctor*, can refer to a person of either sex, you need to know the sex of the person to which the noun refers to determine the gender-appropriate pronoun.

- The doctor gave *her* patients advice. [Doctor is female.]
- The doctor gave *his* patients advice. [Doctor is male.]

When the sex of the noun antecedent is unknown, be sure to follow the guidelines for nonsexist writing in the entry biased language. (*Note*: Some English speakers refer to vehicles and countries as *she*, but contemporary usage favors *it*.)

A pronoun, for example, must agree with its noun antecedent in gender. We must refer to a woman as *she* or *her*, not as *it*; to a man as *he* or *him*, not as *it*; to a barn as *it*, not as *he* or *she*. See also agreement.

general and specific methods of development

General and specific methods of development proceed either from general information to specific details or from specific information to a general conclusion. As with all methods of development, rarely does a writer rely on only one method throughout a document. Most documents blend methods and use combinations of methods.

General to Specific

The general-to-specific development (see Figure G–1) is especially useful for teaching readers about something with which they are not

General statement

Supporting information

Specific details

Subject: Expanding Our Supplier Base for Computer Chips

On the basis of information presented at the supply meeting on April 14, we recommend that the company initiate relationships with computer-chip manufacturers. Several events make such an action necessary.

Our current supplier, Datacom, is experiencing growing pains and is having difficulty shipping the product on time. Specifically, we can expect a reduction of between 800 and 1,000 units per month for the remainder of this fiscal year. The number of units should stabilize at 15,000 units per month thereafter.

Domestic demand for our computers continues to grow. Demand during the current fiscal year is up 500,000 units over the last fiscal year. Our sales projections for the next five years show that demand should peak next year at about 830,000 units given the consumer demand, which will increase exponentially.

Finally, our expansion into Eastern European and Asian markets will require additional shipments of at least 750,000 units per quarter for the remainder of this fiscal year. Sales Department projections put global computer sales at double that rate, or 1,500,000 units per fiscal year, for the next five years.

FIGURE G–1. General-to-Specific Method of Development

familiar because you can begin with generally known information and lead to new and increasingly specific details. This method can also be used to support a general statement with facts or examples that validate the statement. For example, if you begin your writing with the general statement "Companies that diversify are more successful than those that do not," you could follow that statement with examples and statistics that prove to the reader that companies that diversify are, in fact, more successful than companies that do not.

A memo or short report organized entirely in a general-to-specific sequence discusses only one point. All other information in the document supports the general statement, as in Figure G–1 from a memo about locating additional computer-chip suppliers.

Specific to General

Specific-to-general development, as shown in Figure G-2, is especially useful when you wish to persuade skeptical readers of a general principle with an accumulation of specific details and evidence that reach a logical conclusion. It carefully builds its case, often with examples and analogies in addition to facts or statistics, and does not actually make its point until the end. (See also order-of-importance method of development.)

Specific details

Recently, a government agency studied the use of passenger-side air bags in 4,500 accidents involving nearly 7,200 front-seat passengers of the vehicles involved. Nearly all the accidents occurred on routes that had a speed limit of at least 40 mph. Only 20 percent of the adult front-seat passengers were riding in vehicles equipped with passenger-side air bags. Those riding in vehicles not equipped with passenger-side air bags were more than twice as likely to be killed as passengers riding in vehicles that were so equipped.

A conservative estimate is that 40 percent of the adult front-seat passenger-vehicle deaths could be prevented if all vehicles came equipped with passenger-side air bags. Children, however, should always ride in the backseat because other studies have indicated that a child can be killed by the deployment of an air bag. If you are an adult front-seat passenger in an accident, your chances of survival are far greater if the vehicle in which you are riding is equipped with a passenger-side air bag.

General conclusion

FIGURE G–2. Specific-to-General Method of Development

global communication

The prevalence of global communication technology, international trade agreements, and the emergence of Europe as a giant single market means that the ability to communicate with audiences from varied cultural backgrounds is essential. The audiences for such communications include clients, business partners, and colleagues.

Entries such as meetings and résumés in this book are based on U.S. cultural patterns. The treatment of such topics might be very different in other cultures where leadership styles, persuasive strategies, and even legal constraints differ. As illustrated in international correspondence, organizational patterns, forms of courtesy, and ideas about efficiency can vary significantly from culture to culture. What might be seen as direct and efficient in the United States could be seen as blunt and even impolite in other cultures. The reasons behind these differing ways of viewing communication are complex. Those who study cultures have found various ways to measure cultural differences, such as individual versus group orientation, the importance of saving face, and conceptions of time.

Anthropologist Edward T. Hall, a pioneer in cross-cultural research, developed the concept of "contexting" to assess the predominant communication style of a culture.* By "context," Hall means how much or how little an individual assumes another person understands about a subject under discussion. In a very low-context communication, the participants assume they share little knowledge and must communicate in great detail. Low-context cultures tend to assume little prior knowledge on the part of those with whom they communicate; thus thorough documentation is important—written agreements (contracts) are expected, and rules are spelled out in detail.

In a high-context communication, the participants already understand the context and thus do not feel a need to exchange much background information. High-context cultures depend on shared history (or context) to relate to each other. Thus, words and written contracts are not so important, while personal relationships are paramount. Of course, no culture is entirely high or low context; rather, these concepts can be helpful in understanding the predominant communication style of a particular culture.†

Obviously, cultural differences and the reasons behind them are often so subtle that only someone who is very familiar with a culture

*Edward Twitchell Hall and Mildred Reed Hall, *Understanding Cultural Differences: Germans, French and Americans* (Yarmouth, Maine: Intercultural Press, 1990).

†Gerald J. Alred, "Teaching in Germany and the Rhetoric of Culture," *Journal of Business and Technical Communication* 11.3 (July 1997): 353–78.

can explain the effect those differences may have on others from that culture. For that reason, it is best to consult with someone from your intended audience's culture during preparation for writing or for a presentation. Although many resources are available, the following books are useful starting points:

Andrews, Deborah C. *Technical Communication in the Global Community*. 2nd ed. Upper Saddle River, N.J.: Prentice-Hall, 2000.

Scollon, Ron, and Suzanne Wong Scollon. *Intercultural Communication: A Discourse Approach*. 2nd ed. Cambridge, Mass.: Basil Blackwell, 2001.

Varner, Iris, and Linda Beamer. *Intercultural Communication in the Global Workplace*. 3rd ed. New York: McGraw-Hill/Irwin, 2004.

Victor, David A. *International Business Communication*. New York: HarperCollins, 1992.

Writer's Checklist: Communicating Globally

☑ Acknowledge diversity within your organization. Discussing the differing cultures within your company or region will reinforce the idea that people can interpret verbal and nonverbal communications differently.

☑ Invite global and intercultural communication experts to speak at your workplace. Companies in your area may have employees who could be resources for cultural discussions.

☑ Understand that the key to effective communication with global audiences is recognizing that cultural differences, despite the challenges they may present, offer opportunities for growth for both you and your organization.

☑ Consult with someone from your intended audience's culture. Many phrases, gestures, and visual elements are so subtle that only someone who is very familiar with the culture can explain the effect they may have on others from that culture. See also global graphics.

WEB LINK INTERCULTURAL RESOURCES

Intercultural Press is a source of publications aimed at specific cultures as well as a wide variety of subjects from cross-cultural theory to international business. For links to this and other resources for global communication, see <*bedfordstmartins.com/alredtech*> and select *Links for Handbook Entries*.

global graphics

In the global business and technological environment, graphs and other visuals require the same careful attention given to other aspects of global communication. The complex cultural connotations of visuals challenge writers to think beyond their own experience as illustrated in Figures G–3 and G–4.

Symbols, images, and even colors are not free from cultural associations — they depend on context, and context is culturally determined. For instance, in North America, a red cross is commonly used as a symbol for first aid or hospital. In Muslim countries, however, a cross (red or otherwise) represents Christianity, whereas a crescent (usually green) signifies first aid or hospital. A manual for export to Honduras could indicate "caution" by using a picture of a person touching a finger below the eye. In France, however, that gesture means "You can't fool me."

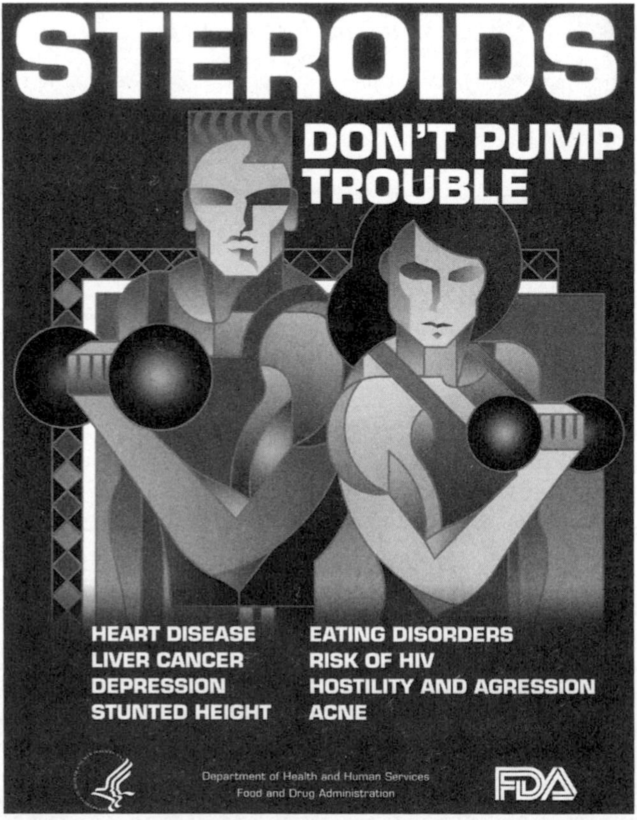

FIGURE G–3. Graphic for a U.S. Audience

FIGURE G–4. Graphic for a Global Audience

Figure G–3 on page 234 shows a graphic warning weight lifters against the use of steroids. The drawings may be appropriate for U.S. audiences and others. However, the image would be highly inappropriate in many cultures where the image of a partially clothed man and woman in close proximity would be contrary to deeply held cultural beliefs and even laws about the public depiction of men and women. Figure G–4, however, depicts weight lifters by using a neutral icon that avoids the connotations associated with more realistic images of people.

These examples suggest why international groups, such as the International Organization for Standardization (ISO), have established agreed-upon symbols, such as those shown in Figure G–5, designed for public signs, guidebooks, and manuals.

Careful attention to the connotations that visual elements may have for a diverse audience makes translations easier, prevents embarrassment, and earns respect for the company and its products and services. See also presentations.

FIGURE G–5. ISO Symbols

Writer's Checklist: Communicating with Global Graphics

- ☑ Consult with someone or test your use of graphics with those from your intended audience's country who can recognize and explain the effect that visual elements will have on readers. See also usability testing.
- ☑ Organize visual information for the intended audience. For example, North Americans tend to read visuals from left to right in clockwise rotation. Middle Eastern cultures read visuals from right to left in counterclockwise rotation.

- ☑ Be sure that the graphics have no unintended religious implications.
- ☑ Carefully consider how you depict people in visuals. Nudity in advertising, for example, is generally acceptable in Europe but much less so in North America and Asia. In some cultures, showing even isolated bare body parts can alienate audiences.
- ☑ Use outlines or neutral abstractions to represent human beings. For example, use stick figures and avoid representing men and women.
- ☑ Examine how you display body positions in signs and visuals. Body positioning can carry unintended cultural meanings very different from your own. For example, some Middle Eastern cultures regard the display of the soles of one's shoes to be disrespectful and offensive.
- ☑ Choose neutral colors (or those you know are appropriate) for your graphics. Generally, black-and-white and gray-and-white illustrations work well. Colors can be problematic. For example, in North America, Europe, and Japan, red indicates danger. In China, however, red symbolizes joy. In Europe and North America, blue generally has a positive connotation; in Japan, blue represents villainy.
- ☑ Check your use of punctuation marks, which are as language specific as symbols. For example, in North America, the question mark generally represents the need for information or the help function in a computer manual or program. In many countries, that symbol has no meaning at all.
- ☑ Create simple visuals and use consistent labels for all visual items. In most cultures, simple shapes with fewer elements are easier to read.
- ☑ Explain the meaning of icons or symbols. Include a glossary to explain technical symbols that cannot be changed, such as company logos.

glossaries

A glossary is an alphabetical list of definitions of specialized terms used in a formal report, a manual, or other long document.

If you are writing a document that will go to readers who are not familiar with specialized or technical terms you use, you may want to include a glossary. If you do, keep the entries concise and be sure they are written in language that all your readers can understand.

- *Amplitude modulation*: Varying the amplitude of a carrier current with an audio-frequency signal.

Arrange the terms alphabetically, with each entry beginning on a new line. The definitions then follow the terms, dictionary style. In a formal report, the glossary appears after the appendix(es) and bibliography, and it begins on a new page.

Including a glossary does not relieve you of the responsibility of defining terms in the text that your reader will not know when those terms are first mentioned.

gobbledygook

Gobbledygook is writing that suffers from an overdose of traits guaranteed to make it stuffy, pretentious, and wordy. Such traits include the overuse of big and mostly abstract words, affectation (especially long variants), buzzwords, inappropriate jargon, clichés, euphemisms, stacked modifiers, and vague words. Gobbledygook is writing that attempts to sound official (officialese), legal (legalese), or scientific; it tries to make a "natural elevation of the geosphere's outer crust" out of a molehill. Consider the following statement from an auto repair release form.

LEGALESE I hereby authorize the above repair work to be done along with the necessary material and hereby grant you and/or your employees permission to operate the car or truck herein described on streets, highways, or elsewhere for the purpose of testing and/or inspection. An express mechanic's lien is hereby acknowledged on above car or truck to secure the amount of repairs thereto.

DIRECT You have my permission to do the repair work listed on this work order and to use the necessary material. You may drive my vehicle to test its performance. I understand that you will keep my vehicle until I have paid for all repairs.

See also clarity, conciseness, and word choice.

good / well

Good is an adjective, and *well* is an adverb.

ADJECTIVE	Janet presented a *good* plan.
ADVERB	The plan was presented *well*.

Well also can be used as an adjective to describe health (a *well* child, *wellness* programs). See also bad/badly.

grammar

Grammar is the systematic description of the way words work together to form a coherent language. In that sense, it is an explanation of the structure of a language. However, grammar is popularly taken to mean the set of rules that governs how a language ought to be spoken and written. In that sense, it refers to the usage conventions of a language.

Those two meanings of grammar—how the language functions and how it ought to function—are easily confused. To clarify the distinction, consider the expression *ain't*. Unless used intentionally to add colloquial flavor, *ain't* is unacceptable because its use is considered nonstandard. Yet taken strictly as a part of speech, the term functions perfectly well as a verb. Whether it appears in a declarative sentence (I ain't going) or an interrogative sentence (Ain't I going?), it conforms to the normal pattern for all verbs in the English language. Although readers may not approve of its use, they cannot argue that it is ungrammatical in such sentences.

To achieve clarity, you need to know both grammar (as a description of the way words work together) and the conventions of usage. Knowing the conventions of usage helps you select the appropriate over the inappropriate word or expression. (See also word choice.) A knowledge of grammar helps you diagnose and correct problems arising from how words and phrases function in relation to one another. For example, knowing that certain words and phrases function to modify other words and phrases gives you a basis for correcting those modifiers that are not doing their job. Understanding dangling modifiers helps you avoid or correct a construction that obscures the intended meaning. In short, an understanding of grammar and its special terminology is valuable chiefly because it enables you to recognize and correct problems so that you can communicate clearly and precisely.

For a complete list of grammar entries, see the Contents by Topic.

WEB LINK GETTING HELP WITH GRAMMAR

For helpful Web sites and grammar exercises, see <*bedfordstmartins.com/alredtech*> and select *Links for Handbook Entries* and *Exercise Central*.

graphs

A graph presents numerical data in visual form and offers several advantages over presenting data within the text or in tables. Trends, movements, distributions, comparisons, and cycles are more readily apparent in graphs than they are in tables. However, although graphs present data in a more comprehensible form than tables do, they are less precise. For that reason, graphs are often accompanied by tables that give exact data. The types of graphs described in this entry include line graphs, bar graphs, pie graphs, and picture graphs. For advice on integrating graphs within text, see visuals; for information about using presentation graphics, see presentations.

Line Graphs

A line graph shows the relationship between two variables or sets of numbers by plotting points in relation to two axes drawn at right angles. The vertical axis usually represents amounts, and the horizontal axis usually represents increments of time. Line graphs that portray more than one set of variables (double-line graphs) allow for comparisons between two sets of statistics for the same period of time. You can emphasize the difference between the two lines by shading the space between them, as shown in Figure G–6.

ETHICS NOTE Be especially careful to proportion the vertical and horizontal scales so that they give a precise presentation of the data that is free of visual distortion. To do otherwise is not only inaccurate but potentially unethical. (See ethics in writing.) In Figure G–7, the graph on

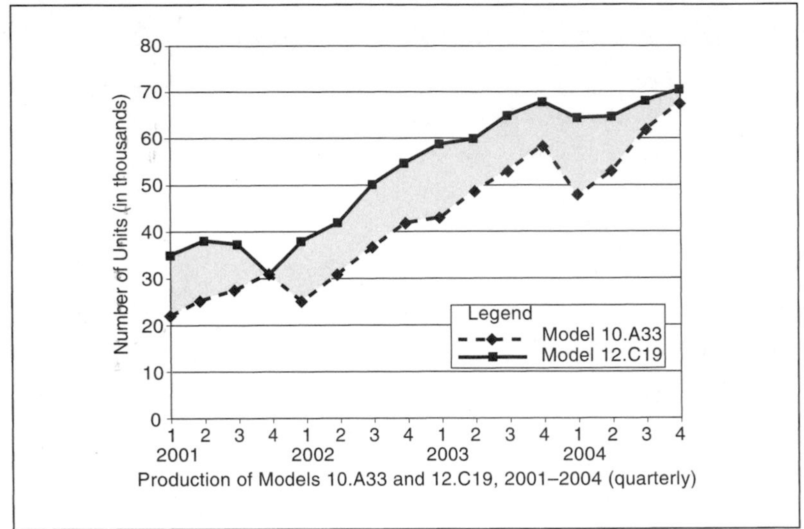

FIGURE G–6. Double-Line Graph (with Shading)

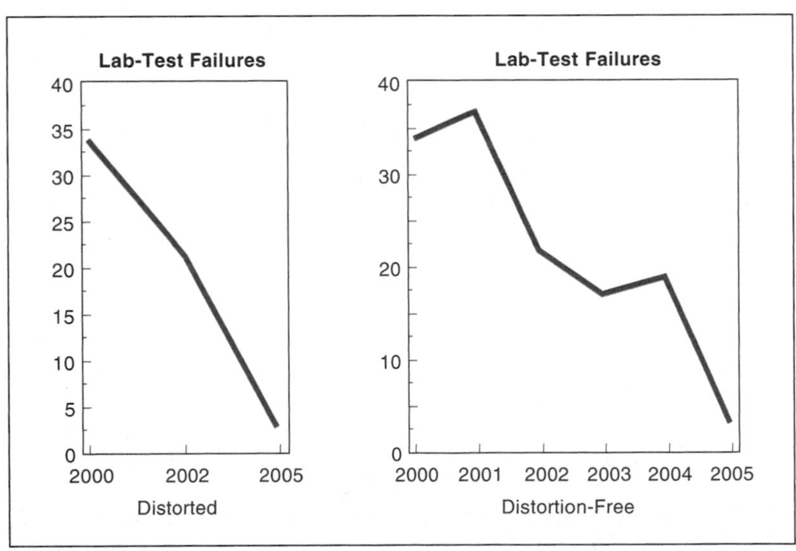

FIGURE G–7. Distorted and Distortion-Free Expressions of Data

the left gives the appearance of a dramatic decrease in lab-test failures because the scale is unevenly compressed, with some of the years selectively omitted. The graph on the right represents the trend more accurately because the years are evenly distributed without omissions. ✦

Bar Graphs

Bar graphs consist of horizontal or vertical bars of equal width, scaled in length to represent some quantity. They are commonly used to show (1) quantities of the same item at different times, (2) quantities of different items at the same time, and (3) quantities of the different parts of an item that make up a whole (in which case, the segments of the bar graph must total 100 percent). The horizontal graph in Figure G–8 shows the quantities of different items for the same period of time.

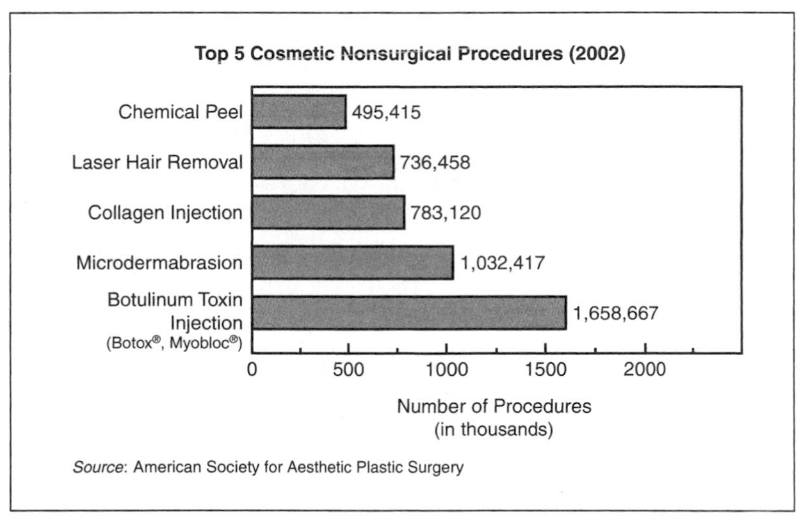

FIGURE G–8. Bar Graph (Quantities of Different Items During a Fixed Period)

Bar graphs can also show the different portions of an item that make up the whole, as shown in Figure G–9. Such a bar graph is divided according to the appropriate proportions of the subcomponents of the item. This type of graph, also called a *column graph* when constructed vertically, can indicate multiple items. Where such items represent parts of a whole, as in Figure G–9, the segments in the bar graph must total 100 percent. Note that in addition to labels, each subdivision

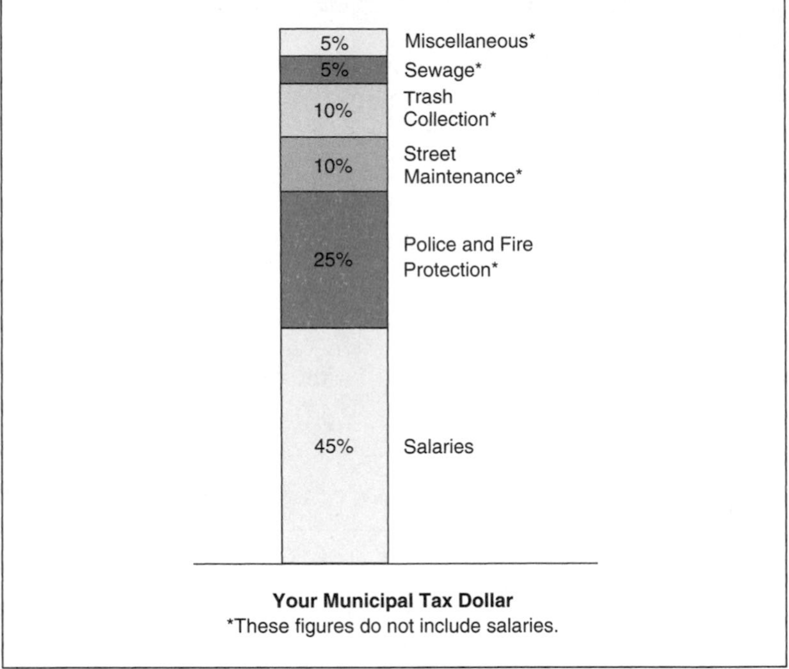

FIGURE G–9. Bar (Column) Graph (Showing the Parts That Make Up the Whole)

of a bar graph must be marked clearly by color, shading, or cross-hatching, with a key or labels that identify the subdivisions represented. Be aware that three-dimensional graphs can make sections seem larger than the amounts they represent.

Pie Graphs

A pie graph presents data as wedge-shaped sections of a circle. The circle equals 100 percent, or the whole, of some quantity, and the wedges represent how the whole is divided. Many times, the data shown in a bar graph could also be depicted in a pie graph. For example, Figure G–9 shows percentages of a whole in bar-graph form. In Figure G–10, the same data are converted into a pie graph, dividing "Your Municipal Tax Dollar" into wedge-shaped sections that represent percentages. Pie graphs also provide a quicker way of presenting information that can be shown in a table; in fact, a table with a more detailed breakdown of the same information often accompanies a pie graph.

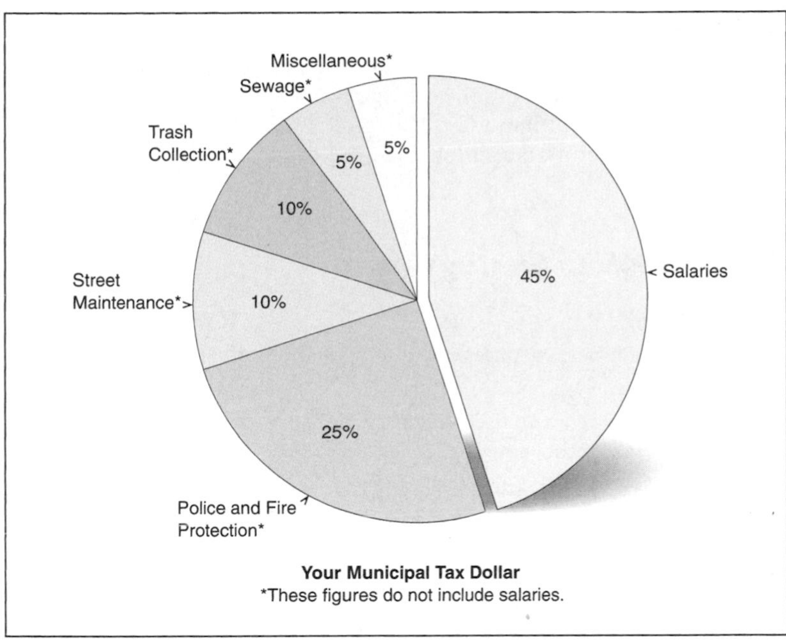

FIGURE G–10. Pie Graph (Showing Percentages of the Whole)

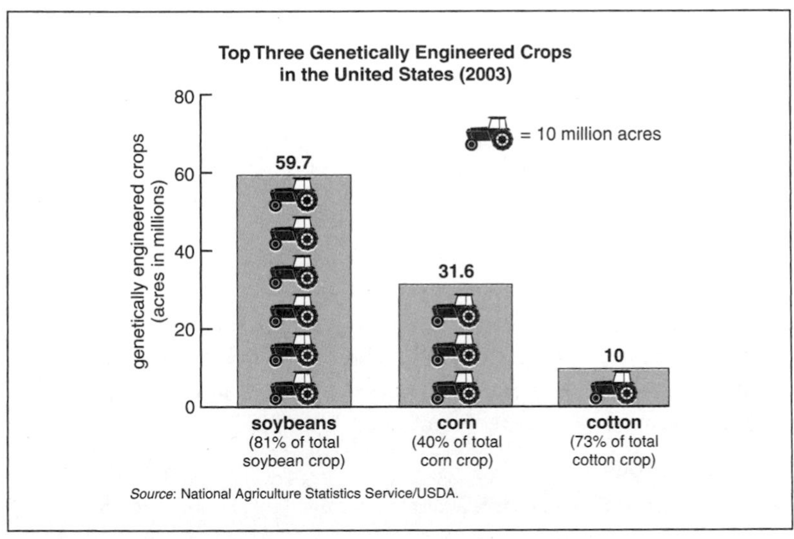

FIGURE G–11. Picture Graph

Picture Graphs

Picture graphs are modified bar graphs that use pictorial symbols of the item portrayed. Each symbol corresponds to a specified quantity of the item, as shown in Figure G–11 on page 243. Note that for precision and clarity, the picture graph includes the total quantity in addition to the symbols.

Writer's Checklist: Creating Graphs

FOR ALL GRAPHS

☑ Use, as needed, a key or legend that lists and defines symbols (see Figure G–6).

☑ Include a source line under the graph at the lower left when the data comes from another source (see Figure G–11).

☑ Place explanatory footnotes directly below the figure caption or label (see Figures G–9 and G–10).

FOR LINE GRAPHS

☑ Indicate the zero point of the graph (the point where the two axes intersect).

☑ Insert a break in the scale if the range of data shown makes it inconvenient to begin at zero.

☑ Divide the vertical axis into equal portions, from the least amount at the bottom (or zero) to the greatest amount at the top.

☑ Divide the horizontal axis into equal units from left to right. If a label is necessary, center it directly beneath the scale.

☑ Make all lettering read horizontally if possible, although the caption or label for the vertical axis is usually positioned vertically (see Figures G–6 and G–11).

FOR BAR GRAPHS

☑ Show either horizontal bars or vertical columns.

☑ Differentiate among the types of data each bar or column or part of a bar or column represents by color, shading, or cross-hatching.

☑ Avoid three-dimensional graphs when they make sections seem larger than the amounts they represent.

FOR PIE GRAPHS

☑ Make sure that the complete circle is equivalent to 100 percent.

☑ Sequence the wedges clockwise from largest to smallest, beginning at the 12 o'clock position, whenever possible.

Writer's Checklist: Creating Graphs (continued)

- ☑ Limit the number of items in the pie graph to avoid clutter and to ensure that the slices are thick enough to be clear.
- ☑ Give each wedge a distinctive color, pattern, shade, or texture.
- ☑ Label each wedge with its percent value and keep all call-outs (labels that identify the wedges) horizontal.
- ☑ Detach a slice, as shown in Figure G–10, if you wish to draw attention to a particular segment of the pie graph.

FOR PICTURE GRAPHS

- ☑ Use picture graphs to add interest to presentations and documents, such as newsletters, that are aimed at wide audiences.
- ☑ Choose symbols that are easily recognizable. See also global graphics.
- ☑ Let each symbol represent a specific number of units.
- ☑ Indicate larger quantities by using more symbols, instead of larger symbols, because relative sizes are difficult to judge accurately.

he / she

The use of either *he* or *she* to refer to both sexes excludes half of the population. (See also biased language.) To avoid this problem, you could use the phrases *he or she* and *his or her*. (Whoever is appointed will find *his or her* task difficult.) However, *he or she* and *his or her* are clumsy when used repeatedly, as are *he/she* and similar constructions. One solution is to reword the sentence to use a plural pronoun; if you do, change the nouns or other pronouns to match the plural form.

- ~~The administrator~~ *Administrators* cannot do ~~his or her job~~ *their jobs* until ~~he or she understands~~ *they understand* the organization's culture.

In other cases, you may be able to avoid using a pronoun altogether.

- Whoever is appointed will find ~~his or her~~ *the* task difficult.

Of course, a pronoun cannot always be omitted without changing the meaning of a sentence.

Another solution is to omit troublesome pronouns by using the imperative mood.

- ~~Everyone must submit his or her~~ *Submit all* expense report*s* by Monday.

headers and footers

A *header* in a report or other document appears at the top of each page, and a *footer* appears at the bottom of each page. Headers or footers may include such elements as the document name, the topic (or subtopic) of a section, the date of the document, the page number, and other identifying information to help readers keep track of where they are in the document. Keep your headers and footers concise—too much information in them can create visual clutter.

Although the information included in headers and footers varies greatly from one organization to the next, the header and footer shown in Figure H–1 are fairly typical. For headers used in letters and memos, see correspondence.

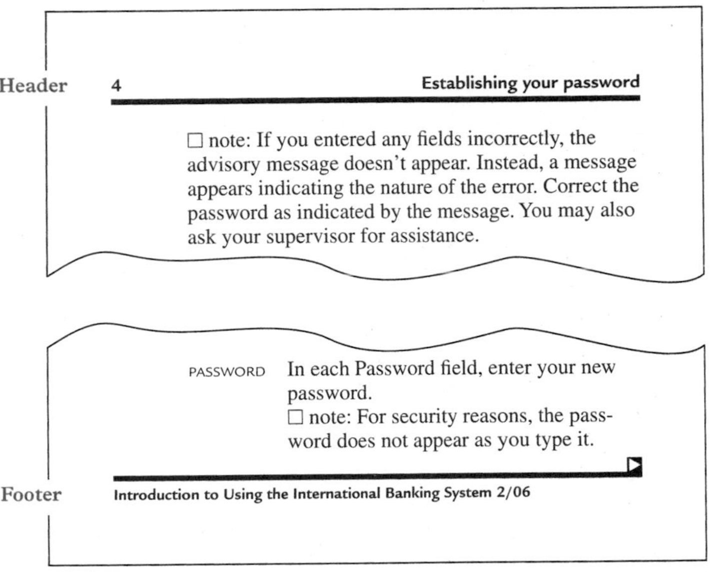

Header
4
Establishing your password
☐ note: If you entered any fields incorrectly, the advisory message doesn't appear. Instead, a message appears indicating the nature of the error. Correct the password as indicated by the message. You may also ask your supervisor for assistance.
PASSWORD
In each Password field, enter your new password.
☐ note: For security reasons, the password does not appear as you type it.
Footer
Introduction to Using the International Banking System 2/06

FIGURE H–1. Header and Footer

headings

Headings (also called *heads*) are titles or subtitles within the body of a document that help readers find information, divide the material into comprehensible segments, highlight the main topics, and signal topic changes. A formal report or proposal may need several levels of headings to indicate major divisions, subdivisions, and even smaller units. However, avoid using more than three levels of headings. See also layout and design.

Headings typically represent the major topics of a document. In a short document, you can use the major divisions of your outline as headings; in a longer document, you may need to use both major and minor divisions.

General Heading Style

No one format for headings is correct. Often a company settles on a standard format, which everyone in the company follows. Sometimes a customer for whom a report or proposal is being prepared requires a particular format. In the absence of specific guidelines, follow the system illustrated in Figure H–2. See also layout and design.

First-level head

ENGINEERING AND MANUFACTURING PLANT LOCATION REPORT

The committee initially considered 30 possible locations for the proposed new engineering and manufacturing plant. Of these, 20 were eliminated almost immediately for one reason or another (unfavorable zoning regulations, inadequate transportation infrastructure, etc.). Of the remaining ten locations, the committee selected for intensive study the three that seemed most promising: Chicago, Minneapolis, and Salt Lake City. We have now visited these three cities, and our observations and recommendations follow.

Second-level head

CHICAGO

Of the three cities, Chicago presently seems to the committee to offer the greatest advantages, although we wish to examine these more carefully before making a final recommendation.

Third-level head

Selected Location

Though not at the geographic center of the United States, Chicago is the demographic center to more than three-quarters of the U.S. population. It is within easy reach of our corporate headquarters in New York. And it is close to several of our most important suppliers of components and raw materials—those, for example, in Columbus, Detroit, and St. Louis. Several factors were considered essential to the location, although some may not have had as great an impact on the selection. . . .

Fourth-level heads

Air Transportation. Chicago has two major airports (O'Hare and Midway) and is contemplating building a third. Both domestic and international air-cargo service are available. . . .

Sea Transportation. Except during the winter months when the Great Lakes are frozen, Chicago is an international seaport. . . .

Rail Transportation. Chicago is served by the following major railroads. . . .

FIGURE H–2. Headings Used in a Document

Decimal Numbering System

Some documents, such as specifications, benefit from the decimal numbering system for ease of cross-referencing sections. The decimal numbering system uses a combination of numbers and decimal points to differentiate among levels of headings. The following example shows the correspondence between different levels of headings and the decimal numbers used:

1. FIRST-LEVEL HEADING
   1.1 Second-level heading
   1.2 Second-level heading
      1.2.1 Third-level heading
      1.2.2 Third-level heading
         1.2.2.1 Fourth-level heading
         1.2.2.2 Fourth-level heading
   1.3 Second-level heading
2. FIRST-LEVEL HEADING

Although decimal headings are indented in an outline or a table of contents, they are flush with the left margins when they function as headings in the body of a report. Every heading starts on a new line, with an extra line of space above and below the heading.

H

Writer's Checklist: Using Headings

- ☑ Use headings to signal a new topic. Use a lower-level heading to indicate a new subtopic within the larger topic.
- ☑ Make headings concise but specific enough to be informative, as in Figure H–2.
- ☑ Avoid too many or too few headings or levels of headings; too many clutter a document, and too few fail to provide recognizable structure.
- ☑ Ensure that headings at the same level are of relatively equal importance and have parallel structure.
- ☑ Subdivide sections only as needed; when you do, try to subdivide them into at least two lower-level headings.
- ☑ Do not leave a heading as the final line of a page. If two lines of text cannot fit below a heading, start the section at the top of the next page.
- ☑ Do not allow a heading to substitute for discussion; the text should read as if the heading were not there.

hyphens

The hyphen (-) serves both to link and to separate words. The hyphen, for example, joins compound words (able-bodied, self-contained, self-esteem) and forms compound numbers from twenty-one through ninety-nine and fractions when they are written out (three-quarters). Two consecutive hyphens indicate a dash.

Hyphens Used with Modifiers

Two- and three-word modifiers that express a single thought are hyphenated when they precede a noun.

- It was a *well-written* report.

However, a modifying phrase is not hyphenated when it follows the noun it modifies.

- The report was *well written.*

If each of the words can modify the noun without the aid of the other modifying word or words, do not use a hyphen.

H

- a *new laser* printer

If the first word is an adverb ending in *-ly*, do not use a hyphen.

- a *privately held* company

A hyphen is always used as part of a letter or number modifier.

- 9-inch, A-frame

In a series of unit modifiers that all have the same term following the hyphen, the term following the hyphen need not be repeated throughout the series; for greater smoothness and brevity, use the term only at the end of the series.

- The third-, fourth-, and fifth-floor laboratories were disinfected.

Hyphens Used with Prefixes and Suffixes

A hyphen is used with a prefix when the root word is a proper noun.

- pre-Columbian, anti-American, post-Newtonian

A hyphen may be used when the prefix ends and the root word begins with the same vowel.

- re-enter, anti-inflammatory

A hyphen is used when *ex-* means "former."

- ex-president, ex-spouse

A hyphen may be used to emphasize a prefix.

- He is *anti-change*.

The suffix *-elect* is hyphenated.

- president-elect

Hyphens and Clarity

The presence or absence of a hyphen can alter the meaning of a sentence.

AMBIGUOUS We need a biological waste management system.

That sentence could mean one of two things: (1) We need a system to manage "biological waste," or (2) We need a "biological" system to manage waste.

CLEAR We need a *biological-waste* management system.

CLEAR We need a biological *waste-management* system.

To avoid confusion, some words and modifiers should always be hyphenated. *Re-cover* does not mean the same thing as *recover*, for example; the same is true of *re-sent* and *resent*, *re-form* and *reform*, *re-sign* and *resign*.

Other Uses of the Hyphen

Hyphens should be used between letters showing how a word is spelled.

- In his e-mail, he misspelled *believed* as b-e-l-e-i-v-e-d.

A hyphen can stand for *to* or *through* between letters and numbers (pages 44-46, the Detroit-Toledo Expressway, A-L and M-Z).

Writer's Checklist: Using Hyphens to Divide Words

☑ Do not divide one-syllable words.

☑ Divide words at syllable breaks, which you can determine with a dictionary.

☑ Do not divide a word if only one letter would remain at the end of a line or if fewer than three letters would start a new line.

☑ Do not divide a word at the end of a page; carry it over to the next page.

☑ If a word already has a hyphen in its spelling, divide the word at the existing hyphen.

☑ Do not use a hyphen to break a URL or an e-mail address at the end of a line because it may confuse readers who could assume that the hyphen is part of the address.

I

idioms

I

An idiom is a group of words that has a special meaning apart from its literal meaning. Someone who "runs for office" in the United States, for example, need not be an athlete. The same candidate would "stand for office" in the United Kingdom. Because such expressions are specific to a culture, nonnative speakers must memorize them.

Idioms are often constructed with prepositions that follow adjectives (*similar to*), nouns (*need for*), and verbs (*approve of*). Some idioms can change meaning slightly with the preposition used, as in *agree to* ("consent") and *agree with* ("in accord"). The following are typical idioms that give nonnative speakers trouble.

call off [cancel]
call on [visit a client]
cross out [draw a line through]
do over [repeat a task]
drop in on [visit unexpectedly]
figure out [solve a problem]
find out [discover information]
get through with [finish]
give up [quit]
hand in [submit]
hand out [distribute]
keep on [continue]
leave out [omit]
look up [research a subject]
put off [postpone]
run into [meet by chance]
run out of [deplete supply]
watch out for [be careful]

Idioms often provide helpful shortcuts. In fact, they can make writing more natural and vigorous. Avoid them, however, if your writing is to be translated into another language or read in other English-speaking countries. Because no system can fully explain such usages, you must check dictionaries or usage guides to interpret their meaning. See also international correspondence and English as a second language.

WEB LINK PREPOSITIONAL IDIOMS

For links to helpful lists of common pairings of prepositions with nouns, verbs, and adjectives, see <*bedfordstmartins.com/alredtech*> and select *Links for Handbook Entries*.

illegal / illicit

If something is *illegal*, it is prohibited by law. If something is *illicit*, it is prohibited by either law or custom. *Illicit* behavior may or may not be *illegal*, but it does violate social convention or moral codes and therefore usually has a clandestine or immoral connotation. (The employee's *illicit* sexual behavior caused a scandal, but the company's attorney concluded that no *illegal* acts were committed.)

illustrations (*see* visuals)

imply / infer

If you *imply* something, you hint or suggest it. (Her e-mail *implied* that the project would be delayed.) If you *infer* something, you reach a conclusion based on evidence or interpretation. (The manager *inferred* from the e-mail that the project would be delayed.)

in / into

In means "inside of"; *into* implies movement from the outside to the inside. (The equipment was *in* the test chamber, so she reached *into* the chamber to adjust it.)

in order to

Most often, *in order to* is a meaningless filler phrase that is dropped into a sentence without thought. See also conciseness.

- ~~In order to~~ *To* start the engine, open the choke and throttle and then press the starter.

However, the phrase *in order to* is sometimes essential to the meaning of a sentence.

- If the vertical scale of a graph line would not normally show the zero point, use a horizontal break in the graph *in order to* include the zero point.

In order to also helps control the pace of a sentence, even when it is not essential to the meaning of the sentence.

- The committee must know the estimated costs *in order to* evaluate the feasibility of the project.

in terms of

When used to indicate a shift from one kind of language or terminology to another, the phrase *in terms of* can be useful.

- *In terms of* gross sales, the year has been relatively successful; however, *in terms of* net income, it has been discouraging.

When simply dropped into a sentence because it easily comes to mind, *in terms of* is meaningless affectation. See also conciseness.

- She was thinking ~~in terms~~ of subcontracting much of the work.

indexing

An index is an alphabetical list of all the major topics and sometimes subtopics in a written work. It cites the pages where each topic can be found and allows readers to find information on particular topics quickly and easily, as shown in Figure I–1. The index always comes at the very end of the work. Many Web sites also provide linked subject indexes to the content of the sites.

The key to compiling a useful index is selectivity. Instead of listing every possible reference to a topic, select references to passages where the topic is discussed fully or where a significant point is made about it.

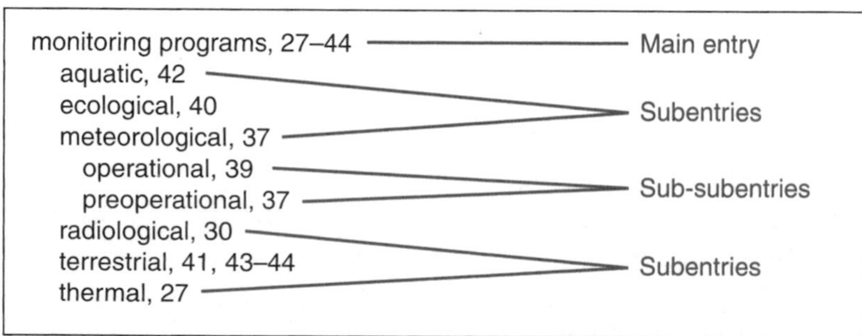

FIGURE I–1. Index Entry (with Main Entry, Subentries, and Sub-subentries)

For index entries like those in Figure I–1, choose key terms that best represent a topic. Key terms are those words or phrases that a reader would most likely look for in an index. For example, the key terms in a reference to the development of legislation about environmental-impact statements would probably be *legislation* and *environmental-impact statement*, not *development*. In selecting terms for index entries, use chapter or section titles only if they include such key terms. For index entries on tables and visuals, use the words from their titles that will function as key terms a reader might seek. Create alphabetical Web site indexes from links to topics in subsites throughout the larger site. Because active Web sites add and remove content continually, their indexes must be updated regularly.

Compiling an Index

Do not attempt to compile an index until the final manuscript is completed because terminology and page numbers will not be accurate before then. The best way to manually compile a list of topics is to read through your written work from the beginning; each time a key term appears in a significant context, note the term and its page number. An index entry can consist solely of a main entry and its page number.

- aquatic monitoring programs, 42

An index entry can also include a main entry, subentries, and even sub-subentries, as shown in Figure I–1. A subentry indicates pages where a specific subcategory or subdivision of the main topic can be found. When you have compiled a list of key terms for the entire work, sort the

DIGITAL TIPS CREATING AN INDEX

Word-processing software can provide a quick and efficient way to create an alphabetical subject index of your document. Following your software's instructions, highlight and code the key terms you wish to index. In Microsoft Word, for example, you can mark a keyword and the software will automatically mark all other instances of the word. The software sorts the coded entries, eliminates duplications, and arranges them alphabetically with their page numbers in a separate section at the end of the document. You can also create headings for each alphabetic grouping of the index (A, B, C, etc.).

Your draft index will still need careful review and revisions, but using the software to create the first draft can certainly save time. For further instructions, see <*bedfordstmartins.com/alredtech*> and select *Digital Tips*, "Creating an Index."

main entries alphabetically, then sort all subentries and sub-subentries alphabetically beneath their main entries. To help indexers with this process, the American Society of Indexers lists software available for indexing at their Web site, <www.asindexing.org>.

Wording Index Entries

The first word of an index entry should be the principal word because the reader will look for topics alphabetically by their main words. Selecting the right word to list first is easier for some topics than for others. For instance, *tips on repairing electrical wire* would not be a suitable index entry because a reader looking for information on electrical wire would not look under the word *tips*. Ordinarily, an entry with two keywords, like *electrical wire*, should be indexed under each word (*electrical wire* and *wire, electrical*). A main index entry should be written as a noun or a noun phrase rather than as an adjective alone or a verb.

I

- electrical wire, 20–22
  grounding, 21
  insulation, 20
  repairing, 22
  size, 21

Cross-Referencing

Cross-references in an index help readers find other related topics in the text. A reader looking up *technical writing*, for example, might find cross-references to *report* or *e-mail*. Cross-references do not include page numbers; they merely direct readers to another main index entry, where they can find page numbers. The two kinds of cross-references are *see* references and *see also* references.

See references are most commonly used with topics that can be identified by several different terms. Listing the topic page numbers by only one of the terms, the indexer then lists the other terms throughout the index as *see* references.

- economic costs. *See* benefit-cost analyses

See references also direct readers to index entries where a topic is listed as a subentry.

- L-shaped fittings. *See* elbows, L-shaped fittings

See also references indicate other entries that include additional information on a topic.

- ecological programs, 40–49. *See also* monitoring programs

Writer's Checklist: Indexing

- ☑ Use lowercase for the first words and all subsequent words of main entries, subentries, and sub-subentries unless they are proper nouns or would otherwise be capitalized. See capitalization.
- ☑ The cross-reference terms *see* and *see also* should appear in italics.
- ☑ Place each subentry in the index on a separate line, indented from its main entry. Indent sub-subentries from the preceding subentry. Indentations allow readers to scan a column quickly for pertinent subentries or sub-subentries.
- ☑ Separate entries from page numbers with commas.
- ☑ Format the index with double columns, as is done in the index to this book.

indiscreet / indiscrete

Indiscreet means "lacking in prudence or sound judgment." (His public discussion of the proposed design was *indiscreet.*) *Indiscrete* means "not divided or divisible into parts." (The separate departments, once combined, become *indiscrete.*) See also discreet/discrete.

individual

Individual is most appropriate when used as an adjective to distinguish a single person from a group. (The *individual* employee's obligations are detailed in the policy manual.) Using *individual* as a noun is an affectation. Use *people* or another appropriate term.

- Several ~~individuals on~~ *members of* the committee did not vote.

ingenious / ingenuous

Ingenious means "marked by cleverness and originality." (Seon Ju's *ingenious* solution led to her promotion.) *Ingenuous* means "straightforward" or "characterized by innocence and simplicity." (The *ingenuous* co-op students bring fresh perspectives to the company.)

inquiries and responses

The purpose of writing an inquiry letter or e-mail is often to obtain answers to specific questions, as in Figure I–2, which shows a college student's requests for specific information from an official at a power company. Inquiries may either benefit the reader (as in requests for information about a product that a company has advertised) or benefit the writer (as in the student's e-mail in Figure I–2). Inquiries that primarily benefit the writer require the use of persuasion and special consideration of your readers' needs. See also correspondence.

From: Kathryn J. Parsons <kjp@fly.ud.edu>
To: Jane E. Metcalf <metcalf@mvpc.org>
Sent: 18 March 2005 09:23:45 -0500 (EDT)
Subject: Inquiry on Heating Systems

Dear Ms. Metcalf:

Could you please send me some information on heating systems for a computerized, energy-efficient house that a team of engineering students at the University of Dayton is designing?

The house, which contains 2,000 square feet of living space (17,600 cubic feet), meets all the requirements in your brochure "Insulating for Efficiency." We need the following information, based on the southern Ohio climate:

1. The proper-size heat pump for such a home.
2. The wattage of the supplemental electrical heating units required.
3. The estimated power consumption and rates for those units for one year.

We will be happy to send you our preliminary design report. If you have questions or suggestions, please contact me at kjp@fly.ud.edu or call 513-229-4598.

Thank you for your help.

Kathryn Parsons

FIGURE I–2. Inquiry

Writing Inquiries

Inquiries need to be specific, clear, and concise in order to receive a prompt, helpful reply. To achieve those goals, keep in mind these guidelines:

- Phrase your questions so that the reader will immediately know the type of information you are seeking, why you need it, and how you will use it.
- If possible, present your questions in a numbered list to make it easy for your reader to respond to them.
- Keep the number of questions to a minimum to improve your chances of receiving a prompt response.
- Offer some inducement for the reader to respond, such as promising to share the results of what you are doing. See also "you" viewpoint.
- Promise to keep responses confidential, if appropriate.

In the closing, thank the reader for taking the time to respond. In addition, make it convenient for the recipient to respond by providing contact information, such as a phone number or an e-mail address, as shown in Figure I–2.

Responding to Inquiries

When you receive an inquiry, determine whether you have both the information and the authority to respond. If you have received an inquiry that you feel you cannot answer, find out who can and forward the inquiry to that person. Notify the writer that you have forwarded the inquiry, as shown in Figure I–3. When an inquiry is forwarded, the person who replies should state in the first paragraph of the response why someone else is answering the original inquiry.

If you are the right person in your organization to respond, answer as promptly as you can, and be sure to answer every question asked, as shown in Figure I–4. How long and how detailed your response should be depends on the nature of the question and the information the writer provides.

inside / inside of

In the phrase *inside of*, the word *of* is redundant and should be omitted.

- The switch is just inside ~~of~~ the door.

From: Jane E. Metcalf <metcalf@mvpc.org>
To: Kathryn J. Parsons <kjp@fly.ud.edu>
Sent: Mon, 21 March 2005 11:42:25 -0500 (EDT)
Subject: RE: Inquiry on Heating Systems

Dear Kathryn Parsons:

Thank you for inquiring about the heating system we recommend for use in homes designed according to the specifications outlined in our brochure "Insulating for Efficiency."

Because I cannot answer your specific questions, I have forwarded your inquiry to Michael Wang, Engineering Assistant in our Development Group. He should be able to answer the questions you have raised. You should be hearing from him shortly.

Best wishes,

Jane E. Metcalf

Jane E. Metcalf, Director of Public Information
Miami Valley Power Company
P.O. Box 1444 ~ Miamitown, OH 45733
Office 513-264-4800 ~ Fax 513-264-4889
Web ~ www.enersaving.com

FIGURE I–3. Response to an Inquiry (Indicating That It Has Been Forwarded)

Using *inside of* to mean "in less time than" is colloquial and should be avoided in writing.

- They were finished ~~inside of~~ *in less than* an hour.

insoluble / insolvable

The words *insoluble* and *insolvable* are sometimes used interchangeably to mean "incapable of being solved." *Insoluble* also means "incapable of being dissolved."

- Until yesterday, the production problem seemed *insolvable*.
- *Insoluble* fiber passes through the intestines largely intact.

From: Michael Wang <mwang@mvpc.org>
To: Kathryn J. Parsons <kjp@fly.ud.edu>
Sent: 28 March 2005 16:09:22 -0500 (EDT)
Subject: RE: Inquiry on Heating Systems

Dear Ms. Parsons:

Jane Metcalf forwarded to me your inquiry of March 18 about the house that your engineering team is designing. I can estimate the heating requirements of a typical home of 17,600 cubic feet as follows:

1. For such a home, we would generally recommend a heat pump capable of delivering 40,000 BTUs, such as our model AL-42 (17 kilowatts).
2. With the AL-42's efficiency, you don't need supplemental heating units.
3. Depending on usage, the AL-42 unit averages between 1,000 and 1,500 kilowatt-hours from December through March. To determine the current rate for such usage, check with Dayton Power and Light Company.

I can give you an answer that would apply specifically to your house based on its particular design (such as number of stories, windows, and entrances). If you send me more details, I will be happy to provide more precise figures for your interesting project.

Sincerely,

Michael Wang

Engineering Assistant
mwang@mvpc.org

FIGURE I–4. Response to an Inquiry

instructions

Instructions that are clear and easy to follow prevent miscommunication and help readers complete tasks effectively and safely. To write effective instructions, you must thoroughly understand the process, system, or device you are describing. Often you must observe someone as he or she completes the task and perform the steps yourself before you begin to write. Finally, keep in mind that the most effective instructions often combine written elements and visual elements that reinforce each other. See also process explanation.

Writing Instructions

Consider your readers' level of knowledge. If all your readers have good backgrounds in the topic, you can use fairly specialized terms. If that is not the case, use plain language or include a glossary for specialized terms that you cannot avoid.

Clear and easy-to-follow instructions are written as commands in the imperative mood, active voice, and (whenever possible) simple present tense.

- *Raise the* ~~The~~ access lid ~~will be raised by the operator.~~ .

Although conciseness is important in instructions, clarity is essential. You can make sentences shorter by leaving out some articles (*a*, *an*, *the*), some pronouns (*you*, *this*, *these*), and some verbs, but such sentences may result in telegraphic style and be harder to understand. For example, the first version of the following instruction for placing a document in a scanner tray is confusing.

CONFUSING	Place document in tray with printed side facing opposite.
CLEAR	Place the document in the document tray with the printed side facing away from you.

One good way to make instructions easy to follow is to divide them into short, simple steps in their proper sequence. Steps can be organized with words (*first*, *next*, *finally*) that indicate time or sequence.

- *First*, determine the problem the customer is having with the computer. *Next*, observe the system in operation. *At that time*, question the operator until you are sure that the problem has been explained completely. *Then* analyze the problem and make any necessary adjustments.

You can also use numbers, as in the following:

- 1. Connect each black cable wire to a brass terminal.
  2. Attach one 4-inch green jumper wire to the back.
  3. Connect the jumper wire to the bare cable wire.

Consider using the numbered- or bulleted-list feature of your word-processing software to create sequenced steps.

Plan ahead for your reader. If the instructions in step 2 will affect a process in step 9, say so in step 2. Sometimes your instructions have to make clear that two operations must be performed simultaneously. Either state that fact in an introduction to the specific instructions or include both operations in one step.

CONFUSING
1. Hold down the CONTROL key.
2. Press the RETURN key before releasing the CONTROL key.

CLEAR
1. While holding down the CONTROL key, press the RETURN key.

If your instructions involve a great many steps, break them into stages, each with a separate heading so that each stage begins again with step 1. Using headings as dividers is especially important if your reader is likely to be performing the operation as he or she reads the instructions.

Illustrating Instructions

Illustrations should be developed together with the text, especially for complex instructions that benefit from visuals that foster clarity and conciseness. Such visuals as drawings, flowcharts, maps, and photographs enable your reader to identify relationships more easily than long explanations. Some instructions, such as those for products sold internationally, use only visuals. The entry global graphics describes visuals that avoid culture-specific connotations and are often used in such instructions.

Consider the layout and design of your instructions to most effectively integrate visuals. Highlight important visuals as well as text by making them stand out from the surrounding text. Consider using boxes and boldface or distinctive headings. Experiment with font style, size, and color to determine which devices are most effective.

The instructions in Figure I–5 on page 264 guide the reader through the steps of streaking a saucer-sized disk of material (called *agar*) used to grow bacteria colonies. The purpose is to thin out the original specimen (the *inoculum*) so that the bacteria will grow in small, isolated colonies.

Warning Readers

Alert your readers to any potentially hazardous materials (or actions) before they reach the step for which the material is needed. Caution readers handling hazardous materials about any requirements for special clothing, tools, equipment, and other safety measures. Highlight warnings, cautions, and precautions to make them stand out visually from the surrounding text. Present warning notices in a box, in all uppercase letters, in large and distinctive fonts, or in color. Experiment

STREAKING AN AGAR PLATE

Distribute the inoculum over the surface of the agar in the following manner:

1. Beginning at one edge of the saucer, thin the inoculum by streaking back and forth over the same area several times, sweeping across the agar surface until approximately one-quarter of the surface has been covered. *Sterilize the loop in an open flame.*
2. Streak at right angles to the originally inoculated area, carrying the inoculum out from the streaked areas onto the sterile surface with only the first stroke of the wire. Cover half of the remaining sterile agar surface. *Sterilize the loop.*
3. Repeat as described in Step 2, covering the remaining sterile agar surface.

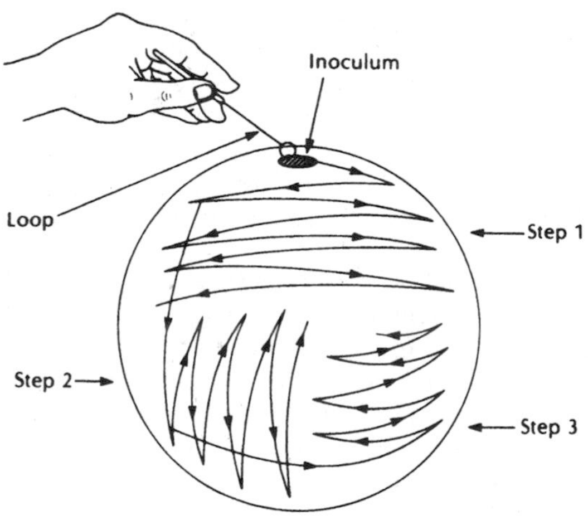

FIGURE I–5. Illustrated Instructions

with font style, size, and color to determine which devices are most effective.

Figure I–6 shows a warning from an instruction manual for the use of a gas grill. Notice that a drawing supports the text of the warning.

Various standards organizations and agencies publish guidelines on the use of terminology, colors, and symbols in instructions and labeling. Two widely influential organizations that publish such guidelines

WARNING

- The assembler/owner is responsible for the assembly, installation, and maintenance of the grill.
- Use the grill outdoors only.
- Do not let children operate or play near your grill.
- Keep the grill area clear and free from materials that burn, gasoline, bottled gas in any form, and other flammable vapors and liquids.
- Do not block holes in bottom and back of grill.
- Visually check burner flames on a regular basis.
- Use the grill in a well-ventilated space. Never use in an enclosed space, carport, garage, porch, patio, or building made of combustible construction, or under overhead construction.
- Do not install your grill in or on recreational vehicles and/or boats.
- Keep grills a distance of 36″, or 3 ft. (approximately 1 m), from buildings to reduce the possibility of fire or heat damage to materials.

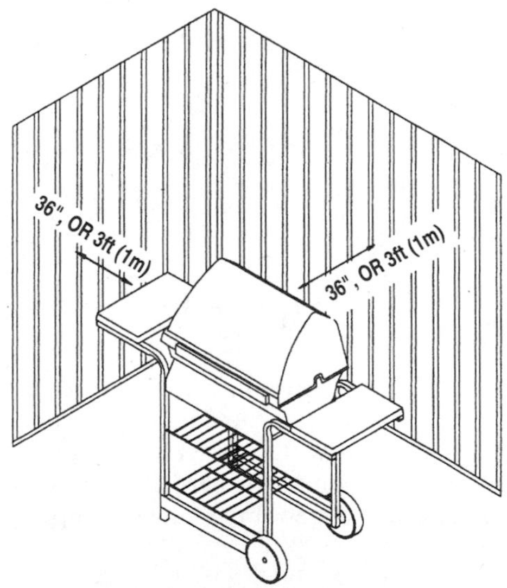

FIGURE I–6. Warning in a Set of Instructions

are the American National Standards Institute (www.ansi.org/) and the International Organization for Standardization (www.iso.org/).

Testing Instructions

To test the accuracy and clarity of your instructions, ask someone who is not familiar with the task to follow your directions. A first-time user can spot missing steps or point out passages that should be worded more clearly. As you observe your tester, note any steps that seem especially confusing and revise accordingly. See usability testing.

Writer's Checklist: Writing Instructions

- ☑ Use the imperative mood and the active voice.
- ☑ Use short sentences and simple present tense as much as possible.
- ☑ Avoid technical terminology and **jargon** that your readers might not know, including undefined **abbreviations**.
- ☑ Do not use elegant variation (two different words for the same thing). See also **affectation**.
- ☑ Eliminate any **ambiguity**.
- ☑ Use effective visuals, and place them properly.
- ☑ Include appropriate warnings and cautions.
- ☑ Verify that measurements, distances, times, and relationships are precise and accurate.
- ☑ Test your instructions by having someone else follow them while you observe.

insure / ensure / assure

Insure, *ensure*, and *assure* all mean "make secure or certain." *Assure* refers to people, and it alone has the connotation of setting a person's mind at rest. (I *assure* you that the equipment will be available.) *Ensure* and *insure* mean "make secure from harm." Only *insure* is widely used in the sense of guaranteeing the value of life or property.

- We need all the data to *ensure* the success of the project.
- We should *insure* the contents of the warehouse.

intensifiers

Intensifiers are adverbs that emphasize degree, such as *very*, *quite*, *rather*, *such*, and *too*. Although intensifiers serve a legitimate and necessary function, unnecessary intensifiers can weaken your writing. Eliminate those that do not make an obvious contribution or replace them with specific details.

- The team was ~~quite~~ happy to learn ~~receive the very good news~~ that it had been awarded a $10,000 ~~rather substantial monetary~~ prize for its design.

Some words (such as *perfect*, *impossible*, and *final*) do not logically permit intensification because, by definition, they do not allow degrees of comparison. Although usage often ignores that logical restriction, to ignore it is, strictly speaking, to defy the basic meanings of such words. See also conciseness and equal/unique/perfect.

interface

An *interface* is a surface that provides a common boundary between two bodies or areas. The bodies or areas may be physical (the *interface* of a piston and a cylinder) or conceptual (the *interface* of mathematics and economics). Do not use *interface* as a substitute for the verb *cooperate*, *interact*, or even *work*. See also affectation and buzzwords.

interjections

An interjection is a word or phrase standing alone or inserted into a sentence to exclaim or to command attention. Grammatically, it has no connection to the sentence. An interjection can be strong (*Hey! Ouch! Wow!*) or mild (*oh*, *well*, *indeed*). A strong interjection is followed by an exclamation mark.

- *Wow!* Profits more than doubled last quarter.

A weak interjection is followed by a comma.

- *Well*, we need to rethink the proposal.

An interjection inserted into a sentence usually requires a comma before it and after it.

- We must, *indeed,* rethink the proposal.

Because they get their main expressive force from sound, interjections are more common in speech than in writing. Use them sparingly.

international correspondence

I

Business correspondence varies among cultures. Organizational patterns, persuasive strategies, forms of courtesy, formality, and ideas about efficiency vary from country to country. For example, in the United States, direct, concise correspondence may demonstrate courtesy by not wasting another person's time. In many Asian cultures, such directness and brevity may seem rude to readers, suggesting that the writer wishes to spend as little time as possible corresponding with the reader. Likewise, where an American writer might consider one brief letter sufficient to communicate a request, a writer in another culture may expect an exchange of three or four longer letters to pave the way for action. The features and communication styles of specific cultures are complex; the entry global communication provides information and resources for cross-cultural study. See also global graphics.

Cultural Differences in Correspondence

When you read correspondence from businesspeople in other cultures or countries, be alert to differences in such features as customary expressions, openings, and closings. Japanese business writers, for example, have traditionally used indirect openings that reflect on the season, compliment the reader's success, and offer hopes for the reader's continued prosperity. Consider as well deeper issues, such as how writers from other cultures express bad news. Japanese writers have also traditionally expressed negative messages, such as refusal letters indirectly to avoid embarrassing the recipient. Such differences are often based on cultural perceptions of time, face-saving, and other expectations.

Cross-Cultural Examples

Figures I–7 and I–8 are a draft and a final version of a letter written by an American businessman to a Japanese businessman. The draft in Figure I–7 does not include enough politeness strategies important to Japanese culture in the opening and closing, and it uses an inappropriately informal salutation with the recipient's first name (*Dear Ichiro:*).

U.S. writer moves too quickly to use of first name.
"Looking forward" is an idiom that may not be clear.
Contraction (*I'm*) is both too informal and may not appear in an English dictionary.
First paragraph gets to business abruptly: add more personal greeting.
Jargon ("transport") is inappropriate.
Abbreviated with U.S. date format: use March 20, 2006.
Inappropriate humor and allusion.
First three sentences use overly simplified style.
The use of "family" is a figure of speech that could be confusing.
Abrupt and informal: add more goodwill to closing.

DRAFT FOR REVIEW

Dear Ichiro:

I'm writing to confirm travel arrangements for your visit to Tucson next month. We at Sun West are looking forward to meeting you and cultivating a successful business relationship between our two companies.

I understand that you will arrive on 3/20/06 on Delta, flight #186 at 1:30 p.m. (itinerary enclosed). When you pick up your baggage at Tucson, go directly toward the "taxi-limo" area where our transport will be holding a card with your name and can take you directly to Lowes Ventana Canyon Resort. The resort has an excellent restaurant—we promise not to take you to the "ptomaine palace" across from our main offices!

We are excited to meet you. We want you to meet our company family. After you meet everyone, you will enjoy a catered breakfast in our conference room. Events include presentations from the president of the company and from departmental directors on several topics.

I have enclosed a guidebook and map of Tucson and material on our company. If you see anything that interests you, let me know, and we will be happy to show you our city and all it has to offer. And if you have any questions about the company before we see you, just e-mail or fax (I don't think regular mail will get to us in time).

Cheers,

Ty Smith
Vice President

FIGURE I–7. Inappropriate International Correspondence (Draft Marked for Revision)

Sun West Corporation, Inc.

2565 North Armadillo
Tucson, AZ 85719
Phone: (602) 555-6677
Fax: (602) 555-6678 sunwest.com

February 28, 2006

Ichiro Katsumi
Investment Director
Toshiba Investment Company
1-29-10 Ichiban-cho
Tokyo 105, Japan

Dear Mr. Katsumi:

I hope that you and your family are well and prospering in the new year. We at Sun West Corporation are very pleased that you will be coming to visit us in Tucson this month. It will be a pleasure to meet you, and we are very gratified and honored that you are interested in investing in our company.

So that we can ensure that your stay will be pleasurable, we have taken care of all of your travel arrangements. You will

- Depart Narita–New Tokyo International Airport on Delta Airlines flight #75 at 1700 on March 20, 2006.
- Arrive at Los Angeles International Airport at 1050 local time and depart for Tucson on Delta flight #186 at 1205.
- Arrive at Tucson International Airport at 1330 local time on March 20 and depart Tucson International Airport on Delta flight #123 at 1845 on March 27.
- Arrive in Salt Lake City, Utah, at 1040 and depart on Delta flight #34 at 1115.
- Arrive in Portland, Oregon, at 1210 local time and depart Portland, Oregon, on Delta flight #254 at 1305.
- Arrive in Tokyo at 1505 local time on March 28.

If you need additional information about your travel plans or information on Sun West Corporation, please call, fax, or e-mail me directly at tsmith@sunwest.com. That way, we will receive your message in time to make the appropriate changes or additions.

FIGURE I–8. Appropriate International Correspondence

Mr. Ichiro Katsumi 2 February 28, 2006

After you arrive in Tucson, a chauffeur from Skyline Limousines will be waiting for you at Gate 12. He or she will be carrying a card with your name, will help you collect your luggage from the baggage claim area, and will then drive you to the Loews Ventana Canyon Resort. This resort is one of the most prestigious in Tucson, with spectacular desert views, high-quality amenities, and one of the best golf courses in the city. The next day, the chauffeur will be back at the Ventana at 0900 to drive you to Sun West Corporation.

We at Sun West Corporation are very excited to meet you and introduce you to all the staff members of our hardworking and growing company. After you meet everyone, you will enjoy a catered breakfast in our conference room. At that time, you will receive a schedule of events planned for the remainder of your trip. Events include presentations from the president of the company and from departmental directors on

- The history of Sun West Corporation
- The uniqueness of our products and current success in the marketplace
- Demographic information and the benefits of being located in Tucson
- The potential for considerable profits for both our companies with your company's investment

We encourage you to read through the enclosed guidebook and map of Tucson. In addition to events planned at Sun West Corporation, you will find many natural wonders and historical sites to see in Tucson and in Arizona in general. If you see any particular event or place that you would like to visit, please let us know. We will be happy to show you our city and all it has to offer.

Again, we are very honored that you will be visiting us, and we look forward to a successful business relationship between our two companies.

Sincerely,

Ty Smith

Ty Smith
Vice President

Enclosures (2)

FIGURE I–8. Appropriate International Correspondence (*continued*)

This draft contains <u>idioms</u> (*looking forward, company family*), <u>jargon</u> (*transport will be holding . . .*), <u>contractions</u> (*I'm, don't*), informal language (*just e-mail or fax, Cheers*), and humor and allusion (*"ptomaine palace" across from our main offices*).

Compare that letter to the one in Figure I–8, which is written in language that is literal and specific. The letter begins with concern about the recipient's family and prosperity because that opening honors traditional Japanese patterns in business correspondence. The letter is free of slang, idioms, and jargon. The sentences are shorter, bulleted lists are used to break up the paragraphs, contractions are not used, and months are spelled out.

When you are writing for international readers, rethink the ingrained habits that define how you express yourself, learn as much as you can about the cultural expectations of others, and focus on politeness strategies that demonstrate your respect for readers. Doing so will help you achieve <u>clarity</u> and mutual understanding with international readers.

WEB LINK GOOGLE'S INTERNATIONAL DIRECTORY

Google's International Business and Trade Directory provides an excellent starting point for searching the Web for information related to customs, communication, and international standards. See <*bedfordstmartins.com/alredtech*> and select *Links for Handbook Entries.*

Writer's Checklist: Writing International Correspondence

- ☑ Observe the guidelines for courtesy, such as those in the *Writer's Checklist: Using Tone to Build Goodwill* in <u>correspondence</u> on page 100.
- ☑ Write clear and complete sentences: Unusual word order or rambling sentences will frustrate and confuse readers. See <u>garbled sentences</u>.
- ☑ Avoid an overly simplified style that may offend or affectation that may confuse the reader. See also <u>English as a second language</u>.
- ☑ Avoid humor, irony, and sarcasm; they are easily misunderstood outside their cultural context.
- ☑ Do not use idioms, jargon, slang expressions, unusual <u>figures of speech</u>, or <u>allusions</u> to events or attitudes particular to American life.
- ☑ Consider whether necessary technical terminology can be found in abbreviated English-language dictionaries; if it cannot, carefully define such terminology.
- ☑ Do not use contractions or <u>abbreviations</u> that may not be clear to international readers.

Writer's Checklist: Writing International Correspondence (continued)

- ☑ Avoid excessive informality, such as using first names too quickly.
- ☑ Write out dates, whether in the month-day-year style (June 11, 2006, not 6/11/06) used in the United States or the day-month-year style (11 June 2006, not 11/6/06) used in many other parts of the world.
- ☑ Specify time zones or refer to international standards, such as Greenwich Mean Time (GMT) or Coordinated Universal Time (UTC).
- ☑ Where possible, use international measurement standards, such as the metric system (18°C, 14 cm, 45 kg, and so on). See also global graphics.
- ☑ Ask someone from your intended audience's culture to review your draft before you complete your final proofreading.

I

Internet research (*see* research)

interviewing for information

The process of interviewing for information can be divided into three parts: (1) determining the proper person to interview, (2) preparing for the interview, and (3) conducting the interview.

Determining the Proper Person to Interview

Many times, your subject or purpose logically points to the proper person to interview for information. If you were writing about using the Web to market a software-development business, you would want to interview someone with extensive experience in Web marketing as well as someone who has built a successful business developing software. The following sources can help you determine the appropriate person to interview: (1) workplace colleagues or faculty in appropriate academic departments, (2) information from the Internet, (3) local chapters of professional societies, and (4) yellow or business pages of the local telephone directory. For advice on finding sources, see research.

Preparing for the Interview

Before the interview, learn as much as possible about the person you are going to interview and the organization for which he or she works. When you contact the prospective interviewee, explain who you are, why you would like an interview, the subject and purpose of the interview, and how much time it will take. You should also let your interviewee know that you will allow him or her to review your draft.

After you have made the appointment, prepare a list of questions to ask your interviewee. Avoid vague, general questions. A question such as "What do you think of the Internet?" is too general to elicit useful information. It is more helpful to ask specific but open-ended questions, such as the following: "Some local professionals in your field are making extensive use of the Internet for helping clients. How has the Internet helped your organization?"

Conducting the Interview

Arrive promptly for the interview and be prepared to guide the discussion. The following *Writer's Checklist: Interviewing Successfully* should help you to do so. During the interview, take only memory-jogging notes that will help you recall the conversation later; do not ask your interviewee to slow down so that you can take detailed notes. As the interview is reaching a close, take a few minutes to skim your notes and ask the interviewee to clarify anything that is ambiguous. Immediately after leaving the interview, use your memory-jogging notes to help you mentally review the interview and record your detailed notes. Do not postpone this step. No matter how good your memory is, you will forget some important points if you do not do this at once. See also note-taking.

Writer's Checklist: Interviewing Successfully

- ☑ Be pleasant but purposeful. You are there to get information, so don't be timid about asking leading questions on the subject.
- ☑ Use the list of questions you have prepared, starting with the less controversial aspects of the topic to get the conversation started and then going on to the more controversial aspects.
- ☑ Let your interviewee do most of the talking. Remember that the interviewee is the expert. See also listening.
- ☑ Be objective. Don't offer your opinions on the subject. You are there to get information, not to debate.
- ☑ Some answers prompt additional questions; ask them as they arise.
- ☑ Don't get sidetracked. If the interviewee strays too far from the subject, ask a specific question to direct the conversation back on track.
- ☑ If you use a tape recorder, do not let it lure you into relaxing so that you neglect to ask crucial questions.
- ☑ After thanking the interviewee, ask permission to contact him or her again to clarify a point or two as you complete your interview notes.

interviewing for a job

A job interview may last 30 minutes, an hour, or several hours. Sometimes, an initial job interview is followed by a series of additional interviews that can last a half or full day. Often, just one or two people conduct a job interview, but, at times, a group of four or more might do so. Job interviews can take place in person, by phone, or by teleconference. Because it is impossible to know exactly what to expect, it is important that you be well prepared. See also application letters, job search, and résumés.

Before the Interview

The interview is not a one-way communication. It presents you with an opportunity to ask questions of your potential employer. In preparation, learn everything you can about the company before the interview. Use the following questions as a guide.

- What kind of organization (e.g., nonprofit, government) is it?
- How diversified is the organization?
- Is it locally owned?
- Does it provide a product or service? If so, what kind?
- How large is the business? How large are its assets?
- Is the owner self-employed? Is the company a subsidiary of a larger operation? Is it expanding?
- How long has it been in business?
- Where will you fit in?

You can obtain information from current employees, the Internet, company publications, and the business section of back issues of local newspapers (available in the library or online). You may be able to learn the company's size, sales volume, product line, credit rating, branch locations, subsidiary companies, new products and services, building programs, and other such information from its annual reports and from publications such as *Moody's Industrials*, *Dun and Bradstreet*, *Standard and Poor's*, and *Thomas' Register*, as well as other business reference sources a librarian might suggest. Ask your interviewer about what you cannot find through your own research. Now is your chance to demonstrate your interest and make certain you are considering a healthy and growing company.

Try to anticipate the questions your interviewer might ask, and prepare your answers in advance. Be sure you understand a question before answering it, and avoid responding too quickly with a rehearsed answer—be prepared to answer in a natural and relaxed manner. Interviewers typically ask the following questions:

- What are your short-term and long-term occupational goals?
- Where do you see yourself five years from now?
- What are your major strengths and weaknesses?
- Do you work better with others or alone?
- How do you spend your free time?
- Describe an accomplishment you are particularly proud of.
- Why are you leaving your current job?
- Why do you want to work for this company?
- Why should I hire you?
- What salary and benefits do you expect?

Many employers use behavioral interviews. Rather than traditional, straightforward questions, the behavioral interview focuses on asking the candidate to provide examples or respond to hypothetical situations. Interviewers who use behavior-based questions are looking for specific examples from your experience. Prepare for the behavioral interview by recollecting challenging situations or problems that were successfully resolved. Examples of behavior-based questions include the following:

- Tell me about a time when you experienced conflict on a team.
- If I were your boss and you disagreed with a decision I made, what would you do?
- How have you used your leadership skills to bring about change?
- Tell me about a time when you failed.

Arrive for your interview on time, or even ten or fifteen minutes early—you may be asked to fill out an application or other paperwork before you meet your interviewer. Always bring extra copies of your résumé and samples of your work (if applicable). If you are asked to complete an application form, read it carefully before you write and proofread it when you are finished. The form provides a written record for company files and indicates to the company how well you follow directions and complete a task.

During the Interview

The interview actually begins before you are seated: What you wear and how you act make a first impression. In general, dress simply and conservatively, avoid extremes in fragrance and cosmetics, and be well-groomed.

Behavior. First, thank the interviewer for his or her time, express your pleasure at meeting him or her, and remain standing until you are offered a seat. Then sit up straight (good posture suggests self-assurance), look directly at the interviewer, and try to appear relaxed and confident. During the interview, you may find yourself feeling a little nervous. Use that nervous energy to your advantage by channeling it into the alertness that you will need to listen and respond effectively. Do not attempt to take extensive notes during the interview, although it is acceptable to jot down a few facts and figures on a small pad. See also listening.

Responses. When you answer questions, do not ramble or stray from the subject. Say only what you must to answer each question properly and then stop, but avoid giving just yes or no answers—they usually do not allow the interviewer to learn enough about you. Some interviewers allow a silence to fall just to see how you will react. The burden of conducting the interview is the interviewer's, not yours—and he or she may interpret your rush to fill a void in the conversation as a sign of insecurity. If such a silence makes you uncomfortable, be ready to ask an intelligent question about the company.

If the interviewer overlooks important points, bring them up. However, let the interviewer mention salary first, if possible. Doing so yourself may indicate that you are more interested in the money than the work. However, make sure you are aware of prevailing salaries and benefits in your field. See salary negotiations.

Interviewers look for a degree of self-confidence and an applicant's understanding of the field, as well as genuine interest in the field, the company, and the job. Ask questions to communicate your interest in the job and company. Interviewers respond favorably to applicants who can communicate and present themselves well.

Conclusion. At the conclusion of the interview, thank the interviewer for his or her time. Indicate that you are interested in the job (if true) and try to get an idea of the company's hiring time frame. Reaffirm friendly contact with a firm handshake.

After the Interview

After you leave the interview, jot down the pertinent information you obtained, as it may be helpful in comparing job offers. As soon as possible following a job interview, send the interviewer a note of thanks in a brief letter or e-mail. Such notes often include the following:

- Your thanks for the interview and to individuals or groups that gave you special help or attention during the interview
- The name of the specific job for which you interviewed
- Your impression that the job is attractive
- Your confidence that you can perform the job well
- An offer to provide further information or answer further questions

Figure I–9 shows a typical example of follow-up correspondence.

If you are offered a job you want, accept the offer verbally and write a brief letter of acceptance as soon as possible—certainly within a week. If you do not want the job, write a refusal letter, as described in acceptance/refusal letters.

From: whaty@exl.gsc.edu (Wilson Hathaway)
To: itsuru@calcutex.com
Sent: 2 June 05 3:49:56 Eastern Daylight Time
Subject: Employment Interview

Dear Mr. Itsuru:

Thank you for the informative and pleasant interview we had yesterday. Please extend my thanks to Mr. Ragins of Human Resources as well.

I came away from our meeting most favorably impressed with Calcutex Industries. I find the position of software designer to be an attractive one and feel confident that my qualifications would enable me to perform the duties to everyone's advantage.

If I can answer any further questions, please let me know.

Sincerely,

Wilson Hathaway

Wilson Hathaway
Gabone State College
P.O. Box 5413 Highton, Ohio 45419
(937) 229-4511 (x341)
Fax (937) 229-4527

FIGURE I–9. Follow-up Correspondence

introductions

DIRECTORY

I

This entry discusses opening strategies for short and routine types of correspondence, such as memos and e-mail, and large writing projects, such as formal reports and major proposals. See also conclusions.

Routine Openings

Not every document needs a fully developed introduction or opening. When your readers are already familiar with your subject, or if what you are writing is short, a brief or routine opening, as shown in the following examples, will provide adequate context.

CORRESPONDENCE

Dear Mr. Ignatowski:
You will be happy to know that we have corrected the error in your bank balance. The new balance shows . . .

PROGRESS REPORT LETTER

Dear Dr. Chang:
To date, 18 of the 20 specimens you submitted for analysis have been examined. Our preliminary analysis indicates . . .

LONGER PROGRESS REPORT

PROGRESS REPORT ON REWIRING THE SPORTS ARENA

The rewiring program at the Sports Arena is proceeding ahead of schedule. Although the costs of certain equipment are higher than our original bid, we expect to complete the project without exceeding our budget because the speedy completion will save labor costs.

Work Completed

As of August 15, we have . . .

E-MAIL

Jane, as I promised in my e-mail yesterday, I've attached the engineering division budget estimates for fiscal year 2006.

Opening Strategies

Opening strategies are aimed at focusing the readers' attention and motivating them to read the entire document.

Objective. In reporting on a project, you might open with a statement of the project's objective so the readers have a basis for judging the results.

- The primary goal of this project was to develop new techniques to solve the problem of waste disposal. Our first step was to investigate . . .

Problem Statement. One way to give readers the perspective of your report is to present a brief account of the problem that led to the study or project being reported.

I

- Several weeks ago a manager noticed a recurring problem in the software developed by Datacom Systems. Specifically, error messages repeatedly appeared when, in fact, no specific trouble. . . . After an extensive investigation, we found that Datacom Systems . . .

For proposals or formal reports, of course, problem statements may be more elaborate and a part of the full-scale introduction, which is discussed later in this entry.

Scope. You may want to present the scope of your document in your opening. By providing the parameters of your material, the limitations of the subject, or the amount of detail to be presented, you enable your readers to determine whether they want or need to read your document.

- This pamphlet provides a review of the requirements for obtaining an FAA pilot's license. It is not intended as a textbook to prepare you for the examination itself; rather, it outlines the steps you need to take and the costs involved.

Background. The background or history of a subject may be interesting and lend perspective and insight to a subject. Consider the following example from a newsletter describing the process of oil drilling:

- From the bamboo poles the Chinese used when the pyramids were young to today's giant rigs drilling in hundreds of feet of water, there has been considerable progress in the search for oil. But whether in ancient China or a modern city, underwater or on a mountaintop, the objective of drilling has always been the same—to manufacture a hole in the ground, inch by inch.

Summary. You can provide a summary opening by describing in abbreviated form the results, conclusions, or recommendations of your article or report. Be concise: Do not begin a summary by writing "This report summarizes. . . ."

CHANGE This report summarizes the advantages offered by the photon as a means of examining the structural features of the atom.

TO As a means of examining the structure of the atom, the photon offers several advantages.

Interesting Detail. Often an interesting detail will gain the readers' attention and arouse their curiosity. Readers of a white paper for a manufacturer of telescopes and scientific instruments, for example, may be persuaded to invest if they believe that the company is developing innovative, cutting-edge products.

- The rings of Saturn have puzzled astronomers ever since they were discovered by Galileo in 1610 using the first telescope. Recently, even more rings have been discovered. . . .
  Our company's Scientific Instrument Division designs and manufactures research-quality, computer-controlled telescopes that promise to solve the puzzles of Saturn's rings by enabling scientists to use multicolor differential photometry to determine the rings' origins and compositions.

Definition. Although a definition can be useful as an opening, do not define something with which the reader is familiar or provide a definition that is obviously a contrived opening (such as "Webster defines *technology* as . . ."). A definition should be used as an opening only if it offers insight into what follows.

- *Risk* is often a loosely defined term. In this report, risk refers to a qualitative combination of the probability of an event and the severity of the consequences of that event. In fact, . . .

Anecdote. An anecdote can be used to attract and build interest in a subject that may otherwise be mundane; however, this strategy is best suited to longer documents and presentations.

- In his poem "The Calf Path," Sam Walter Foss tells of a wandering, wobbly calf trying to find its way home at night through the lonesome woods. It made a crooked path, which was taken up the next day by a lone dog. Then "a bellwether sheep pursued the trail over vale and steep, drawing behind him the flock, too, as all good bellwethers do." At last the path became a country road; then a lane that bent and turned and turned again. The lane became a village street, and at last the main street of a flourishing city. The poet ends by saying, "A hundred thousand men were led by a calf, near three centuries dead."

Medical researchers today often follow a "calf path" because they pursue only well-established lines of investigation rather than. . . .

Quotation. Occasionally, you can use a quotation to stimulate interest in your subject. To be effective, however, the quotation must be pertinent—not some loosely related remark selected from a book of quotations.

- According to Deborah Andrews, "technical communicators in the twenty-first century must reach audiences and collaborate across borders of culture, language, and technology" [B3]. One way of accomplishing that goal is to make sure our training includes cross-cultural experiences that provide . . .

Forecast. Sometimes you can use a forecast of a new development or trend to arouse the reader's interest.

- In the not-too-distant future, we may be able to use a hand-held medical diagnostic device similar to those in science fiction to assess the complete physical condition of accident victims. This project and others are now being developed at The Seldi Group, Inc.

Persuasive Hook. While all opening strategies contain persuasive elements, the hook uses persuasion most overtly. A brochure touting the newest innovation in tax-preparation software might address readers as follows:

- Welcome to the newest way to do your taxes! TaxPro EZ ends the headache of last-minute tax preparation with its unique Web-Link feature.

Full-Scale Introductions

The purpose of a full-scale introduction is to give readers enough general information about the subject to enable them to understand the details in the body of the document. An introduction should accomplish any or all of the following:

- *State the subject.* Provide background information, such as definition, history, or theory, to provide context for your readers.

- *State the purpose.* Make your readers aware of why the document exists and whether the material provides a new perspective or clarifies an existing perspective.
- *State the scope.* Tell readers the amount of detail you plan to cover.
- *Preview the development of the subject.* Especially in a longer document, outline how you plan to develop the subject. Providing such information allows readers to anticipate how the subject will be presented and helps them evaluate your conclusions or recommendations.

Consider writing an opening or introduction last. Many writers find that it is only when they have drafted the body of the document that they have a full enough perspective on the subject to introduce it adequately.

Manuals and Specifications

You may need to write one kind of introduction for reports, academic papers, or trade journal articles and a different kind for manuals or specifications. When writing an introduction for a manual or set of specifications, identify the topic and its primary purpose or function in the first sentence or two. Be specific, but do not go into elaborate detail. Your introduction sets the stage for the entire document, and it should provide readers with a broad frame of reference and an understanding of the overall topic. Then the reader is ready for technical details in the body of the document.

How technical your introduction should be depends on your readers: What are their technical backgrounds? What kind of information are they seeking in the manual or specification? The topic should be introduced with a specific audience in mind—a computer user, for example, has different interests in an application program and a different technical vocabulary than a programmer. Whether you need to provide explanations or definitions of terminology will depend on your intended audience. The following example is written for readers who understand such terms as "constructor" and "software modules."

- The System Constructor is a program that can be used to create operating systems for a specific range of microcomputer systems. The constructor selects requested operating software modules from an existing file of software modules and combines those modules with a previously compiled application program to create a functional operating system designed for a specific hardware configuration. It selects the requested software modules, establishes the necessary linkage between the modules, and generates the control tables for the system according to parameters specified at run time.

You may encounter a dilemma that is common in technical writing: Although you cannot explain topic A until you have explained topic B, you cannot explain topic B before explaining topic A. The solution is to explain both topics in broad, general terms in the introduction, as in the following paragraph that introduces "source units" and "destination units":

- The NEAT/3 programming language, which treats all peripheral units as file storage units, allows your program to perform data input or output operations depending on the specific unit. Peripheral units from which your program can only input data are referred to as *source units*; those to which your program can only output data are referred to as *destination units*.

Then, when you need to write a detailed explanation of topic A (*source units*), you will be able to do so because your reader will know just enough about both topics to be able to understand your detailed explanation.

investigative reports

Investigative reports may be written for a variety of reasons—most often in response to a request for information. You might be asked, for instance, to determine how your Web site compares to those of competing companies in your industry or to learn the level of satisfaction among your customers. An investigative report gives a precise analysis of a topic and offers conclusions, as is the case of the investigative report shown in Figure I–10.

Open the report with a statement of its primary and (if any) secondary purposes, then define the scope of your investigation. If the report includes a survey of opinions, indicate the number of people surveyed, income categories, occupations, and other identifying information. (See also questionnaires.) Include any information that is pertinent in defining the extent of the investigation. Then report your findings and, if necessary, discuss their significance in the report with your conclusions.

Sometimes the person requesting the investigative report may ask you to make recommendations as a result of your findings. In that case, the report may be referred to as a *recommendation report*. See also feasibility reports.

Memo

To: Noreen Rinaldo, Training Manager
From: Charles Lapinski, Senior Instructor *CL*
Date: February 7, 2006
Subject: Adler's Basic English Program

As requested, I have investigated Adler Medical Instruments' (AMI's) Basic English Program to determine whether we might adopt a similar program.

The purpose of AMI's program is to teach medical technologists outside the United States who do not speak or read English to understand procedures written in a special 800-word vocabulary called *Basic English.* This program eliminates the need for AMI to translate its documentation into a number of different languages. The Basic English Program does not attempt to teach the medical technologists to be fluent in English but, rather, to recognize the 800 basic words that appear in Adler's documentation.

Course Analysis

The course teaches technologists a basic medical vocabulary in English; it does not provide training in medical terminology. Students must already know, in their own language, the meaning of medical vocabulary (e.g., the meaning of the word *hemostat*). Students must also have basic knowledge of their specialty, must be able to identify a part in an illustrated parts book, must have used AMI products for at least one year, and must be able to read and write in their own language.

Students are given an instruction manual, an illustrated book of equipment with parts and their English names, and pocket references containing the 800 words of the Basic English vocabulary plus the English names of parts. Students can write the corresponding word in their language beside the English word and then use the pocket reference as a bilingual dictionary. The course consists of 30 two-hour lessons, each lesson introducing approximately 27 words. No effort is made to teach pronunciation; the course teaches only recognition of the 800 words, which include 450 nouns, 70 verbs, 180 adjectives and adverbs, and 100 articles, prepositions, conjunctions, and pronouns.

Course Success

The 800-word vocabulary enables the writers of documentation to provide medical technologists with any information that might be required because the subject areas are strictly limited to usage, troubleshooting, safety, and operation of AMI medical equipment. All nonessential words (*apple*, *father*, *mountain*, and so on) are eliminated, as are most synonyms (for example, *under* appears, but *beneath* does not).

Conclusions and Recommendations

AMI's program appears to be quite successful, and a similar approach could also be appropriate for us. I see two possible ways in which we could use some or all of the elements of AMI's program: (1) in the preparation of our student manuals or (2) as AMI uses the program.

I think it would be unnecessary to use the Basic English methods in the preparation of manuals for *all* of our students. Most of our students are English speakers to whom an unrestricted vocabulary presents no problem.

As for our initiating a program similar to AMI's, we could create our own version of the Basic English vocabulary and write our instructional materials in it. Because our product lines are much broader than AMI's, however, we would need to create illustrated parts books for each of the different product lines.

FIGURE I–10. Investigative Report

italics

Italics is a style of type used to denote emphasis and to distinguish foreign expressions, book titles, and certain other elements. *This sentence is printed in italics.* Italic type is signaled by underlining in manuscripts submitted for publication or where italic font is not available (see also e-mail). You may need to italicize words that require special emphasis in a sentence. (Contrary to projections, sales have *not* improved.) Do not overuse italics for emphasis, however. (This will hurt *you* more than *me*.)

Foreign Words and Phrases

Foreign words and phrases are italicized (*bonjour*, *guten tag*, the sign said *"Se habla español"*). Foreign words that have been fully assimilated into English need not be italicized (cliché, etiquette, vis-à-vis, de facto, résumé). When in doubt about the treatment of a word, consult a current dictionary. See also foreign words in English.

Titles

Italicize the titles of separately published documents, such as books, periodicals, newspapers, pamphlets, brochures, legal cases, movies, and television programs.

- The book *Turning Workplace Conflicts into Collaboration* was reviewed in the *New York Times*.

Abbreviations of such titles are italicized if their spelled-out forms would be italicized.

- The *NYT* is one of the nation's oldest newspapers.

Italicize the titles of compact discs, videotapes, plays, long poems, paintings, sculptures, and long musical works.

CD-ROM	*Computer Security Tutorial on CD-ROM*
PLAY	Arthur Miller's *Death of a Salesman*
LONG POEM	T. S. Eliot's *The Wasteland*
MUSICAL WORK	Gershwin's *Porgy and Bess*

Use quotation marks for parts of publications, such as chapters of books and articles or sections within periodicals.

Proper Names

The names of ships, trains, and aircraft (but not the companies or governments that own them) are italicized (U.S. aircraft carrier *Inde-*

pendence, U.S. space shuttle *Endeavour*). Craft that are known by model or serial designations are not italicized (DC-7, Boeing 747).

Words, Letters, and Figures

Words, letters, and figures discussed as such are italicized.

- The word *inflammable* is often misinterpreted.
- The *S* and *6* keys on my keyboard do not function.

Subheads

Subheads in a report are sometimes italicized.

- *Training Managers.* We are leading the way in developing first-line managers who not only are professionally competent but . . .

See also headings and layout and design.

its / it's

Its is a possessive pronoun and does not use an apostrophe. *It's* is a contraction of *it is*.

- *It's* essential that the lab maintain *its* quality control.

See also expletives and possessive case.

J

jargon

Jargon is a highly specialized slang that is unique to an occupational or a professional group. Jargon is at first understood only by insiders; over time, it may become known more widely. For example, computer programmers coined the term *debugging* to describe the discovery and correction of errors in software. If all your readers are members of a particular occupational group, jargon may provide an efficient means of communicating. However, if you have any doubt that your entire audience is part of such a group, avoid using jargon. See also affectation, functional shift, and gobbledygook.

job descriptions

Most large companies and many small ones use formal job descriptions to specify the duties of and requirements for many of the jobs in the firm. Job descriptions fulfill several important functions: They provide information on which equitable salary scales can be based; they help management determine whether all functions within a company are adequately supported; and they let both prospective and current employees know exactly what is expected of them. Together, all the job descriptions in a firm present a picture of the organization's structure.

Sometimes middle managers are given the task of writing job descriptions for their employees. In many organizations, though, employees are required to draft their own job descriptions, which supervisors then check and approve.

Although job-description formats vary from organization to organization, they commonly contain the following sections:

- The *accountability* section identifies, by title only, the person to whom the employee reports.

- The *scope of responsibilities* section provides an overview of the primary and secondary functions of the job and states, if applicable, who reports to the employee.
- The *specific duties* section gives a detailed account of the specific duties of the job as concisely as possible.
- The *personal requirements* section lists the required or preferred education, training, experience, and licensing for the job.

The job description shown in Figure J–1 on page 290 is typical. It never mentions the person holding the job described; it focuses, instead, on the job and the qualifications required to fill the position.

Writer's Checklist: Writing Job Descriptions

- ☑ Before attempting to write your job description, list all the different tasks you do in a week or a month. Otherwise, you will almost certainly leave out some of your duties.
- ☑ Focus on content. Remember that you are describing your job, not yourself.
- ☑ List your duties in decreasing order of importance. Knowing how your various duties rank in importance makes it easier to set valid job qualifications.
- ☑ Begin each statement of a duty with a verb and be specific. Write "Greet and assist customers" rather than "Take care of customers."
- ☑ Review existing job descriptions that are considered well written.

job search

Whether you are applying for your first job or want to change careers entirely, begin by assessing your skills, interests, and abilities, perhaps through brainstorming.* Next, consider your career goals and values. For instance, do you prefer working independently or collaboratively? Do you enjoy public settings? meeting people? How important are career stability and location? Finally, ask yourself what you would most like to be doing in the immediate future, in two years, and in five years.

Once you have reflected and brainstormed about the job that is right for you, a number of sources can help you locate the job you

*A good source for stimulating your thinking is the most recent edition of *What Color Is Your Parachute? A Practical Manual for Job-Hunters & Career-Changers* by Richard Nelson Bolles, published by Ten Speed Press.

Manager, Technical Publications
Dakota Electrical Corporation

Accountability

Reports directly to the Vice President, Customer Service.

Scope of Responsibilities

The Manager of Technical Publications plans, coordinates, and supervises the design and development of technical publications and documentation required to support the sale, installation, and maintenance of Dakota products. The manager is responsible for the administration and morale of the staff. The supervisor for instruction manuals and the supervisor for parts manuals report directly to the manager.

Specific Duties

- Directs an organization currently comprising 20 people (including two supervisors), over 75 percent of whom are writing professionals and graphic artists
- Screens, selects, and hires qualified applicants for the department
- Prepares a formal orientation program to familiarize writing trainees with the production of reproducible copy and graphic arts
- Evaluates the performance of and determines the salary adjustments for all department employees
- Plans documentation to support new and existing products
- Subcontracts publications and acts as a purchasing agent when needed
- Offers editorial advice to supervisors
- Develops and manages an annual budget for the Technical Publications Department
- Cooperates with the Engineering, Parts, and Service Departments to provide the necessary repair and spare parts manuals upon the introduction of new equipment
- Serves as a liaison between technical specialists, the publications staff, and field engineers
- Recommends new and appropriate uses for the department within the company
- Keeps up with new technologies in printing, typesetting, art, and graphics and uses them to the advantage of Dakota Electrical Corporation where applicable

Requirements

- B.A. in technical communication
- Minimum of three years' professional writing experience and a general knowledge of graphics, production, and Web design
- Minimum of two years' management experience with a knowledge of the general principles of management
- Strong interpersonal skills

J

FIGURE J–1. Job Description

want. Of course, you should not rely on any one of these sources exclusively:

- Networking
- Campus career services
- Web resources
- Advertisements
- Trade and professional journal listings
- Private (or temporary) employment agencies
- Letters of inquiry

Keep a file during your job search of dated job ads, copies of application letters and résumés, and the names of important contacts. This collection can serve as a future resource and reminder. See also interviewing for a job and salary negotiations.

J

Networking

Networking involves communicating with people who might provide useful advice or may know of potential jobs in your interest areas. They may include people already working in your chosen field, contacts in professional organizations, professors, family members, or friends. Use your contacts to expand your network of contacts. Consider that of all open positions, an estimated 80 percent are filled through networking.*

Campus Career Services

A visit to a college career-development center is another good way to begin your job search. Government, business, and industry recruiters often visit campus career offices to interview prospective employees; recruiters also keep career counselors aware of their companies' current employment needs and submit job descriptions to them. Not only can career counselors help you select a career, they can also put you in touch with the best, most current resources—identifying where to begin your search and saving you time. Career-development centers often hold workshops on résumé preparation and offer other job-finding resources on their Web sites.

Web Resources

Using the Web can enhance your job search in a number of ways. First, you can consult sites that give advice about careers, job seeking, and résumé preparation. Second, you can learn about businesses and

*From *JobStar Central*, an online job-search guide hosted by the *Wall Street Journal* at <http://jobstar.org>.

organizations that may hire employees in your area by visiting their Web sites. Such sites often list job openings, provide instructions for applicants, and offer other information, such as employee benefits. Third, you can learn about jobs in your field and post your résumé for prospective employers at employment databases, such as Monster.com or America's Job Bank. Fourth, you can post your résumé at your personal Web site. Although posting your résumé at an employment database will undoubtedly attract more potential employers, including your résumé at your own site has benefits. For example, you might provide a link to your site in e-mail correspondence or provide your Web site's URL in an inquiry letter to a prospective employer. If you use a personal Web site, however, it should contain only material that would be of interest to prospective employers, such as examples of your work, awards, and other professional items.

Employment specialists suggest that you spend time on the Web in the evening or early morning so that you can focus on in-person contacts during working hours.

 WEB LINK **FINDING A JOB**

For job-hunting tips, sample documents, and links to the sites mentioned in this entry, see <*bedfordstmartins.com/alredtech*> and select *Finding an Internship or Job* and *Links for Handbook Entries*.

Advertisements

Many employers advertise in the classified sections of newspapers and on their own Web sites. For the widest selection of help-wanted listings, look in the Sunday editions or the help-wanted Web pages of local and big-city newspapers. Use the search options they provide or the general strategies for database searches discussed in the entry research.

A clinical medical technologist seeking a job, for example, might find the specialty listed under "Medical Technologist" or "Clinical Laboratory Technologist." Depending on a hospital's or pathologist's needs, the listing could be even more specific, such as "Blood Bank Technologist" or "Hematology Technologist."

As you read the ads, take notes on salary ranges, job locations, job duties and responsibilities, and even the terminology used in the ads to describe the work. A knowledge of keywords and key expressions that are generally used to describe a particular type of work can be helpful when you prepare your résumé and letters of application.

Trade and Professional Journal Listings

In many industries, associations publish periodicals of interest to people working in the industry. Such periodicals (print and online) often contain job listings. To learn about the trade or professional associations for your occupation, consult resources on the Web, such as Google's Directory of Professional Organizations or online resources offered by your library or campus career office. You may also consult the following references at a library: *Encyclopedia of Associations*, *Encyclopedia of Business Information Sources*, and *National Directory of Employment Services*.

Private (or Temporary) Employment Agencies

Private employment agencies are profit-making organizations that are in business to help people find jobs—for a fee. Reputable agencies provide you with job leads, help you organize your job search, and supply information on companies doing the hiring. A staffing agency, or temporary placement agency, could match you with an appropriate temporary or permanent job in your field. Temporary work for an organization for which you might want to work permanently is an excellent way to build your network while continuing your job search.

Choose an employment or a temporary-placement agency carefully. Some are well established and reputable; others are not. Check with your local Better Business Bureau and your college career office before you sign an agreement with a private employment agency. Further, be sure you understand who is paying the agency's fee. Often the employer pays the agency's fee; however, if you have to pay, make sure you know exactly how much. As with any written agreement, read the fine print carefully.

Letters of Inquiry

If you would like to work for a particular firm, write and ask whether it has any openings for people with your qualifications. Normally, you can send the letter to the department head, the director of human resources, or both; for a small firm, however, write to the head of the firm.

Other Sources

Local, state, and federal government agencies offer many employment services. Local government agencies are listed in telephone and Web directories under the name of your city, county, or state.

journal articles (*see* **trade journal articles**)

K

kind of / sort of

The phrases *kind of* and *sort of* should be used only to refer to a class or type of things.

- He described a special *kind of* mirror called a quantum switch.

Do not use *kind of* or *sort of* to mean "rather," "somewhat," or "somehow." That usage can lead to vagueness; it is better to be specific.

K

VAGUE	It was *kind of* a bad year for the company.
SPECIFIC	The company's profits fell 10 percent last year.

know-how

The informal term *know-how*, meaning "special competence or knowledge," should be avoided in formal writing style.

- The applicant has impressive technical ~~know-how.~~ skill.

L

laboratory reports

A laboratory report communicates information acquired from laboratory testing or a major investigation. It should begin by stating the reason that a laboratory investigation was conducted; it should also list the equipment and methods used during the test, the problems encountered, the results and conclusions reached, and any recommendations. The example in Figure L–1 on pages 296–97 shows sections from a typical laboratory report.

A laboratory report emphasizes the equipment and procedures used in the investigation because those two factors can be critical in determining the accuracy of the data and even replicating the procedure if necessary. Although this emphasis often requires the use of the passive voice, you should present the results of the laboratory investigation clearly and precisely. If your report requires graphs or tables, integrate them into your report as described in the entry visuals. For reporting routine or simple laboratory tests, see test reports.

lay / lie

Lay is a transitive verb—a verb that requires a direct object to complete its meaning—that means "place" or "put."

- We will *lay* the foundation of the building one section at a time.

The past-tense form of *lay* is *laid.*

- We *laid* the first section of the foundation last month.

The perfect-tense form of *lay* is also *laid.*

- Since June, we *have laid* all but two sections of the foundation.

Lay is frequently confused with *lie,* which is an intransitive verb—a verb that does not require an object to complete its meaning—that means "recline" or "remain."

PCB Exposure from Oil Combustion

Wayne County Firefighters Association

Submitted to:
Mr. Philip Landowe
President, Wayne County Firefighters Association
Wandell, IN 45602

Submitted by:
Analytical Laboratories, Incorporated
1220 Pfeiffer Parkway
Indianapolis, IN 46223
February 28, 2006

INTRODUCTION

Waste oil used to train firefighters was suspected by the Wayne County Firefighters Association of containing polychlorinated biphenyls (PCBs). According to information provided by Mr. Philip Landowe, President of the Association, it has been standard practice in training firefighters to burn 20–100 gallons of oil in a diked area of approximately 25–50 m^3. Firefighters would then extinguish the fire at close range. Exposure would last several minutes, and the exercise would be repeated two or three times each day for one week.

Oil samples were collected from three holding tanks near the training area in Englewood Park on November 14, 2005. To determine potential firefighter exposure to PCBs, bulk oil analyses were conducted on each of the samples. In addition, the oil was heated and burned to determine the degree to which PCB is volatized from the oil, thus increasing the potential for firefighter exposure via inhalation.

TESTING PROCEDURES

Bulk oil samples were diluted with hexane, put through a cleanup step, and analyzed in electron-capture gas chromatography. The oil from the underground tank that contained PCBs was then exposed to temperatures of 1008°C without ignition and 2008°C with ignition. Air was passed over the enclosed sample during heating, and volatized PCB was trapped in an absorbing medium. The absorbing medium was then extracted and analyzed for PCB released from the sample.

RESULTS

Bulk oil analyses are presented in Table 1. Only the sample from the underground tank contained detectable amounts of PCB. Aroclor 1260, containing 60 percent chlorine, was found to be present in this sample at 18 mg. Concentrations of 50 mg PCB in oil are considered hazardous. Stringent storage and disposal techniques are required for oil with PCB concentrations at these levels.

TABLE 1. Bulk Oil Analyses

Source	*Sample #*	*PCB Content (mg/g)*
Underground tank	6062	18*
(11' deep)		
Circle tank	6063a	<1
(3' deep)	6063b	<1
Square pool	6064a	<1
(3' deep)	6064b	<1

*Aroclor 1260 is the PCB type. This sample was taken for volatilization study.

FIGURE L–1. Laboratory Report

DISCUSSION AND CONCLUSIONS

At a concentration of 18 mg/g, 100 gallons of oil would contain approximately 5.5 g of PCB. Of the 5.5 g of PCB, about 0.3 g would be released to the atmosphere under the worst conditions.

The American Conference of Governmental Industrial Hygienists has established a threshold limit value (TLV)* of 0.5 mg/m^3 air for a PCB containing 54 percent C1 as a time-weighted average over an 8-hour work shift and has stipulated that exposure over a 15-minute period should not exceed 1 mg/m^3. The 0.3 g of released PCB would have to be diluted to 600 m^3 air to result in a concentration of 0.5 mg/m^3 or less. Because the combustion of oil lasted several minutes, a dilution to more than 600 m^3 is likely; thus, exposure would be less than 0.5 mg/m^3.

In summary, because exposure to this oil was limited and because PCB concentrations in the oil were low, it is unlikely that exposure from inhalation would be sufficient to cause adverse health effects. However, we cannot rule out the possibility that excessive exposure may have occurred under certain circumstances, based on factors such as excessive skin contact and the possibility that oil with a higher-level PCB concentration could have been used earlier. The practice of using this oil should be terminated.

*The safe average concentration that most individuals can be exposed to in an 8-hour day.

FIGURE L–1. Laboratory Report (*continued*)

- Injured employees should *lie* down and remain still until the EMTs arrive.

The past-tense form of *lie* is *lay*. This form causes the confusion between *lie* and *lay*.

- The injured employee *lay* still for approximately five minutes.

The perfect-tense form of *lie* is *lain*.

- The injured employee *had lain* still for approximately five minutes before the EMTs arrived.

layout and design

DIRECTORY

The layout and design of a document can make even the most complex information accessible and give readers a favorable impression of the writer and the organization. To accomplish those goals, a design should help readers find information easily; offer a simple and uncluttered presentation; and highlight structure, hierarchy, and order. The design must also fit the purpose of the document and its context. For example, if clients are paying a high price for consulting services, they may expect a sophisticated, polished design; if employees inside an organization expect management to be frugal, they may accept—even expect—an economical and standard company design.

Effective design is based on visual simplicity and harmony and can be achieved with careful selection of typography, page-design elements, and appropriate visuals, as well as thoughtful layout of text and visual components on a page.

Typography

Typography refers to the style and arrangement of type on a page. A complete set of all the letters, numbers, and symbols available in one typeface (or style) is called a *font*. The letters in a typeface have a number of distinctive characteristics, as shown in Figure L–2.

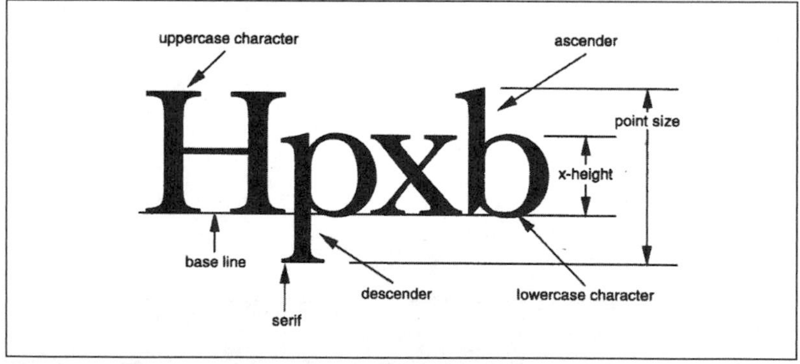

FIGURE L–2. Primary Components of Letter Characters

Typeface and Type Size. For most on-the-job writing, select a typeface primarily for its legibility. Avoid typefaces that may distract readers. Instead, choose popular typefaces with which readers are familiar, such as Times Roman, Garamond, or Gill Sans. Avoid using more than two typefaces in the text of a document. For certain documents, however, such as newsletters, you may wish to use distinctively different typefaces to provide contrast among various elements like headlines, headings, inset quotes, and sidebars. Always experiment before making final decisions, and keep in mind your audience.

One way typefaces are characterized is by the presence or absence of serifs. Serif typefaces have projections, as shown in Figure L–2; sans serif styles do not. (*Sans* is French for "without.") The text of this book is set in Plantin, a serif typeface. Although sans serif type has a modern look, serif type is easier to read, especially in the smaller sizes. Sans serif, however, works well for headings (like the entry titles in this book) and for Web sites and other documents read on-screen.

Ideal font sizes for the main text of paper documents range from 10 to 12 points. However, for some elements or documents, you may wish to select typeface sizes that are smaller (as in footnotes) or larger (as in headlines for brochures). See Figure L–3 for a visual comparison of type sizes in a serif typeface.

6 pt. This size might be used for dating a source.
8 pt. This size might be used for footnotes.
10 pt. This size might be used for figure captions.
12 pt. This size might be used for main text.
14 pt. This size might be used for headings.

FIGURE L–3. Type Sizes (6- to 14-Point)

Your readers and the distance from which they will read a document should help determine type size. For example, instructions that will rest on a table at which the reader stands require a larger typeface than a document that will be read up close. For presentations and writing for the Web, preview your document to see the effectiveness of your choice of point sizes and typefaces.

Type Style and Emphasis. One method of achieving emphasis through typography is to use capital letters. HOWEVER, LONG STRETCHES OF ALL UPPERCASE LETTERS ARE DIFFICULT TO READ. (See also e-mail.) Use all uppercase letters only in short spans, such as in headings. Likewise, use italics sparingly because *continuous italic type reduces legibility and thus slows readers.* Of course, italics are useful if your aim is to slow readers, as in cautions and warnings. **Boldface**, used in moderation, may be the best cuing device because it is visually different yet retains the customary shapes of letters and numbers.

Page-Design Elements

Thoughtfully used design elements can not only provide emphasis but also give a document visual logic by highlighting organization. Consistency and moderation are important—use the same technique to highlight a particular feature throughout your document and be careful not to overuse any single technique. The following typical elements can be used to make your document accessible and effective: justification, headings, headers and footers, lists, columns, white space, and color. Some of these elements are illustrated in Figure L–4.

L

Justification. Left-justified (ragged-right) margins are generally easier to read than full-justified margins, especially for text using standard margins on 8½″ × 11″ pages. Left-justified is also better if full justification causes your word-processing software to insert irregular spaces between words, producing unwanted white space or unevenness in blocks of text. Full-justified text is more appropriate for publications aimed at a broad readership that expects a more formal, polished appearance. Full justification is also useful with multiple-column formats because the spaces between the columns (called *alleys*) need the definition that full justification provides.

Headings. Headings reveal the organization of a document and help readers decide which sections they need to read. You should provide typographic contrast between headings and the base text with either a different typeface or a different style (bold, italic, caps, etc.). Headings are often effective in **boldface** or in a sans serif typeface that contrasts with a base text in a serif typeface.

Headers and Footers. A header in a report, letter, or other document appears at the top of each page, and a footer appears at the bottom of each page. Document pages may have headers or footers or both that include such elements as the topic or subtopic of a section, identifying number, the date the document was written, page number, and the document name. Keep your headers and footers concise—too much information in them can create visual clutter. For headers used in letters and memos, see correspondence.

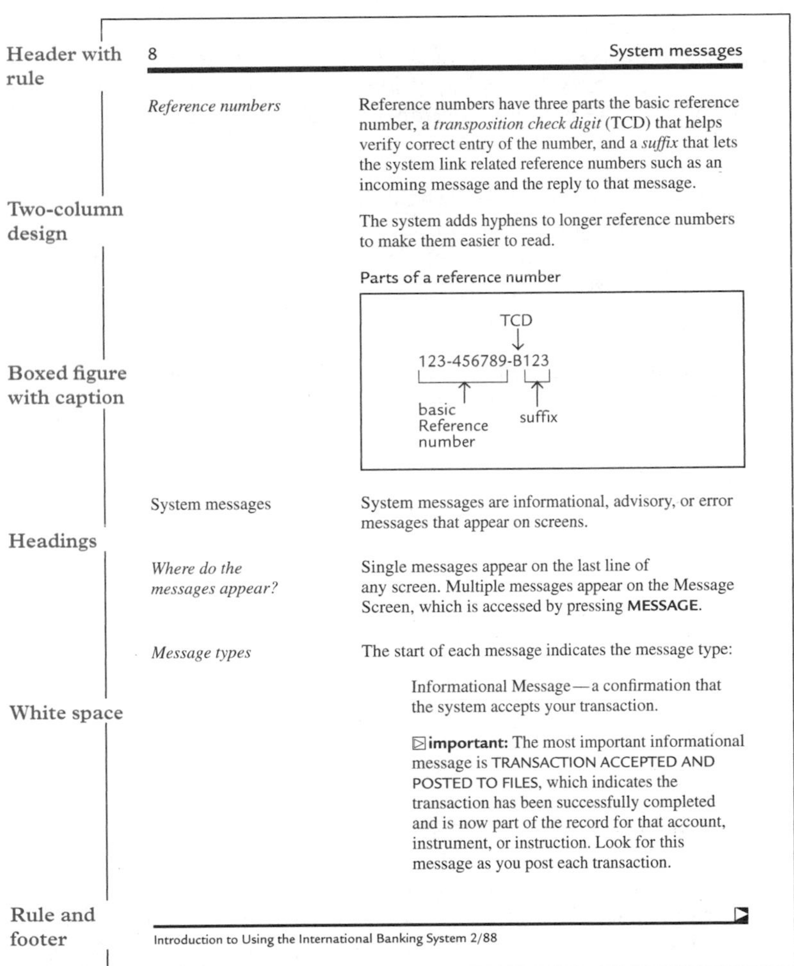

8 System messages

Reference numbers

Reference numbers have three parts the basic reference number, a *transposition check digit* (TCD) that helps verify correct entry of the number, and a *suffix* that lets the system link related reference numbers such as an incoming message and the reply to that message.

The system adds hyphens to longer reference numbers to make them easier to read.

Parts of a reference number

System messages

System messages are informational, advisory, or error messages that appear on screens.

Where do the messages appear?

Single messages appear on the last line of any screen. Multiple messages appear on the Message Screen, which is accessed by pressing **MESSAGE**.

Message types

The start of each message indicates the message type:

Informational Message—a confirmation that the system accepts your transaction.

▷**important:** The most important informational message is TRANSACTION ACCEPTED AND POSTED TO FILES, which indicates the transaction has been successfully completed and is now part of the record for that account, instrument, or instruction. Look for this message as you post each transaction.

Introduction to Using the International Banking System 2/88

FIGURE L–4. Sample Page Illustrating Layout and Design Choices

Lists. Lists are an effective way to highlight words, phrases, and short sentences. Lists are particularly useful for presenting certain types of information, such as steps in sequence, materials or parts needed, concluding points, and recommendations.

Columns. As you design pages, consider how columns may improve the readability of your document. A single-column format works well with larger typefaces, double spacing, and left-justified margins.

For smaller typefaces and single-spaced lines, the two-column structure keeps text columns narrow enough so that readers need not scan back and forth across the width of the entire page for every line. A word on a line by itself at the end of a column is called an *orphan*. A single word carried over to the top of a column or page is called a *widow*. Avoid both.

White Space. White space visually frames information and breaks it into manageable chunks. For example, white space between paragraphs helps readers see the information in each paragraph as a unit. White space between sections can also serve as a visual cue to signal that one section is ending and another is beginning.

Color. Color and screening (shaded areas on a page) can distinguish one part of a document from another or unify a series of documents. They can set off sections within a document, highlight examples, or emphasize warnings. In tables, screening can highlight column titles or sets of data to which you want to draw the reader's attention.

Visuals

Readers notice visuals before they notice text, and they notice larger visuals before they notice smaller ones. Thus, the size of an illustration suggests its relative importance. For newsletter articles and publications aimed at wide audiences, consider especially the proportion of the illustration to the text. Magazine designers often use the three-fifths rule: Page layout is more dramatic and appealing when the major element (photograph, drawing, or other illustration) occupies three-fifths rather than half the available space. The same principle can be used to enhance the visual appeal of a report.

Illustrations can be gathered in one place (for example, at the end of a report), but placing them in the text closer to their accompanying explanations makes them more effective. Using illustrations in the text also provides visual relief. For advice on the placement of illustrations, see the *Writer's Checklist: Creating and Integrating Visuals* (pages 562–63).

Icons. Visuals that are especially useful for Web design are icons, which are pictorial representations of concepts. Commonly used icons include the national flags used for Web pages to symbolize different language versions of a document. To be effective, icons must be simple and easily recognized or defined, as shown in the entry global graphics.

Captions. Captions are titles that highlight or describe visuals or blocks of text. Captions often appear below or above figures and tables and in the left or right margins next to blocks of text.

Rules. Rules are vertical or horizontal lines used to create boxes for visuals or divide one area of the page from another. They can effectively highlight elements and make information more accessible. Rules and boxes, for example, set off visuals from surrounding explanations or highlight warning statements from the steps in instructions.

Page Layout and Thumbnails

Page layout involves combining typography, design elements, and visuals on a page to make a coherent whole. The flexibility of your design is affected by your design software, your method of printing the document, and your budget.

Before you spend time positioning actual text and visuals on a page, especially for documents like brochures, you may want to create a thumbnail sketch, in which blocks indicate the placement of elements. You can go further by roughly assembling all the thumbnail pages to show the size, shape, form, and general style of a large document. Such a mock-up, called a *dummy*, allows you to see how a finished document will look.

WEB LINK **DESIGNING DOCUMENTS**

Word-processing programs offer many options for improving the layout-and-design elements of your document. For a tutorial on these options, see <*bedfordstmartins.com/alredtech*> and select *Tutorials*, "Designing Documents with a Word Processor." For step-by-step instructions for setting margins, alignment, columns, and other design elements, select *Digital Tips*, "Laying Out a Page."

lend / loan

Both *lend* and *loan* can be used as verbs, but *lend* is more common. (You can *lend* [or *loan*] them the money if you wish.) Unlike *lend*, *loan* can be a noun. (The bank approved our *loan*.)

letters (*see* **correspondence** and **e-mail**)

liable / libel

Liable means "legally subject to" or "responsible for."

- Employers are held *liable* for their employees' actions.

Libel refers to "anything circulated in writing or pictures that injures someone's good reputation." When someone's reputation is injured in speech, the term is *slander.*

- When an editorial accused our board of directors of bribery, the board sued the newspaper for *libel.*
- If the mayor supports the bribery charge in her speech tonight, we will accuse her of *slander.*

In workplace writing, *liable* should retain its legal meaning. Where a condition of probability is intended, use *likely.*

- Rita is ~~liable~~ *likely* to be promoted.

library research (*see* **research**)

like / as

To avoid confusion between *like* and *as*, remember that *like* is a preposition and *as* (or *as if*) is a conjunction. Use *like* with a noun or pronoun that is not followed by a verb.

- The new supervisor behaves *like* a novice.

Use *as* before clauses, which contain verbs.

- He responded *as* we expected he would.
- It seemed *as if* the presentation would never end.

Like and *as* are used in comparisons: *Like* is used in elliptical constructions that omit the verb, and *as* is used when the verb is retained.

- He adapted to the new system *like* a duck to water.
- He adapted to the new system *as* a duck adapts to water.

listening

Effective listening enables the listener to understand the directions of an instructor, the message in a speaker's presentation, the goals of a manager, and the needs and wants of customers. Above all, it lays the foundation for cooperation. Productive communication occurs when both the speaker and the listener focus clearly on the content of the message and attempt to eliminate as much interference as possible.

Fallacies About Listening

Most people assume that because they can hear they know how to listen. In fact, *hearing* is passive, whereas *listening* is active. Hearing voices in a crowd or a ringing telephone requires no analysis and no active involvement. We hear such sounds without choosing to listen to them—we have no choice but to hear them. Listening, however, requires taking action, interpreting the message, and assessing its worth. Listening also requires that you consider the context of messages and the differences in meaning that may be the result of differences in the speaker's and the listener's occupation, education, culture, sex, race, or other factors. See also biased language, connotation/denotation, English as a second language, global communication, idioms, international correspondence, and jargon.

L

Active Listening

To listen actively, you should (1) make a conscious decision to listen actively, (2) define your purpose for listening, (3) take specific actions to listen more efficiently, and (4) adapt to the situation.

Step 1: Make a Conscious Decision. The first step to active listening is simply making up your mind to do so. Active listening requires a conscious effort, something that does not come naturally. To listen actively, the well-known precept is true: "Seek first to understand and *then* to be understood."*

Step 2: Define Your Purpose. Knowing why you are listening can go a long way toward managing the most common listening problems: drifting attention, formulating your response while the speaker is still talking, and interrupting the speaker. To help you define your purpose for listening, ask yourself these questions:

*Stephen R. Covey, *The 7 Habits of Highly Effective People: Powerful Lessons in Personal Change*, 15th ed. (New York: Free Press, 2004).

- What kind of information do I hope to get from this exchange, and how will I use it?
- What kind of message do I want to send while I am listening? (Do I want to portray understanding, determination, flexibility, competence, or patience?)
- What factors—boredom, daydreaming, anger, impatience—might interfere with listening during the interaction? How can I keep these factors from placing a barrier between the speaker and me?

Step 3: Take Specific Actions. Becoming an active listener requires a willingness to become a responder rather than a reactor. A *responder* is a listener who slows down the communication to be certain that he or she is accurately receiving the message sent by the speaker. A *reactor* simply says the first thing that comes to mind, without checking to make sure that he or she accurately understands the message. Take the following actions to help you become a responder and not a reactor.

- Make a conscious effort to be impartial when evaluating a message. For example, do not dismiss a message because you dislike the speaker or are distracted by the speaker's appearance, mannerisms, or accent.
- Slow down the communication by asking for more information or by paraphrasing the message received before you offer your thoughts. Paraphrasing lets the speaker know you are listening, gives the speaker an opportunity to clear up any misunderstanding, and keeps you focused.
- Listen with empathy by putting yourself in the speaker's position. When people feel they are being listened to empathetically, they tend to respond with appreciation and cooperation, thereby improving the communication.
- To help you stay focused on what the speaker is saying, take notes while you are listening. Note-taking not only communicates your attentiveness to the speaker, it also reinforces the message and helps you remember it.

Step 4: Adapt to the Situation. The requirements of active listening differ from one situation to another. For example, when you are listening to a lecture, you may be listening only for specific information. However, if you are on a team project that depends on everyone's contribution, you need to listen at the highest level so that you can gather information as well as pick up on nuances the other speakers may be communicating. See also collaborative writing.

lists

Lists can save readers time by allowing them to see at a glance specific items, questions, or directions. Lists also help readers by breaking up complex statements and by allowing key ideas to stand out, as in Figure L–5.

Before we agree to hold the engineering conference at the Brent Hotel, we should make sure the hotel facilities provide the following:

- Business center with phones, faxes, Internet access, and copying services for the conference committee
- Ground-floor exhibit area large enough for thirty 8-foot-by-15-foot booths
- Eight meeting rooms to accommodate 25 people each
- Internet access, projection screens for presentations, and overhead projectors in each room
- Ballroom and dining facilities for 250 people

To confirm that the Brent Hotel is our best choice, we should tour the facilities during our stay in Kansas City.

FIGURE L–5. Bulleted List in a Paragraph

Writer's Checklist: Using Lists

☑ List only comparable items and use parallel structure.

☑ Use only words, phrases, or short sentences of the same general length.

☑ Provide context by introducing each list, typically with a complete sentence followed by a colon.

☑ Ensure coherence by providing for adequate transition before and after a list.

☑ Use bullets, as opposed to numbers, when rank or sequence is not important.

☑ Do not overuse lists or include too many items, as in pages or presentation slides that contain only lists.

literature reviews

A literature review is a summary of the relevant literature (printed and electronic) available on a particular subject over a specified period of time. For example, a literature review might describe significant material published in the past five years on a technique for improving emergency medical diagnostic procedures, or it might describe all reports written in the past ten years on efforts to improve safety procedures at a particular company. A literature review tells readers what has been published on a particular subject and gives them an idea of what material they should read in full.

Some trade journal articles or theses begin with a brief literature review to bring the reader up to date on current research in the field. The writer then uses the review as background for his or her own discussion of the subject. Figure L–6 shows a two-paragraph literature review that serves as an introduction for a medical article on facial clefts in newborns. A literature review also could be a whole document in

L

Assessment of attractiveness of newborn infants with unrepaired facial clefts by different groups of individuals was studied to investigate the validity of a surgeons' rule of thumb system for this based upon cleft severity.

Recent studies have attempted to construct objective rating scales to quantify the severity of facial disfigurement in children and young people (Eliason et al., 1991; Poole et al., 1991; Roberts-Harry et al., 1992; Tobiasen and Hiebert, 1993). The medical community increasingly recognizes the need for a holistic approach to congenital disfigurement, including such aspects as parents' reactions, feelings, and attitudes (Bull and Rumsey, 1988). Results of these studies suggest that there may be a mistaken assumption of consensus between plastic surgeons and people unfamiliar with facial clefts in how each group respectively rates attractiveness of facial appearance. The studies all refer, however, to repaired facial clefts and other anomalies; none has assessed unrepaired facial clefts in neonates. Tobiasen (1991) has concluded that it is premature to suggest that surgeons rate impairment differently from others. Surgeons commonly use a rule of thumb system to classify severity on the basis of dichotomous categorizations (i.e., bilateral or unilateral clefting along the axis of symmetry; complete or incomplete extension to the nasal alar, and with or without palate involvement). The proposed system used by plastic surgeons and derived from Freedlander et al. (1990) is shown in Table 1.

Source: Slade, Pauline. "Relationships Between Cleft Severity and Attractiveness of Newborns with Unrepaired Clefts." *Cleft Palate–Craniofacial Journal*, vol. 32, no. 4 (July 1995): 318–322.

FIGURE L–6. Literature Review

itself. Fully developed literature reviews are good starting points for detailed research.

To prepare a literature review, you must first research published material on your topic. Because your readers may begin their research based on your literature review, carefully and accurately cite all bibliographic information. As you review each source, note the scope of the book or article and judge its value to the reader. Save all printouts of computer-assisted searches—you may want to incorporate the sources in a bibliography. See also documenting sources.

Begin a literature review by defining the area to be covered and the types of works to be reviewed. For example, a literature review may be limited to articles and reports and not include any books. You can arrange your discussion chronologically, beginning with a description of the earliest relevant literature and progressing to the most recent (or vice versa). You can also subdivide the topic, discussing works in various subcategories of the topic.

Related to literature reviews are annotated bibliographies, which also give readers information about published material. Unlike literature reviews, however, an annotated bibliography cites each bibliographic item and then describes it in a single block of text. The description (or *annotation*) may include the purpose of the article or book, its scope, the main topics covered, its historical importance, and anything else the writer feels the reader should know.

logic errors

Logic is essential to convincing readers that your conclusions are valid. Errors in logic can undermine the point you are trying to communicate and your credibility. (See also persuasion.) Typical logic errors include *lack of reason, sweeping generalizations, non sequiturs, false cause, biased or suppressed evidence, fact versus opinion,* and *loaded arguments.*

Lack of Reason

When a statement is contrary to the reader's common sense, that statement is not reasonable. If, for example, you stated, "New York is a small town," your reader might immediately question your statement. However, if you stated, "Although New York's population is over eight million, it is composed of neighborhoods that function as small towns," your reader could probably accept the statement as reasonable.

Sweeping Generalizations

Sweeping generalizations are statements that are too broad or all-inclusive to be supportable; they generally enlarge an observation about a small group to refer to an entire population. A flat statement, such as "Engineers are poor writers," ignores any possibility that some engineers may be good writers. Using such generalizations weakens your credibility.

Non Sequiturs

A non sequitur is a statement that does not logically follow a previous statement.

- I cleared off my desk, and the report is due today.

The missing link in these statements is that the writer cleared his or her desk to make space for materials to help finish the report today. In your own writing, be careful that you do not allow logical gaps that produce non sequiturs.

False Cause

A false cause (also called *post hoc, ergo propter hoc*) refers to the logical fallacy that because one event followed another event, the first somehow caused the second.

L

- I didn't bring my umbrella today. No wonder it is now raining.
- Because we now perform more testing at our Sherwood facility, fewer of the collected samples show contaminants.
  [The Sherwood facility may not have any relationship to fewer contaminated samples.]

Such errors in reasoning can happen when the writer hastily concludes that two events are related without examining the causal connections between them.

Biased or Suppressed Evidence

ETHICS NOTE A conclusion reached as a result of biased or suppressed evidence—self-serving data, questionable sources, purposely omitted or incomplete facts—is both illogical and unethical. Suppose you are preparing a report on the acceptance of a new policy among employees. If you distribute questionnaires only to those who think the policy is effective, the resulting evidence will be biased. If you purposely ignore employees who do not believe the policy is effective, you will be suppressing evidence. Intentionally ignoring relevant data that might not support your position not only produces inaccurate results but is unethical. See also ethics in writing. ✦

Fact Versus Opinion

Distinguish between fact and opinion. Facts include verifiable data or statements, whereas opinions are personal conclusions that may or may not be based on facts. For example, it is verifiable that distilled water boils at 100°C; that it tastes better or worse than tap water is an opinion. Distinguish the facts from your opinions in your writing so that your readers can draw their own conclusions.

Loaded Arguments

When you include an opinion in a statement and then reach conclusions that are based on that statement, you are loading the argument. Consider the following opening for a memo:

- I have several suggestions to improve the poorly written policy manual. First, we should change . . .

Unless everyone agrees that the manual is poorly written, readers may reject a writer's entire message because they disagree with this loaded premise. Do not load arguments in your writing; conclusions reached with loaded statements are weak and can produce negative reactions in readers who detect the loading.

WEB LINK **UNDERSTANDING AN ARGUMENT**

Dr. Frank Edler of Metropolitan Community College in Omaha, Nebraska, offers a tutorial in recognizing the logical components of an argument and thinking critically. See <*bedfordstmartins.com/alredtech*> and select *Links for Handbook Entries*.

loose / lose

Loose is an adjective meaning "not fastened" or "unrestrained." (He found a *loose* wire.) *Lose* is a verb meaning "be deprived of" or "fail to win." (I hope we do not *lose* the contract.)

lowercase and uppercase letters

(*see* **capitalization**)

malapropisms

A malapropism is a word that sounds similar to the one intended but is ludicrously wrong in the context.

INCORRECT	Our employees are no longer *sedimentary* now that we have a fitness center.
CORRECT	Our employees are no longer *sedentary* now that we have a fitness center.

Intentional malapropisms are sometimes used in humorous writing; unintentional malapropisms can embarrass a writer. See also figures of speech.

M

manuals

Manuals—whether printed or electronic—help customers and technical specialists use and maintain products. The manuals are usually written by professional technical writers, although in smaller companies, engineers and technicians may write them. See also instructions and review the entries process explanation and technical writing style.

Types of Manuals

Following are descriptions of typical technical manuals.

User Manuals. User manuals are aimed at skilled or unskilled users of equipment and provide instructions for the setup, operation, and maintenance of a product. User manuals also typically include safety precautions and troubleshooting charts and guides.

Tutorials. Tutorials are self-study guides for users of a product or system. Either packaged with user manuals or provided electronically, tutorials walk novice users through the operation of a product or system.

Training Manuals. Training manuals are used to train individuals in some procedure or skill, such as operating equipment, flying an airplane, or processing an insurance claim. In many technical and vocational fields, training manuals are the primary teaching devices. Training manuals are often accompanied by DVDs, CD-ROMs, or other audiovisual material.

Operators' Manuals. Written for skilled operators of construction, manufacturing, computer, or military equipment, operators' manuals contain essential instructions and safety warnings. They are often published in a convenient format that allows operators to use them at a work site.

Service Manuals. Service manuals help trained technicians repair equipment or systems, usually at the customer's location. Such manuals often contain troubleshooting guides for locating technical problems.

Special-Purpose Manuals. A number of manual types, including programmer reference manuals, overhaul manuals, handling and setup manuals, and safety manuals, have very specialized and limited uses.

Designing and Writing Effective Manuals

Identify and Write for Your Audience. Will you be writing for novice users, intermediate users, or experts? Or will you be instructing a combination of users with different levels of technical knowledge and experience? Identify your audience before you begin writing. Depending on your audience, you will make the following decisions:

- Which details to include (fewer for experts, more for novices)
- What level of technical vocabulary to use (necessary technical terminology for experts, plain language for novices)
- Whether to include a summary list of steps (experts will probably prefer using this summary list and not the entire manual; once novices and intermediate users gain more skill, they will also prefer referring just to this summary list)

Design your manual so that readers can use the equipment, software, or machinery while they are also reading your instructions.

- Readers should be able to easily find particular sections and instructions. Use generous amounts of white space between sections and actions, and use headings, subheadings, and words that readers will find familiar.

- Write with precision and accuracy so that readers can perform the procedures easily.
- Use clearly drawn and labeled visuals, in addition to written text, to show readers exactly what equipment, online screens, or other items in front of them should look like.

Provide an Overview. An overview at the beginning of a manual should explain the overall purpose of the procedure, how the procedure can be useful to the reader, and any cautions or warnings the reader should know about before starting. If readers know the purpose of the procedure, along with its specific workplace applications, they will be more likely to learn, recall the material that follows, and pay close attention to that material.

Create Major Sections. Divide any procedure into separate goals, create major sections to cover those goals, and state them in the section headings. If your manual has chapters, you can divide each chapter into specific subsections. In a manual about designing a Web page, an introductory chapter might include these subsections: (1) obtaining Web space, (2) viewing the new space, (3) using the Pico Editor to create text.

Indicate the Goals of Actions. Within each section, explain *why* readers must follow each step or each related set of steps. The conventional way to indicate a goal is to use either the infinitive form of a verb ("*To turn* on the machine") or the gerund form ("*Turning* on the machine"). Whichever verb form you choose, use it consistently throughout your manual each time you want to indicate a goal. In the following example, the manual writer decided to indicate the goal of a set of actions (Copying Files) with a boldfaced, upper- and lowercase, flush-left heading using the gerund form of the verb.

Copying Files

Action 1 At the Filer, type F (for filecopy)

Action 2 Type the file's old volume and filename in the "VOLUME:FILENAME" format, then hit <ENTER>.

Use the Imperative Verb Form for Actions. The conventional way to indicate an action is by using the imperative form of verbs (for example, "*Type* your name"; "*Press* the Control Key"). Use the imperative form consistently each time you want to indicate an action. In the example about copying files, the manual writer designated actions by placing subheads (Action 1, Action 2, and so on) in the left margin and imperative verbs ("type" and "hit") in the instructions to the right of the subheads.

Use Simple and Direct Verbs. Simple and direct verbs are most meaningful, especially for novice readers. Avoid jargon and terms known only to intermediate readers or experts, unless you know that they constitute the entire reading audience.

POOR VERB CHOICE	BETTER VERB CHOICE
Attempt	Try
Depress	Press
Discontinue	Stop
Display	Show
Employ	Use
Enumerate	Count
Execute	Do
Observe	Watch
Segregate	Divide

Indicate the Response of Actions. When appropriate, indicate the expected response of an action to reassure readers that they are performing the procedure correctly (for example, "A blinking light will appear" or "You will see a red triangle"). Use the form you choose consistently throughout a set of instructions to designate a response. In the following example, you can see how the response is indented farther to the right than the action, begins with a subhead ("Response"), and is indicated by a full sentence.

Action 1	At the Filer, type F (for filecopy)
Response	The system asks for the old volume and filename of the file you want to copy: **Filecopy what file?**

M

Writer's Checklist: Preparing Manuals

- ☑ Determine the best medium for your manual: Web, online document, CD-ROM, spiral binding, loose-leaf binding, and so on. See also layout and design.
- ☑ Pay attention to organization and outlining because complex products and systems need well-organized manuals to be useful to readers.
- ☑ Use a consistent format for each part of a manual: headings, subheadings, goals, actions, responses, cautions, warnings, and tips.
- ☑ Use standardized symbols, as described in global graphics, especially for international readers or where regulations require them.

Writer's Checklist: Preparing Manuals (continued)

- ☑ Use visuals—such as screen shots, schematics, exploded-view drawings, flowcharts, photographs, and tables—placing them where they would most benefit readers.
- ☑ Include indexes to help readers find information.
- ☑ Use warning statements and standard symbols for potential dangers in instructions.
- ☑ Conduct usability testing on manual drafts so that you can detect errors and other problems readers are likely to have.
- ☑ Have manuals reviewed by your peers as well as by technical and legal experts to ensure that the manuals are helpful and accurate. See also revision and proofreading.

maps

M

Maps are often used to show specific geographic features (roads, mountains, rivers, and the like). They can also illustrate geographic distributions of population, climate patterns, corporate branch offices, and so forth. The map in Figure M–1, from an environmental assessment, shows the overlapping geographic areas served by three electric utilities in Missouri, Iowa, and Illinois. Note that the map contains a figure number and title, scale of distances, key (or legend), compass, and distinctive highlighting for emphasis. Maps are often used in reports, proposals, brochures, and other documents in which readers need to know the location or geographic orientation of buildings and other facilities.

Writer's Checklist: Creating and Using Maps

- ☑ Follow the general guidelines discussed in visuals for placement of maps.
- ☑ Label each map clearly, and assign each map a figure number if it is one of a number of illustrations.
- ☑ Clearly identify all boundaries in the map. Eliminate unnecessary boundaries.
- ☑ Eliminate unnecessary information that may clutter a map. For example, if the purpose of the map is to show population centers, do not include mountain elevations, rivers, or other physical features.
- ☑ Include a scale of miles/kilometers or feet/meters to give your readers an indication of the map's proportions.
- ☑ Indicate which direction is north with an arrow or a compass symbol.

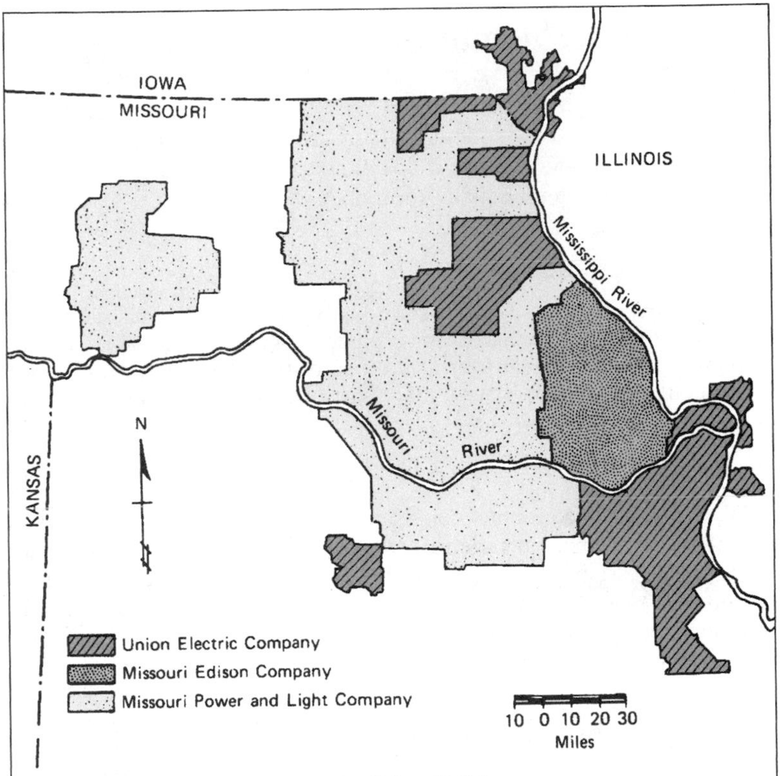

Figure 12. Location of Service Areas of Three Utilities
Source: The U.S. Nuclear Regulatory Commission

FIGURE M–1. Map

Writer's Checklist: Creating and Using Maps (continued)

☑ Emphasize key features by using color, shading, dots, cross hatching, or other appropriate symbols.

☑ Include a key, or legend, that explains what the different colors, shadings, or symbols represent.

WEB LINK MAPS AND MAPPING INFORMATION

The University of Iowa Libraries offer a useful site with links to online maps and other types of mapping and cartographic resources. See <*bedfordstmartins.com/alredtech*> and select *Links for Handbook Entries*.

mathematical equations

You can accurately prepare material with mathematical equations and make it easy to read by following consistent standards throughout a document. Unless you need to follow a specific style manual or specifications, the following guidelines should serve you well.

Formatting Equations

Set short and simple equations, such as $x(y) = y^2 + 3y + 2$, as part of the running text rather than displaying them on separate lines, as long as an equation does not appear at the beginning of a sentence. If a document contains multiple equations, identify them with numbers, as the following example shows:

$$x(y) = y^2 + 3y + 2 \tag{1}$$

Number displayed equations consecutively throughout the work. Place the equation number, in parentheses, at the right margin of the same line as the equation (or of the first line if the equation runs longer than one line). Leave at least four spaces between the equation and the equation number. Refer to displayed equations by number, for example, as "Equation 1" or "Eq. 1."

M

Positioning Displayed Equations

Equations that are set off from the text need to be surrounded by space. Triple-space between displayed equations and normal text. Double-space between one equation and another and between the lines of multiline equations. Count space above the equation from the uppermost character in the equation; count space below from the lowermost character.

Type displayed equations either at the left margin or indented five spaces from the left margin, depending on their length. When a series of short equations is displayed in sequence, align them on the equal signs.

$$\begin{aligned}
p(x,y) &= \sin(x + y) && (2)\\
p(x,y) &= \sin x \cos y + \cos x \sin y && (3)\\
p(x_0, y_0) &= \sin x_0 \cos y_0 + \cos x_0 \sin y_0 && (4)\\
q(x,y) &= \cos(x + y) && (5)\\
&= \cos x \cos y - \sin x \sin y &&\\
q(x_0, y_0) &= \cos x_0 \cos y_0 - \sin x_0 \sin y_0 && (6)
\end{aligned}$$

Break an equation that requires two lines at the equal sign, carrying the equal sign over to the second portion of the equation.

$$\begin{aligned} &{}_0\!\int^1 (f_n - \tfrac{u}{r} f_n)^2\, r\, dr + 2n\, {}_0\!\int^1 f_n f_n dr \\ &= {}_0\!\int^1 (f_n - \tfrac{u}{r} f_n)^2\, r\, dr + n f_n^2(1) \end{aligned} \tag{7}$$

If you cannot break an equation at the equal sign, break it at a plus or minus sign that is not in parentheses or brackets. Bring the plus or minus sign to the next line of the equation, which should be positioned to end near the right margin of the equation.

$$\begin{aligned} ø(x, y, z) = (x^2 + y^2 + z^2)^{1/2}\,(x - y + z)\,(x + y - z)^2 \\ - [f(x, y, z) - 3x^2] \end{aligned} \tag{8}$$

The next best place to break an equation is between parentheses or brackets that indicate multiplication of two major elements.

For equations that require more than two lines, start the first line at the left margin, end the last line at the right margin (or four spaces to the left of the equation's number), and center intermediate lines between the margins. Whenever possible, break equations at operational signs, parentheses, or brackets.

Omit punctuation after displayed equations, even when they end a sentence and even when a key list defining terms follows (for example, P = pressure, psf; V = volume, cu ft; T = temperature, °C). Punctuation may be used before an equation, however, depending on the grammatical construction.

> The term $(n)_r$ may be written in a more familiar way by using the following algebraic device:
>
> $$(n)_r = \frac{(n)(n-1)(n-2)\ldots(n-r+1)(n-r)(n-r-1)\ldots 3\cdot 2\cdot 1}{(n-r)(n-r-1)\ldots 3\cdot 2\cdot 1} \tag{9}$$
>
> $$= \frac{n!}{(n-r)!}$$

M

maybe / may be

Maybe (one word) is an adverb meaning "perhaps." (*Maybe* the legal staff can resolve this issue.) *May be* (two words) is a verb phrase. (It *may be* necessary to hire a specialist.)

media / medium

Media is the plural of *medium* and should always be used with a plural verb.

- Many communication *media are* available today.
- The Internet *is* a multifaceted *medium.*

meetings

Meetings allow people to share information and collaborate to produce better results than exchanges of e-mail messages or other means would allow. (See also selecting the medium.) Like a presentation, a successful meeting requires planning and preparation.

Planning a Meeting

For a meeting to be successful, determine the focus of the meeting, decide who should attend, and choose the best time and place to hold it. Prepare an agenda for the meeting and determine who should take the minutes.

Determine the Purpose of the Meeting. The first step in planning a meeting is to focus on the desired outcome. Ask yourself the following question to help you determine the purpose of the meeting: What should participants know, believe, do, or be able to do as a result of attending the meeting?

Once you have focused your desired outcome, use the information to write a purpose statement for the meeting that answers the questions *what* and *why*.

M

- The purpose of this meeting is to gather ideas from the technical maintenance staff [*what*] in order to create a new aircraft maintenance schedule [*why*].

Decide Who Should Attend. Schedule a meeting for a time when all or most of the key people can be present. If a meeting must be held without some key participants, ask those people prior to the meeting for their contributions or invite them to participate by speakerphone. Of course, the meeting minutes should be distributed to everyone, including significant nonattendees.

Choose the Meeting Time. The time of day and the length of the meeting can influence its outcome. Consider the following when you are planning a meeting:

- People need Monday morning to focus on the week's work after the weekend.
- People need Friday afternoon to complete tasks that must be finished before the weekend.
- Long meetings should include adequate breaks to allow participants to check their messages, make phone calls, and refresh themselves.

- Meetings held during the last 15 minutes of the day will be quick, but few people will remember what happened.
- Speakerphone participants may need consideration for their time zones.

Choose the Meeting Location. Having a meeting on your own premises can give you an advantage: You feel more comfortable, which, along with your guests' newness to their surroundings, may help you get an edge. Holding the meeting on someone else's premises, however, can signal cooperation. For balance, especially when people are meeting for the first time or are discussing sensitive issues, having a meeting at a neutral site may be the best solution. No one gains an advantage in off-site meetings, and attendees often feel freer to participate.

Establish the Agenda. A tool for focusing the group, the agenda is an outline of what the meeting will address. Always prepare an agenda for a meeting, even if it is only an informal list of main topics. Ideally, the agenda should be distributed to attendees a day or two before the meeting. For a longer meeting in which participants are required to make a presentation, try to distribute the agenda a week or more in advance.

The agenda should list the attendees, the meeting time and place, and the topics you plan to discuss. If the meeting includes presentations, list the time allotted for each speaker. Finally, indicate an approximate length for the meeting so that participants can plan the rest of their day. Figure M–2 shows a typical agenda.

Design Meeting Agenda

Purpose:	To get creative ideas for the new software
Date:	May 12, 2006
Place:	Conference Room E
Time:	9:30 a.m.–11:00 a.m.
Attendees:	New Products Manager, Software Engineering Manager and Designers, Technical Publications Manager, Technical Training Manager

Topic	Presenter	Time
The Software	Bob Arbuckle	9:30–9:45
The Campaign	Maria Lopez	9:45–10:00
The Design Strategy	Mary Winifred	10:00–10:15
Discussion	Led by Dave Grimes	10:15–11:00

FIGURE M–2. Meeting Agenda

If the agenda is distributed in advance of the meeting, it should be accompanied by a memo or an e-mail informing people of the following:

- The purpose of the meeting
- The date and place
- The meeting start and stop times
- The names of the people invited
- Instructions on how to prepare

Figure M–3 shows a cover memo sent as an e-mail to accompany an agenda.

From: Elrod Lauter <elauter@mmsoftware.com>
To: Software Engineering Manager, New Products Manager, Technical Publication Manager, Technical Training Manager
Sent: Thur, 05 May 2006 13:30:12 EST
Subject: Design Meeting (May 12, 2006)
Attachments: Meeting Agenda.doc (29 KB)

Purpose of the Meeting

The purpose of this meeting is to get your ideas for the next generation of software.

Date, Time, and Location

Date: May 12, 2006
Time: 9:30 a.m.–11:00 a.m.
Place: Conference Room E (go to the ground floor, take a right off the elevator, third door on the left)

Attendees

Those addressed above

Meeting Preparation

Everyone should be prepared to offer suggestions on the following topics:

- Features of the new software
- Techniques for designing the software
- Customer profile of potential buyers
- FAQs—questions customers may ask

Agenda

Please see the attached document.

FIGURE M–3. E-mail to Accompany an Agenda

Assign the Minute-Taking. Delegate the minute-taking to someone other than the leader. The minute-taker should record major decisions made and tasks assigned. To avoid misunderstandings, the minute-taker must record each assignment, the person responsible for it, and the date on which it is due.

For a standing committee, it is best to rotate the responsibility of taking minutes. See also minutes of meetings and note-taking.

Conducting the Meeting

Assign someone to write on a board or project a computer image of information that needs to be viewed by everyone present.

During the meeting, keep to your agenda; however, allow room for differing views, and foster an environment in which participants listen respectfully to one another. Create a productive environment.

- Consider the feelings, thoughts, ideas, and needs of others—do not let your own agenda blind you to other points of view.
- Help other participants feel valued and respected by listening to them and responding to what they say.
- Respond positively to the comments of others whenever possible.
- Consider communication styles and approaches that are different from your own, particularly those from other cultures. See also global communication.

Deal with Conflict. Despite your best efforts, conflict is inevitable. However, conflict is potentially valuable; when managed positively, it can stimulate creative thinking by challenging complacency and showing ways to achieve goals more efficiently or economically. See collaborative writing.

Members of any group are likely to vary greatly in their personalities and attitudes, and you may encounter people who approach meetings differently. Consider the following tactics for the interruptive, negative, rambling, overly quiet, and territorial personality types.

- The *interruptive person* rarely lets anyone finish a sentence and may intimidate the group's quieter members. Tell that person in a firm but nonhostile tone to let the others finish in the interest of getting everyone's input. By addressing the issue directly, you signal to the group the importance of putting common goals first.
- The *negative person* has difficulty accepting change and often considers a new idea or project from a negative point of view. Such negativity, if left unchecked, can demoralize the group and suppress enthusiasm for new ideas. If the negative person brings up a valid point, however, ask for the group's suggestions to remedy the

issue being raised. If the negative person's reactions are not valid or are outside the agenda, state the necessity of staying focused on the agenda and perhaps recommend a separate meeting to address those issues.

- The *rambling person* cannot collect his or her thoughts quickly enough to verbalize them succinctly. Restate or clarify this person's ideas. Try to strike a balance between providing your own interpretation and drawing out the person's intended meaning.
- The *overly quiet person* may be timid or may just be deep in thought. Ask for this person's thoughts, being careful not to embarrass the person. In some cases, you can have a quiet person jot down his or her thoughts and give them to you later.
- The *territorial person* fiercely defends his or her group against real or perceived threats and may refuse to cooperate with members of other departments, companies, and so on. Point out that although such concerns may be valid, everyone is working toward the same overall goal and that goal should take precedence.

Close the Meeting. Just before closing the meeting, review all decisions and assignments. Paraphrase each to help the group focus on what they have agreed to do and to ensure that the minutes will be complete and accurate. Now is the time to raise questions and clarify any misunderstandings. Set a date by which everyone at the meeting can expect to receive copies of the minutes. Finally, thank everyone for participating, and close the meeting on a positive note.

Writer's Checklist: Planning and Conducting Meetings

- ☑ Develop a purpose statement for the meeting to focus your planning.
- ☑ Invite only those essential to fulfilling the purpose of the meeting.
- ☑ Select a time and place convenient to all those attending.
- ☑ Create an agenda and distribute it a day or two before the meeting.
- ☑ Assign someone to take meeting minutes.
- ☑ Ensure that the minutes record key decisions; assignments; due dates; and the date, time, and location of the follow-up meeting.
- ☑ Follow the agenda to keep everyone focused.
- ☑ Respect the views of others and how they are expressed.
- ☑ Review strategies in this entry for handling conflict and attendees whose style of expression may prevent getting everyone's best thinking.
- ☑ Close the meeting by reviewing key decisions and assignments.

memos

Memos are used within organizations to report results, instruct employees, announce policies, disseminate information, and delegate responsibilities. Whether sent on paper, as e-mails, or as attachments to e-mails, memos provide a record of decisions made and actions taken. They also can play a key role in the management of many organizations because managers use memos to inform and motivate employees. See also selecting the medium.

Writing Memos

Many of the principles discussed in correspondence and the "Five Steps to Successful Writing" apply to writing memos.

Development and Protocol. To produce an effective memo, outline it first, even if you simply jot down points to be covered and then order them logically. With careful preparation, your memos will be both concise and adequately developed. Adequate development of your thoughts is crucial to the clarity of your message, as the following example indicates.

ABRUPT Be more careful on the loading dock.

DEVELOPED To prevent accidents on the loading dock, follow these procedures:
1. Check . . .
2. Load only . . .
3. Replace . . .

Although the abrupt version is concise, it is not as clear and specific as the developed revision. Do not assume your readers will know what you mean. Readers who are in a hurry may misinterpret a vague memo. See also conciseness.

Although such practices vary, be alert to the protocol of addressing and sending memos in your organization. Consider who should receive a memo and in what order—senior managers, for example, take precedence over junior managers. If rank does not apply, alphabetizing recipients by last name is safe. Who should be copied on a memo is a sensitive matter, and you need to learn both the formal and informal practices in an organization. The best general advice, however, is to copy only those who need to see the information.

Openings. Although methods of development vary, a memo normally begins with a statement of its main point. Consider the following example:

- Because of the recent hacker attack on our Web site, I recommend that we no longer post the e-mail addresses of our laboratory employees. This practice . . .

When your reader is not familiar with the subject or with the background of a problem, provide an introductory paragraph before stating the main point of the message. Doing so is especially important in memos that will serve as records of crucial information. Generally, longer or complex subjects benefit most from more thorough introductions. However, even when you are writing a short message about a familiar subject, remind readers of the context. In the following example, words that provide context are shown in *italics*.

- *As Maria Lopez recommended,* I reviewed the office reorganization plan. I like most of the features; however, . . .

Do not state the main point first when (1) the reader is likely to be highly skeptical or (2) you are disagreeing with your superiors. In such cases, a more persuasive tactic is to state the problem first, then present the specific points supporting your final recommendation. See also persuasion and refusal letters.

M

Writing Style and Tone. Whether your memo is formal or informal depends entirely on your readers and your purpose. A message to a coworker who is also a friend is likely to be informal, while an internal proposal to several readers or to someone two or three levels higher in your organization is likely to be more formal. Consider the following versions of a statement:

TO AN EQUAL	I can't go along with the plan because I think it has serious logistical problems. First, . . .
TO A SUPERIOR	The logistics of moving the department may pose serious problems. First, . . .

A message that gives instructions to a subordinate should also be relatively formal, impersonal, and direct, unless you are trying to reassure or praise. When writing to subordinates, remember that *managing* does not mean *dictating.* If you are too formal, sprinkling your writing with fancy words, you may seem stuffy and pompous. In fact, affectation may both irritate and baffle readers, cause a loss of time, and produce costly errors. Consider the unintended secondary messages the following notice conveys:

- It has been decided that the office will be open the day after Thanksgiving.

"It has been decided" not only sounds impersonal but also communicates an authoritarian, management-versus-employee tone. The passive voice also suggests that the decision-maker does not want to say "I have decided" and thus be identified. One solution is to remove the first part of the sentence.

- The office will be open the day after Thanksgiving.

The best solution, however, would be to suggest both that there is a good reason for the decision and that employees are privy to (if not a part of) the decision-making process.

- Because we must meet the December 15 deadline for submitting the Bradley Foundation proposal, the office will be open the day after Thanksgiving.

By subordinating the bad news (the need to work on that day), the writer focuses on the reasoning behind the decision to work. Employees may not necessarily like the message, but they will at least understand that the decision is not arbitrary and is tied to an important deadline.

M

Lists and Headings. Lists can give impact to important points by making it easier for your reader to quickly grasp information. Be careful, however, not to overuse lists. A message that consists almost entirely of lists is difficult to understand because it forces the reader to connect the separate and disjointed items on the page. Further, lists lose their effectiveness when they are overused.

Headings are another attention-getting device, particularly in long memos. They divide material into manageable segments, call attention to main topics, and signal a shift in topic. Readers can scan the headings and read only the section or sections appropriate to their needs.

Closings. A memo closing can accomplish many important tasks, such as building positive relationships with readers, encouraging colleagues and employees, and letting recipients know what you will do or what you expect of them.

- I will discuss the problem with the marketing consultant and let you know by Monday what we are able to change.

Although routine statements are sometimes unavoidable (Thanks again for your help), make your closing work for you by providing specific prompts to which the reader can respond. See also conclusions.

- If you would like further information, such as a copy of the questionnaire we used, please let me know.

Format and Design

Memo formats and conventions vary greatly. Although no single standard exists, Figure M–4 shows a typical $8^{1}/_{2}'' \times 11''$ format with a printed company name. When memos require more than one page, use a second page header, as shown in Figure C–13 on page 112.

Subject Lines. Subject lines announce the topic; because they also aid filing and later retrieval, they must be specific and accurate.

VAGUE	Subject: Tuition Reimbursement
VAGUE	Subject: Time-Management Seminar
SPECIFIC	Subject: Tuition Reimbursement for Time-Management Seminar

Capitalize all major words in a subject line, except articles, prepositions, and conjunctions with fewer than five letters unless they are the first or last words. Remember that the subject line should not substitute for an opening that provides context for the message.

Signature. If you are sending a printed memo, the final step is signing or initialing it, a practice that lets readers know that you approve of its contents. Where you sign or initial the memo depends on the practice of your organization. Figure M–4 shows a typical placement of initials.

WEB LINK **WRITING MEMOS**

For links to articles about when to write memos and tips for following organization protocol, see <*bedfordstmartins.com/alredtech*> and select *Links for Handbook Entries.*

methods of development

A logical method of development satisfies the readers' need for shape and structure in a document, whether it is an e-mail, a report, or a Web

SOFTWARE DESIGN SERVICES
MEMORANDUM

TO: Barbara Smith, Software Design Manager
FROM: Hannah Kaufman, Vice President HK
DATE: April 14, 2006
SUBJECT: Schedule for ACM Electronics Software

ACM Electronics has asked us to prepare a comprehensive proposal for its Milwaukee operation by August 8, 2006. We have worked with electronics firms in the past, so this proposal should be relatively easy to prepare. My guess is that the job will take nearly two months. I would like you to prepare a tentative schedule for the project.

Additional Personnel

In preparing the schedule, check the status of the following:

- Schedules of all staff designers
- Availability of freelance designers
- Availability of freelance programmers

Ordinarily, we would not need to depend on outside personnel; however, because our bid for the *Wall Street Journal* special project is still under consideration, we could be short of staff in June and July. Further, we have to consider vacations that have already been approved.

Scheduling Deadlines

Please give me target deadlines by April 18. A successful proposal will give us a good chance to obtain the contract to do ACM Electronics' new expansion into international operations as well.

I know your staff can do the job.

cc: Ted Harris, President
Fred Moore, Marketing Director

FIGURE M–4. Typical Memo Format

M

page. It helps you as a writer move smoothly and logically from the introduction to a conclusion.

Choose the method or, as is often the case, a combination of methods that best suits your subject, your readers, and your purpose. Following are the most common methods, each of which is discussed in further detail in its own entry.

- Cause-and-effect method of development begins with either the cause or the effect of an event. This approach can be used to develop a report that offers a solution to a problem, beginning with the problem and moving on to the solution or vice versa.
- Chronological method of development emphasizes the time element of a sequence, as in a trouble report that traces events as they occurred in time.
- Comparison method of development is useful when writing about a new topic that is in many ways similar to another topic that is more familiar to your readers.
- Definition method of development extends definitions with additional details, examples, comparisons, or other explanatory devices. See also defining terms.
- Division-and-classification method of development either separates a whole into component parts and discusses each part separately (*division*) or groups parts into categories that clarify the relationship of the parts (*classification*).
- General and specific methods of development proceed either from general information to specific details or from specific information to a general conclusion.
- Order-of-importance method of development presents information in either decreasing order of importance, as in a proposal that begins with the most important point, or increasing order of importance, as in a presentation that ends with the most important point.
- Sequential method of development emphasizes the order of elements in a process and is particularly useful when writing step-by-step instructions.
- Spatial method of development describes the physical appearance of an object or area (such as a room) from top to bottom, inside to outside, front to back, and so on.

Rarely does a writer rely on only one of these methods. In fact, documents often blend methods of development, and most large documents use combinations of methods. For example, in describing the organization of a computer manufacturer, you might use elements from three methods of development. You could divide the larger topic (the manufacturer) into departments (division and classification), arrange the departments by their order of importance within the organization (order of importance), and present their manufacturing operations in the order they occur (sequential). As this example illustrates, when outlining a document, you may base your major division on one primary method of development appropriate to your purpose and then subordinate other methods to it.

minutes of meetings

Organizations and committees keep official records of their meetings; such records are known as *minutes*. Because minutes are often used to settle disputes, they must be accurate, complete, and clear. When approved, minutes of meetings are official and can be used as evidence in legal proceedings. An example of minutes is shown in Figure M–5.

Keep your minutes brief and to the point. Except for recording motions, which must be transcribed word for word, summarize what occurs and paraphrase discussions. To keep the minutes concise, follow a set format, and use headings for each major point discussed. See also note-taking.

Avoid abstractions and generalities; always be specific. Refer to everyone in the same way—a lack of consistency in titles or names may suggest a deference to one person at the expense of another. Avoid adjectives and adverbs that suggest good or bad qualities, as in "Mr. Sturgess's *capable* assistant read the *comprehensive* report to the subcommittee." Minutes should be objective and impartial.

NORTH TAMPA MEDICAL CENTER

Minutes of the Monthly Meeting
Medical Audit Committee

DATE: June 26, 2006

PRESENT: G. Miller (Chair), C. Bloom, J. Dades, K. Gilley, D. Ingoglia (Secretary), S. Ramirez

ABSENT: D. Rowan, C. Tsien, C. Voronski, R. Fautier, R. Wolf

Dr. Gail Miller called the meeting to order at 12:45 p.m. Dr. David Ingoglia made a motion that the June 2, 2006, minutes be approved as distributed. The motion was seconded and passed.

The committee discussed and took action on the following topics.

(1) TOPIC: Meeting Time

Discussion: The most convenient time for the committee to meet.
Action taken: The committee decided to meet on the fourth Tuesday of every month, at 12:30 p.m.

FIGURE M–5. Minutes of a Meeting

If a member of the committee is to follow up on something and report back to the committee at its next meeting, clearly state the person's name and the responsibility he or she has accepted.

Writer's Checklist: Preparing Minutes of Meetings

Include the following in meeting minutes:

- ☑ The name of the group or committee holding the meeting
- ☑ The topic of the meeting
- ☑ The kind of meeting (a regular meeting or a special meeting called to discuss a specific subject or problem)
- ☑ The number of members present and, for committees or boards of ten or fewer members, their names
- ☑ The place, time, and date of the meeting
- ☑ A statement that the chair and the secretary were present or the names of any substitutes
- ☑ A statement that the minutes of the previous meeting were approved or revised
- ☑ A list of any reports that were read and approved
- ☑ All the main motions that were made, with statements as to whether they were carried, defeated, or tabled (vote postponed), and the names of those who made and seconded the motions (motions that were withdrawn are not mentioned)
- ☑ A full description of resolutions that were adopted and a simple statement of any that were rejected
- ☑ A record of all ballots with the number of votes cast for and against resolutions
- ☑ The time the meeting was adjourned (officially ended) and the place, time, and date of the next meeting
- ☑ The recording secretary's signature and typed name and, if desired, the signature of the chairperson

mixed constructions

A mixed construction is a sentence in which the elements do not sensibly fit together. The problem may be a grammar error, a logic error, or both.

- Because the copier wouldn't start, ~~explains why~~ we had to call a technician.

The original sentence mixes a subordinate clause (*Because the copier wouldn't start*) with a verb (*explains*) that attempts to incorrectly use the subordinate clause as its subject. The revision correctly uses the pronoun *we* as the subject of the main clause. See also sentence construction.

modifiers

Modifiers are words, phrases, or clauses that expand, limit, or make otherwise more specific the meaning of other elements in a sentence. Although we can create sentences without modifiers, we often need the detail and clarification they provide.

WITHOUT MODIFIERS	Production decreased.
WITH MODIFIERS	*Automobile* production decreased *rapidly.*

Most modifiers function as adjectives or adverbs. Adjectives describe qualities or impose boundaries on the words they modify.

- *noisy* machinery, *ten* automobiles, *this* printer, *an* animal

An adverb modifies an adjective, another adverb, a verb, or an entire clause.

- Under test conditions, the brake pad showed *much* less wear than it did under actual conditions.
  [The adverb *much* modifies the adjective *less.*]
- The redesigned brake pad lasted *much* longer.
  [The adverb *much* modifies another adverb, *longer.*]
- The wrecking ball hit the wall of the building *hard.*
  [The adverb *hard* modifies the verb *hit.*]
- *Surprisingly,* the motor failed even after all the durability and performance tests it had passed.
  [The adverb *surprisingly* modifies an entire clause: *the motor failed.*]

Adverbs are intensifiers when they increase the impact of adjectives (*very* fine, *too* high) or adverbs (*very* slowly, *rather* quickly). Be cautious using intensifiers; their overuse can lead to vagueness and hence to inaccuracies.

Stacked (Jammed) Modifiers

Stacked modifiers are strings of modifiers preceding <u>nouns</u> that make writing unclear or difficult to read.

- Your *staffing-level authorization reassessment* plan should result in a major improvement.

The noun *plan* is preceded by three long modifiers, a string that forces the reader to slow down to interpret its meaning. Stacked modifiers are often the result of an overuse of <u>buzzwords</u> or <u>jargon</u>. See how breaking up the stacked modifiers makes the example easier to read.

- Your plan for reassessing the staffing-level authorizations should result in a major improvement.

Misplaced Modifiers

A modifier is misplaced when it modifies the wrong word or phrase. A misplaced modifier can cause <u>ambiguity</u>.

- We *almost* lost all of the parts.
  [The parts were *almost* lost but were not.]
- We lost *almost* all of the parts.
  [Most of the parts were in fact lost.]

To avoid ambiguity, place modifiers as close as possible to the words they are intended to modify. Likewise, place phrases near the words they modify. Note the two meanings possible when the phrase is shifted in the following sentences:

- The equipment *without the accessories* sold the best.
  [Different types of equipment were available, some with and some without accessories.]
- The equipment sold the best *without the accessories*.
  [One type of equipment was available, and the accessories were optional.]

Place clauses as close as possible to the words they modify.

REMOTE We sent the brochure to several local firms *that had four-color art.*

CLOSE We sent the brochure *that had four-color art* to several local firms.

Squinting Modifiers

A squinting modifier is one that can be interpreted as modifying either of two sentence elements simultaneously, thereby confusing readers about which is intended. See also dangling modifiers.

- We agreed *on the next day* to make the adjustments. [Did they agree *to make the adjustments on the next day*? Or *on the next day*, did they agree to make the adjustments?]

A squinting modifier can sometimes be corrected simply by changing its position, but often it is better to rewrite the sentence.

- We agreed that *on the next day* we would make the adjustments. [The adjustments were to be made on the next day.]
- *On the next day*, we agreed that we would make the adjustments. [The agreement was made on the next day.]

mood

The grammatical term *mood* refers to the verb functions that indicate whether the verb is intended to make a statement or ask a question, give a command, or express a hypothetical possibility.

The *indicative mood* states a fact, gives an opinion, or asks a question.

- The setting *is* correct.
- *Is* the setting correct?

The *imperative mood* expresses a command, suggestion, request, or plea. In the imperative mood, the implied subject *you* is not expressed. (*Install* the system today.)

The *subjunctive mood* expresses something that is contrary to fact, conditional, hypothetical, or purely imaginative; it can also express a wish, a doubt, or a possibility. In the subjunctive mood, *were* is used instead of *was* in clauses that speculate about the present or future, and the base form (*be*) is used following certain verbs, such as *propose*, *request*, or *insist*. See also progressive tense.

- If we *were* to finish the tests today, we would be ahead of schedule.
- The senior researcher insisted that she [I, you, we, they] *be* in charge of the project.

The most common use of the subjunctive mood is to express clearly that the writer considers a condition to be contrary to fact. If the condition is not considered to be contrary to fact, use the indicative mood.

SUBJUNCTIVE If I *were* president of the firm, I would change several hiring policies.

INDICATIVE Although I *am* president of the firm, I don't control every aspect of its policies.

TIPS FOR DETERMINING MOOD

In written and especially in spoken English, the tendency increasingly is to use the indicative mood where the subjunctive traditionally has been used. Note the differences between traditional and contemporary usage in the following examples.

Traditional (formal) use of the subjunctive mood

- I wish he *were* here now.
- If I *were* going to the conference, I would room with him.
- I requested that she *show* up on time.

Contemporary (informal) use of the indicative mood

- I wish he *was* here now.
- If I *was* going to the conference, I would room with him.
- I requested that she *shows* up on time.

In formal technical writing, it is best to use the more traditional expressions.

M

Ms. / Miss / Mrs.

Ms. is widely used in business and public life to address or refer to a woman, especially if her marital status is either unknown or irrelevant to the context. Traditionally, *Miss* is used to refer to an unmarried woman, and *Mrs.* is used to refer to a married woman. Some women may indicate a preference for *Ms.*, *Miss*, or *Mrs.*, which you should honor. If a woman has an academic or a professional title, use the appropriate form of address (*Doctor*, *Professor*, *Captain*) instead of *Ms.*, *Miss*, or *Mrs.* See also biased language.

mutual / common

Common is used when two or more persons (or things) share something or possess it jointly.

- We have a *common* desire to make the program succeed.
- The fore and aft guidance assemblies have a *common* power source.

Mutual may also mean "shared" (*mutual* friend, of *mutual* benefit), but it usually implies something given and received reciprocally and is used with reference to only two persons or parties.

- Melek respects Roth, and from my observations the respect is *mutual*.
  [Roth also respects Melek.]

N

narration

Narration is the presentation of a series of events in a prescribed (often chronological) sequence. Much narrative writing explains how something happened: a laboratory study, a site visit, an accident, the decisions in an important meeting. See also trip reports and trouble reports.

Effective narration rests on two key writing techniques: the careful, accurate sequencing of events and a consistent point of view on the part of the narrator. Narrative sequence and essential shifts in the sequence are signaled in three ways: chronology (clock and calendar time), transitional words pertaining to time (*before*, *after*, *next*, *first*, *while*, *then*), and verb tenses that indicate whether something has happened (past tense) or is under way (present tense). The point of view indicates the writer's relation to the information being narrated as reflected in the use of person. Narration usually expresses a first- or third-person point of view. First-person narration indicates that the writer is a participant, and third-person narration indicates that the writer is writing about what happened to someone or something else.

The narrative shown in Figure N–1 reconstructs the final hours of the flight of a small aircraft that attempted to land at the airport in Hailey, Idaho. Instead, the plane crashed into the side of a mountain, killing the pilot and the copilot. Because of the disastrous outcome of the flight, the investigators needed to "tell the story" in detail so that any lessons learned could be made available to other fliers. To do that, they had to recount the events as closely as possible. Investigators sequenced the events as precisely as possible from the beginning of the flight until the crash site was located by referencing verified clock times and approximating those that could not be verified. The verb tenses throughout indicate a past action: *departed*, *canceled*, *cleared*, *called*, *recorded.*

Although narration often exists in combination with other forms of discourse, once a narrative is under way, it should not be interrupted by lengthy explanations or analyses. Explain only what is necessary for readers to follow the events.

History of the Flight*

At 0613 m.s.t.[1] on January 3, 1983, N805C, a Canadair Challenger owned and operated by the A. E. Staley Company departed Decatur, Illinois, on a flight to Friedman Memorial Airport, Hailey, Idaho. The route of the flight was via Capitol, Illinois; Omaha, Nebraska; Scotts Bluff, Nebraska; Riverton, Wyoming; Idaho Falls, Idaho, direct 43°30′ north latitude, 114°17′ west longitude.

The en route portion of the flight was uneventful, and about 35 nmi [nautical miles] east of Idaho Falls N805C was cleared by the Salt Lake City, Utah, Air Route Traffic Control Center (ARTCC) to descend from 39,000 feet to 22,000 feet. N805C descended to 22,000 feet and the flightcrew then requested a descent to 17,000 feet. About 35 nmi east of Sun Valley Airport, after being cleared, N805C descended to 17,000 feet. About 0901, N805C's flightcrew canceled their flight plan and, shortly thereafter, changed the transponder from 1311, the assigned discrete code, to 1200, the visual flight rules (VFR) code. At 0901:07, the data analysis reduction tool (DART) radar data showed a 1311 beacon code at 17,000 feet about 11 nmi east of the Sun Valley Airport. At 0901:37, the DART radar data showed a 1200 VFR transponder beacon code with no altitude readout about 2 nmi west of the 1311 beacon code that was recorded at 0901:07.

At 0904:10, DART radar data recorded a 1200 code target at 13,500 feet almost directly over the Sun Valley Airport. According to an employee of the airport's fixed base operator, N805C's flightcrew called on the airport's UNICOM[2] frequency and requested a landing advisory and asked if a food order had been placed. The flightcrew then stated that there would "be a quick turn," and placed a fuel request. This was the last transmission heard from N805C.

About 1030, the chief pilot of the A. E. Staley Manufacturing Company, who was to board N805C at Sun Valley, arrived at the airport. Since the airplane was overdue, he instituted inquiries to several nearby airports to determine where the airplane had landed. At 1300, he asked air traffic control to make a full communications search. At 1400, after being told that the airplane had not been found, he requested an air search. While waiting for search and rescue teams to arrive, the chief pilot rented an airplane and at about 1700, found the accident site. The impact site, elevation about 6,510 feet, was about 2.2 nmi north of Sun Valley Airport at coordinates 43°32′ 50″ N latitude, 114°17′ 35″ W longitude.

**Source*: National Transportation Safety Board, "Aircraft Accident Report: A. E. Staley Manufacturing Co., Inc., Canadair Challenger CL-600, N805C, Hailey, Idaho, January 3, 1983." National Technical Information Service, Springfield, Va., 1983.

[1]All times, unless otherwise noted, are mountain standard time based on the 24-hour clock.
[2]UNICOM. The nongovernment air/ground radio communications facility, which may provide airport advisory information at certain airports. The Sun Valley UNICOM did not record, nor was it required to record or log, the time of radio communications.

FIGURE N–1. Narration

nature

Nature, when used to mean "kind" or "sort," is vague. Avoid this usage in your writing. Say exactly what you mean.

- The ~~nature of~~ *exclusionary clause in* the contract caused the problem.

needless to say

Although the phrase *needless to say* sometimes occurs in speech and writing, you can either eliminate the phrase or comment on a statement to follow with a more descriptive word.

- ~~Needless to say,~~ *Understandably,* departmental cutbacks have meant decreased efficiency.

newsletter articles

If your organization publishes a newsletter, you may be asked to contribute an article on a subject in your area of expertise. In fact, an article is a good way to promote your work or your department.

Before you begin to write, consider the traditional *who*, *what*, *where*, *when*, and *why* of journalism (*Who* did it? *What* was done? *Where* was it done? *When* was it done? *Why* was it done?) and then add *how*, which may be of as much interest to your colleagues as any of the five *w*'s.

Next, determine whether the company has an official policy or position on your subject. If so, adhere to it as you prepare your article. If there is no company policy, determine as nearly as you can management's attitude toward your subject.

Gather several fairly recent issues of the newsletter and study the style and tone of the writing and the approach used for various kinds of subjects. Understand those perspectives before you begin to work on your own article. Ask yourself the following questions about your subject: What is its significance to the organization? What is its significance to my coworkers? The answers to those questions should help you establish the style, tone, and approach for your article and also heavily influence your conclusion. See also context.

Research for a newsletter article frequently consists of interviewing for information. Interview everyone concerned with your subject. Get all available information and all points of view. Be sure to give maximum credit to the maximum number of people by quoting statements from those involved in projects and naming those who have developed initiatives. See also quotations.

Figure N–2 shows an article written for *Connection*, a newsletter produced by Ken Cook Company and distributed to current and prospective clients. This article describes an innovative approach, called

Connection

Ken Cook Co.
Partners in Product Documentation

www.kencook.com
June 2000

NEW PARTS MANUAL CONCEPT FOR LIFE FITNESS

Vincent A. Hebein, Manager of Logistics and Product Manager of Customer Support Services, of Life Fitness had a problem. He needed to create a catalog for Customer Support Services that would raise the bar for product support catalogs in the fitness and sports accessories industry.

Ken Cook Co. had the answer. Instead of using confusing line and exploded view drawings to illustrate replacement parts for Life Fitness equipment, why not use a "see through" photograph to illustrate the parts placement on the machine? Warren Metzger, Senior Technical Writer, and Dave Duecker and Pete Zielinski, Documentation Design Specialists, of Ken Cook Co. put the process together.

After some experimentation, we found a way of "peeling off" enough of the outside covers to visualize the components inside. You need a photograph of the machine with the outside covers on and another one with the outside covers off without moving the camera. Peter used Adobe® Photoshop® to bring both images together and reveal just the right amount of inside detail to get the desired effect. Dave used QuarkXPress™ to design the page layout and index the closeup photograph of each component to the illustration of each machine prepared by Peter. All pages and the cover were designed to complement the Life Fitness product catalog released earlier this year.

Metzger directed the photography at the new Customer Support Services facility in Franklin Park, Illinois. Tracy Pecs of Customer Support Services chose the replacement parts we needed to illustrate and made sure that we had the correct parts available to photograph. Aaron Sztuk of Life Fitness coordinated the effort of removing the shrouds and outside covers of each machine.

"This catalog is really a showcase," according to Jeff Mayer, Sales and Marketing Executive of Ken Cook Co. "Many applications lend themselves to this process from automobiles to outboards," said Mayer. Brett Kehoe, Account Executive, and Beth Bergeson, Coordinator, at Ken Cook Co. helped Mayer manage the project and keep it on course.

Life Fitness, a division of Brunswick Corporation, is the global leader in designing and manufacturing a full line of reliable, high-quality fitness equipment for commercial and consumer use. Its cardiovascular and strength-training products, including the renowned Lifecycle exercise bike, are used in health, fitness and wellness facilities as well as in homes worldwide. The company is headquartered near Chicago and distributes its equipment in more than 80 countries.

Life Fitness is the largest commercial fitness equipment provider in the world with the No. 1 market share in this segment. The Lifecycle exercise bike is currently the No. 1 bike in commercial facilities and has set the standard for efficient, easy-to-use workout programs since its introduction 30 years ago in 1968. Likewise, Life Fitness treadmills, introduced in 1991, are the No. 1 treadmills in commercial facilities. The company is also a leading manufacturer of cross-trainers, stairclimbers and strength-training equipment.

FEATURES

FIGURE N–2. Newsletter Article

"see through" photography, which was used for the parts catalog of a fitness equipment manufacturer. Notice the inset photograph of a stair-climber, which is intended to draw readers to the article. By describing the innovative see through approach as well as the preeminent reputation of the fitness equipment manufacturer, the article positions Ken Cook Company at the leading edge of product support publications.

Writer's Checklist: Writing Newsletter Articles

Writing a newsletter article requires a more journalistic approach than writing a report. Because newsletters are usually not required reading, you do not have a captive audience. See also emphasis.

☑ Write an intriguing title to catch the readers' attention; rhetorical questions often work well.

☑ Include eye-catching photographs or visuals that entice your audience to read the lead paragraph of your introduction. See also layout and design.

☑ Fashion a lead, or first paragraph, that will encourage further reading. The first paragraph generally makes the transition from the title to the substantive body of the article.

☑ Offer a well-developed presentation of your subject to hold the readers' interest all the way to the end of the article.

☑ Write a conclusion that emphasizes the significance of your subject to your audience and stresses the points you want your readers to retain.

☑ Follow the steps listed in the Checklist of the Writing Process as you prepare your newsletter article.

newsletters

Newsletters are publications that are designed to inform and to create and sustain interest and membership in an organization. They can also be used to sell products and services. The two main types of newsletters are organizational newsletters and subscription newsletters.

Organizational newsletters like the one shown in Figure N–3 are sent to employees or members of an association to keep them informed about issues regarding their company or group, such as the development of new products or policies, or the accomplishments of individuals or teams. Stories in organizational newsletters can also urge members to take a specific action. For example, a health club's newsletter

Connection

Ken Cook Co.

Partners in Product Documentation

www.kencook.com

April 2003

KEN COOK CO. A FORD TIER 1 SUPPLIER

Ford Motor Company awarded Ken Cook Co. a Tier 1 Supplier status in December, 2002. This prestigious qualification allows the company to expand services to automobile manufacturers under the Ford umbrella. Tier 1 status was awarded because of a 20 year relationship with Volvo Cars of North America, a member of Ford's Premier Automotive Group.

"I'm a strong supporter of Ken Cook Co. and am pleased to see them become a Tier 1 supplier. KCC is an invaluable partner to Volvo and this action secures the integrity of our relationship," says Barry Morse, service operations specialist for Volvo Cars.

Over the 20 year period, Ken Cook Co. has provided technical documentation services to the Rockleigh, NJ automobile manufacturer. Comprehensive services include dealer support programs, data management, publication development and replication, call center operations, online order fulfillment, warehousing and custom distributions of technical documentation.

"We've been very successful in helping Volvo support their dealer and service network. We are currently in the final stages of completing a web-based project to address EPA and CARB requirements affecting the automotive industry," said Bruce Radtke, Ken Cook Co. account executive responsible for managing the Volvo account. The website, which will make service related information available on a subscription basis to both Volvo dealer and independent repair shops, was launched April 1st, 2003.

"I'm a strong supporter of Ken Cook Co. and am pleased to see them become a Tier 1 supplier. KCC is an invaluable partner to Volvo and this action secures the integrity of our relationship."

Barry Morse,
service operations specialist
for Volvo Cars of North America

Visit the Volvo Cars service support website at www.volvotechinfo.com.

"Achieving the designation of a Tier 1 supplier is a significant milestone for our company and a testament to the strong relationship we have built with Volvo," says Ken Cook, president of Ken Cook Co. "With a successful business model already in place, our vision is to share our expertise in technical information management and provide other Ford companies with the programs and services Volvo has found indispensable." Other companies within the Ford Premier Automotive Group include Jaguar, Land Rover and Aston Martin.

VOLVO

FEATURES

FIGURE N–3. Company Newsletter (Front Page)

could describe new equipment being installed, its purpose, and how to use it.

Subscription newsletters are designed to attract and build a readership interested in buying specific products or services or in learning more about investing or financial matters. Subscribers are buying information, and they expect a certain level of value for their money. For example, a person with experience in the stock market could create a financial newsletter and charge subscribers a monthly fee for the investing advice in that newsletter; a person who collects movie memorabilia could create an online newsletter that includes stories about ways to find and sell rare movie posters.

Before you begin to develop a newsletter, decide on its specific purpose and the specific readers you will be targeting; then make sure the newsletter's appearance and editorial choices create a sense of identification among the readership. Because newsletters often involve different individuals who work on design, content, and project management, see collaborative writing. See also persuasion and promotional writing. If you are asked to contribute an article to a newsletter, see newsletter articles.

You will need to acquire a mailing list (names and addresses of your readers), and you will need to decide on the most strategic way to get the newsletter to the readers, whether through interoffice mail, the post office, or online. Because it can be time-consuming and technically problematic to send out hundreds or thousands of online newsletters by yourself, you may also need to subscribe to a list-hosting service.

As you develop and edit articles, you need to research the topics and interview relevant sources. As you research trade journals, business and technology magazines, newspapers, or the Internet, find specific angles for the articles that will appeal to your select audience. Attempt to provide content that they will not read elsewhere. Other strategies include interviewing and profiling customers or your employees. Be aware that your fact-checking needs to be meticulous. Unlike the general readership of a newspaper, newsletter readers are often specialists in their fields. Because newsletters are often distributed to branches and clients abroad, see global communication and global graphics. See also interviewing for information.

As shown in Figures N–2 and N–3, a newsletter's format should be simple and consistent, yet visually appealing to your readership. Use the active voice and a conversational tone. Boldface names of customers or association members included in your articles to identify them and bring attention to their contributions. Use headings and bullets to break up the text and make the newsletter easy to read. Keep your sentences simple and paragraphs short. See conciseness and layout and design.

Using word-processing or desktop-publishing software, create newspaper columns and one or two visuals per page that complement the text. On the front page, identify the organization, include the date, volume and issue numbers, and a contents box. Depending on the length and quantity of the text and photographs or other visuals to be included, newsletters are usually 8½″ × 11″ or 17″ × 22″ pages folded in half. If your budget is generous and the scale of your project is substantial, consider using a professional printer to produce your newsletter.

nominalizations

A nominalization is a weak verb (*make, do, conduct, perform*) combined with a noun, when the verb form of the noun would communicate the same idea more directly and concisely.

- The quality assurance team will ~~perform an evaluation of~~ *evaluate* the new software.

If you use nominalizations solely to make your writing sound more formal, the result will be affectation. You may occasionally have an appropriate use for a nominalization. For example, you might use a nominalization to slow the pace of your writing. See also conciseness and technical writing style.

none

None can be considered either a singular or a plural pronoun, depending on the context. See also agreement.

- *None* of the material *has* been ordered.
  [Always use a singular verb with a singular noun, in this case, "material."]
- *None* of the clients *has* been called yet.
  [Use a singular verb even with a plural noun (*clients*) if the intended emphasis is on the idea of *not one*.]
- *None* of the clients *have* been called yet.
  [Use a plural verb if you intend *none* to refer to all clients.]

For emphasis, substitute *no one* or *not one* for *none* and use a singular verb.

- We paid the retail price for three of the machines, ~~none~~ *not one* of which was worth the money.

nor / or

Nor always follows *neither* in sentences with continuing negation. (They will *neither* support *nor* approve the plan.) Likewise, *or* follows *either* in sentences. (The firm will accept *either* a short-term *or* a long-term loan.)

Two or more singular subjects joined by *or* or *nor* usually take a singular verb. However, when one subject is singular and one is plural, the verb agrees with the subject nearer to it. See also conjunctions and parallel structure.

SINGULAR	Neither the *architect* nor the *client was* happy with the design.
PLURAL	Neither the *architect* nor the *clients were* happy with the design.
SINGULAR	Neither the *architects* nor the *client was* happy with the design.

note-taking

The purpose of note-taking is to summarize and record information you extract during research. (For taking notes at a meeting, see minutes of meetings.) The challenge in taking notes is to condense someone else's thoughts into your own words without distorting the original thinking. As you extract information, let your knowledge of the audience and the purpose of your writing guide you.

✪ ETHICS NOTE Resist copying your source word for word as you take notes; instead, paraphrase the author's idea or concept. You must do more than just change a few words in the original passage; otherwise, you will be guilty of plagiarism. See also paraphrasing. ✦

On occasion, when your source concisely sums up a great deal of information or points to a trend important to your subject, you are justified in directly quoting the source and incorporating it into your document. If you use a direct quote, enclose the material in quotation marks in your notes. In your finished writing, provide the source of your quotation. As a general rule, you will rarely need to quote anything longer than a paragraph. See also documenting sources.

When taking notes on abstract ideas, as opposed to factual data, do not sacrifice clarity for brevity—notes expressing concepts can lose their meaning if they are too brief. The critical test is whether a week later you understand the note and can recall the significant ideas of the passage. Consider the information in the following paragraph:

> Long before the existence of bacteria was suspected, techniques were in use for combating their influence in, for instance, the decomposition of meat. Salt and heat were known to be effective, and these do in fact kill bacteria or prevent them from multiplying. Salt acts by the osmotic effect of extracting water from the bacte-

rial cell fluid. Bacteria are less easily destroyed by osmotic action than are animal cells because their cell walls are constructed in a totally different way, which makes them much less permeable.

The paragraph says essentially three things:

1. Before the discovery of bacteria, salt and heat were used to combat the effects of bacteria.
2. Salt kills bacteria by extracting water from their cells by osmosis, hence its use in curing meat.
3. Bacteria are less affected by the osmotic effect of salt than are animal cells, because bacterial cell walls are less permeable.

If your readers' needs and your objective involved tracing the origin of the bacterial theory of disease, you might want to note that salt was traditionally used to kill bacteria long before people realized what caused meat to spoil. It might not be necessary to your topic to say anything about the relative permeability of bacterial cell walls.

You should record notes in a way you find efficient. Some find various shareware note and index programs useful. However, jotting notes on 3″ × 5″ index cards is often more flexible, and the cards are especially useful for outlining complex and long-term projects.*

Writer's Checklist: Taking Notes

- ☑ Ask yourself the following questions: What information do you need to fulfill your purpose? What are the needs of your readers?
- ☑ Record only the most important ideas and concepts. Be sure to record all vital names, dates, and definitions.
- ☑ When in doubt about whether to take a note, consider the difficulty of finding the source again should you want it later.
- ☑ Use quotations when sources summarize a great deal of information or point to important trends.
- ☑ Ensure proper credit: Record the author, title, publisher, place, page number, URL, and date of publication or retrieval. (On subsequent notes from the same source, include only the author and page number.)
- ☑ Use your own shorthand and record notes in a way that you find efficient, whether in an electronic document or on index cards.
- ☑ Photocopy or download pages and highlight passages that you intend to quote.
- ☑ Check your notes for accuracy against the original material before moving on to another source.

*Peter Walsh, *How to Organize Just About Everything* (New York: Free Press, 2004).

nouns

DIRECTORY

A noun names a person, place, thing, concept, action, or quality.

Types of Nouns

The two basic types of nouns are proper nouns and common nouns. *Proper nouns*, which are capitalized, name specific people, places, and things (H. G. Wells, Boston, United Nations, Nobel Prize). See also capitalization.

Common nouns, which are not capitalized unless they begin sentences, name general classes or categories of persons, places, things, concepts, actions, and qualities (writer, city, organization, award). Common nouns include collective nouns, concrete nouns, abstract nouns, count nouns, and mass nouns.

N

Collective nouns are common nouns that indicate a group or collection. They are plural in meaning but singular in form (audience, jury, brigade, staff, committee). (See the subsection Collective Nouns on page 349 for advice on using singular or plural forms with collective nouns.)

Concrete nouns are common nouns used to identify those things that can be discerned by the five senses (paper, keyboard, glue, nail, grease).

Abstract nouns are common nouns that refer to things that cannot be discerned by the five senses (loyalty, pride, valor, peace, devotion).

Count nouns are concrete nouns that identify things that can be separated into countable units (desks, envelopes, printers, pencils, books).

Mass nouns are concrete nouns that identify things that cannot be easily separated into countable units (water, air, electricity, oil, cement). See also English as a second language.

Noun Functions

Nouns function as subjects of verbs, direct and indirect objects of verbs and prepositions, subjective and objective complements, or appositives.

- The *metal* failed during the test. [subject]
- The bricklayer cemented the *blocks* efficiently. [direct object of a verb]
- The state presented our *department* a safety award. [indirect object]
- The event occurred within the *year*. [object of a preposition]
- A dynamo is a *generator*. [subjective complement]
- The regional manager was appointed *chairperson*. [objective complement]
- George Thomas, the *treasurer*, gave his report last. [appositive]

Words normally used as nouns can also be used as adjectives and adverbs.

- It is *company* policy. [adjective]
- He went *home*. [adverb]

Collective Nouns

When a collective noun refers to a group as a whole, it takes a singular verb and pronoun.

- The staff *was* divided on the issue and could not reach *its* decision until May 15.

N

When a collective noun refers to individuals within a group, it takes a plural verb and pronoun.

- The staff *returned* to *their* offices after the conference.

A better way to emphasize the individuals on the staff would be to use the phrase *the staff members*.

- The staff members *returned* to *their* offices after the conference.

Treat organization names and titles as singular.

- LRM Associates *has* grown 30 percent in the last three years; *it* will move to a new facility in January.

Plural Nouns

Most nouns form the plural by adding *-s*.

- dolphin/dolphins, pencil/pencils

Nouns ending in *ch*, *s*, *sh*, *x*, and *z* form the plural by adding *-es*.

- search/searches, glass/glasses, wish/wishes, six/sixes, buzz/buzzes

Nouns that end in a consonant plus *y* form the plural by changing the *y* to *-ies*.

- delivery/deliveries

Some nouns ending in *o* add *-es* to form the plural, but others add only *-s*.

- tomato/tomatoes, dynamo/dynamos

Some nouns ending in *f* or *fe* add *-s* to form the plural; others change the *f* or *fe* to *-ves*.

- cliff/cliffs, fife/fifes, hoof/hooves, knife/knives

Some nouns require an internal change to form the plural.

- woman/women, man/men, mouse/mice, goose/geese

Some nouns do not change in the plural form.

- many *fish*, several *deer*, fifty *sheep*

Hyphenated and open compound nouns form the plural in the main word.

- sons-in-law, high schools, editors in chief

Compound nouns written as one word add *-s* to the end.

- two *tablespoonfuls*

If you are unsure of the proper usage, check a dictionary. See possessive case for a discussion of how nouns form possessives.

number (grammar)

Number is the grammatical property of nouns, pronouns, and verbs that signifies whether one thing (singular) or more than one (plural) is being referred to. (See also agreement.) Nouns normally form the plural by simply adding *-s* or *-es* to their singular forms.

- *Partners* in successful *businesses* are not always personal friends.

Some nouns require an internal change to form the plural.

- woman/women, man/men, goose/geese, mouse/mice

All pronouns except *you* change internally to form the plural.

- I/we, he/they, she/they, it/they

By adding *-s* or *-es*, most verbs show the singular of the third person, present tense, indicative mood.

- he *stands*, she *works*, it *goes*

The verb *be* normally changes form to indicate the plural.

SINGULAR I *am* ready to begin work.
PLURAL We *are* ready to begin work.

numbers

DIRECTORY

The standards for using numbers vary; however, unless you are following an organizational or a professional style manual, observe the following guidelines.

Numerals or Words

Write numbers from zero to ten as words, and write numbers above ten as numerals.

- I rehearsed my presentation *three* times.
- The association added *150* new members.

Spell out numbers that begin a sentence, however, even if they would otherwise be written as figures.

- *One hundred and fifty* people attended the meeting.

If spelling out such a number seems awkward, rewrite the sentence so that the number does not appear at the beginning.

Spell out approximate and round numbers.

- We've had *over a thousand* requests this month.

In most writing, spell out small ordinal numbers, which express degree or sequence (first, second; but 27th, 42nd), when they are single words (our nineteenth year) or when they modify a century (the twenty-first century). However, avoid ordinal numbers in dates (use March 30 or 30 March, not March 30th).

Plurals

Indicate the plural of numerals by adding *-s* (7s, the late 1990s). Form the plural of a written number (like any noun) by adding *-s* or *-es* or by dropping the *y* and adding *-ies* (elevens, sixes, twenties). See also apostrophes.

Measurements

Express units of measurement as numerals (3 miles, 45 cubic feet, 9 meters). When numbers run together in the same phrase, write one as a numeral and the other as a word.

- The order was for ~~12~~ *twelve* 6-foot tables.

Generally give percentages as numerals and write out the word *percent*, except when the number is in a table. (Approximately *85 percent* of the land has been sold.)

Fractions

Express fractions as numerals when they are written with whole numbers (27½ inches, 4¼ miles). Spell out fractions when they are expressed without a whole number (one-fourth, seven-eighths). Always write decimal numbers as numerals (5.21 meters).

Money

In general, use numerals to express exact or approximate amounts of money.

- We need to charge *$28.95* per unit.
- The new system costs *$60,000.*

Use words to express indefinite amounts of money.

- The printing system may cost *several thousand dollars.*

 TIPS FOR PUNCTUATING NUMBERS

Some rules for punctuating numbers in English are summarized as follows.

Use a comma to separate numbers with four or more digits into groups of three, starting from the right (5,289,112,001 atoms).

Do not use a comma in years, house numbers, zip codes, and page numbers.

- June 2005
- 92401 East Alameda Drive
- The zip code is 91601.
- Page 1204

Use a period to represent the decimal point (4.2 percent; $3,742,097.43). See also global communication and global graphics.

Use numerals and words for rounded amounts of money over one million dollars.

- The contract is worth *$6.8 million.*

Use numerals for more complex or exact amounts.

- The corporation paid *$2,452,500* in taxes last year.

For amounts under a dollar, ordinarily use numerals and the word *cents* (The pens cost *50 cents* each), unless other figures that require dollar signs appear in the same sentence.

- The business-card holders cost *$10.49* each, the pens cost *$.50* each, and the pencil-cup holders cost *$6.49* each.

Time

Divide hours and minutes with colons when *a.m.* or *p.m.* follows (7:30 a.m., 11:30 p.m.). Do not use colons with the 24-hour system (0730, 2330). Spelled out time is not followed by *a.m.* or *p.m.* (seven o'clock in the evening).

Dates

In the United States dates are usually written in a month-day-year sequence (August 11, 2006). Never use the strictly numerical form for dates (8/11/06) because the date is not immediately clear, especially in international correspondence.

Addresses

Spell out numbered streets from one to ten unless space is at a premium (East Tenth Street). Write building numbers as numerals. The only exception is the building number *one* (One East Monument Street). Write highway numbers as numerals (U.S. 40, Ohio 271, I-94).

Documents

In manuscripts, page numbers are written as numerals, but chapter and volume numbers may appear as numerals or words (Page 37, Chapter 2 or Chapter Two, Volume 1 or Volume One). Express figure and table numbers as numerals (Figure 4, Table 3).

Do not follow a word representing a number with a numeral in parentheses that represents the same number. Doing so is redundant.

- Send five ~~(5)~~ copies of the report.

O

objects

The three kinds of objects are direct objects, indirect objects, and objects of prepositions. All objects are nouns or noun equivalents: pronouns, verbals, and noun phrases and clauses. See also complements.

Direct Object

A direct object answers the question *what?* or *whom?* about a verb and its subject.

- We sent a *full report.* [We sent *what?*]
- Bill telephoned the *client.* [Bill telephoned *whom?*]

Indirect Object

O

An indirect object is a noun or noun equivalent that occurs with a direct object after certain kinds of transitive verbs, such as *give*, *wish*, *cause*, and *tell.* The indirect object answers the question *to whom or what?* or *for whom or what?* The indirect object always precedes the direct object.

- We sent the *general manager* a full report.
  [*Report* is the direct object; the indirect object, *general manager*, answers the question, "We sent a full report *to whom?*"]

Object of a Preposition

The object of a preposition is a noun or pronoun that is introduced by a preposition, forming a prepositional phrase.

- At the *meeting*, the district managers approved the contract.
  [*Meeting* is the object, and *at the meeting* is the prepositional phrase.]

OK / okay

The expression *okay* (also spelled *OK*) is common in informal writing, but it should be avoided in reports and most correspondence.

- Mr. Sturgess ~~gave his okay to~~ *approved* the project.

on / onto / upon

On is normally used as a preposition meaning "attached to" or "located at." (Install the shelf *on* the north wall.) *Onto* implies movement to a position on or movement up and on. (The commuters surged *onto* the platform.)

Similarly, *on* stresses a position of rest (A book lay *on* the table), and *upon* emphasizes movement or a condition (Final payment will be made *upon* completion).

one

When used as an indefinite pronoun, *one* may help you avoid repeating a noun. (We need a new plan, not an old *one*.) *One* is often redundant in phrases in which it restates the noun, and it may take the proper emphasis away from the adjective.

- The training program was not ~~a unique one.~~ *unique.*

One can also be used in place of a noun or personal pronoun in a statement. (*One* cannot ignore *one's* physical condition.) Using *one* in that way is formal and impersonal; in any but the most formal writing, you should address your reader directly and personally as *you*. (*You* cannot ignore *your* physical condition.) See also point of view.

one of those . . . who

A dependent clause beginning with *who* or *that* and preceded by *one of those* takes a plural verb.

- She is *one of those* managers *who are* concerned about their writing.
- This is *one of those* policies *that make* no sense when you examine them closely.

In those two examples, *who* and *that* refer to plural antecedents (*managers* and *policies*) and thus take plural verbs (*are* and *make*). See also agreement.

If the phrase *one of those* is preceded by *the only*, however, the verb should be singular.

- She is *the only one of those* managers *who is* concerned about her writing.
  [The verb is singular because its subject, *who*, refers to a singular antecedent, *one*. If the sentence were reversed, it would read, "Of those managers, she is *the only one who is* concerned about her writing."]
- This is *the only one of those* policies *that makes* no sense when you examine it closely.
  [If the sentence were reversed, it would read, "Of those policies, this is *the only one that makes* no sense when you examine it closely."]

only

The word *only* should be placed immediately before the word or phrase it modifies. See also modifiers.

- We ~~only~~ lack ^only^ financial backing.

Be careful with the placement of *only* because it can change the meaning of a sentence.

- *Only* he said that he was tired.
  [He alone said that he was tired.]
- He *only* said that he was tired.
  [He actually was not tired, although he said he was.]
- He said *only* that he was tired.
  [He said nothing except that he was tired.]
- He said that he was *only* tired.
  [He said that he was nothing except tired.]

O

order-of-importance method of development

The order-of-importance method of development is a particularly effective and common organizing strategy. This method can use one of two ordering strategies—decreasing order, which is often best for written documents, and increasing order, which is especially effective for oral presentations.

Decreasing Order

Decreasing order begins with the most important fact or point, then moves to the next most important, and so on, ending with the least important. This order is especially appropriate for a memo or other correspondence addressed to a busy decision-maker (see Figure O–1), who may be able to reach a decision after considering only the most important points. In a report addressed to various readers, some of whom may be interested in only the major points and others who may need all the information, decreasing order may be ideal for your purpose.

The advantages of decreasing order are that (1) it gets the reader's attention immediately by presenting the most important point first,

Memorandum

To: Tawana Shaw, Director, Human Resources Department
From: Frank W. Nemitz, Chief, Product Marketing *FWN*
Date: November 20, 2006
Subject: Selection of Manager of the Technical Writing Department

I have assessed the candidates for Manager of the Technical Writing Department as you requested, and the following are my evaluations.

Top-Ranked Candidate
The most qualified candidate is Michelle Bryant, acting manager of the department. In her 12 years in the department, Ms. Bryant has gained wide experience in all facets of its operations. She has maintained a consistently high production record and has demonstrated the skills and knowledge required for the supervisory duties she is now handling. She has continually been rated "outstanding" in all categories in her job-performance appraisals. However, her supervisory experience is limited to her present three-month tenure as acting manager of the department, and she lacks the college degree required by the job description.

Second-Ranked Candidate
Michael Bastick, Graphics Coordinator, my second choice, also has strong potential for the position. An able administrator, he has been with the company for seven years. Further, he is enrolled in a management-training course at Metro State University's downtown campus. I have ranked him second because he lacks supervisory experience and because his most recent work as a technical writer has been limited.

Third-Ranked Candidate
Jane Fine, my third-ranked candidate, has shown herself to be an exceptionally skilled writer in her three years with the Public Relations Department. Despite her obvious potential, she does not yet have the breadth of experience in technical writing that is required to manage the Technical Writing Department. Jane Fine also lacks on-the-job supervisory experience.

FIGURE O–1. Decreasing Order-of-Importance Method of Development

(2) it makes a strong initial impression, and (3) it ensures that even the most hurried reader will not miss the most important point.

Increasing Order

Increasing order begins from the least important point or fact, then progresses to the next more important, and builds finally to the most important or strongest point.

Increasing order of importance is effective for writing in which (1) you want to save your strongest points until the end or (2) you need to build the ideas point by point to an important conclusion (see Figure O–2). Many oral presentations benefit especially from increasing

From: Harry Mathews <matthews@appliedsciences.com>
To: Sun-Hee Kim <kim@appliedsciences.com>
Sent: Friday, May 19, 2006 9:29 AM
Subject: Recruiting Qualified Electronics Technicians

As our company continues to expand, and with the planned opening of the Lakeland Facility late next year, we need to increase and refocus our recruiting program to keep our company staffed with qualified electronics technicians. Below is my analysis of possible recruiting options.

Technical School Recruitment
Over the past three years, we have relied on our in-house internship program and on local and regional technical school graduates to fill these positions. Although our in-house internship program provides a qualified pool of employees, technical school enrollments in the area have in the past provided candidates who are already trained. Each year, however, fewer technical school graduates are being produced, and even the most vigorous Career Day recruiting has yielded disappointing results.

Military Veteran Recruitment
In the past, we relied heavily on the recruitment of skilled veterans from all branches of the military. This source of qualified applicants all but disappeared when the military offered attractive reenlistment bonuses for skilled technicians in uniform. As a result, we need to become aggressive in our attempts to reach this group through advertising. I would like to meet with you soon to discuss the details of a more dynamic recruiting program for skilled technicians leaving the military.

I am certain that with the right recruitment campaign, we can find the skilled employees essential to our expanding role in electronics products and consulting.

FIGURE O–2. Increasing Order-of-Importance Method of Development

order because it leaves the audience with the strongest points freshest in their minds. The disadvantage of increasing order, especially for written documents, is that it begins weakly, and the reader or listener may become impatient or distracted before reaching your main point. In the example given in Figure O–2, the writer begins with the least productive source of applicants and builds up to the most productive source.

organization

Organization is essential to the success of any writing project, from a formal report to a Web design or an effective presentation. Good organization is achieved by outlining and by using a logical and appropriate method of development that suits your subject, your readers, and your purpose.

During the organization stage of the writing process, you must consider a layout and design that will be helpful to your reader and a format appropriate to your subject and purpose. If you intend to include visuals with your writing, consider them as you create your outline, especially if they need to be prepared by someone else while you are writing and revising the draft. See also "Five Steps to Successful Writing."

organizational charts

An organizational chart shows how the various components of an organization are related to one another. This type of visual is useful when you want to give your readers an overview of an organization or to display the lines of authority within it, as in Figure O–3.

The title of each organizational component (office, section, division) is placed in a separate box. The boxes are then linked to a central authority, as shown in Figure O–3. If your readers need the information, include the name of the person and position title in each box.

As with all illustrations, place the organizational chart as close as possible to but not preceding the text that refers to it.

outlining

An outline is the skeleton of the document you are going to write; at the least, it should list the main topics and subtopics of your subject in a logical method of development.

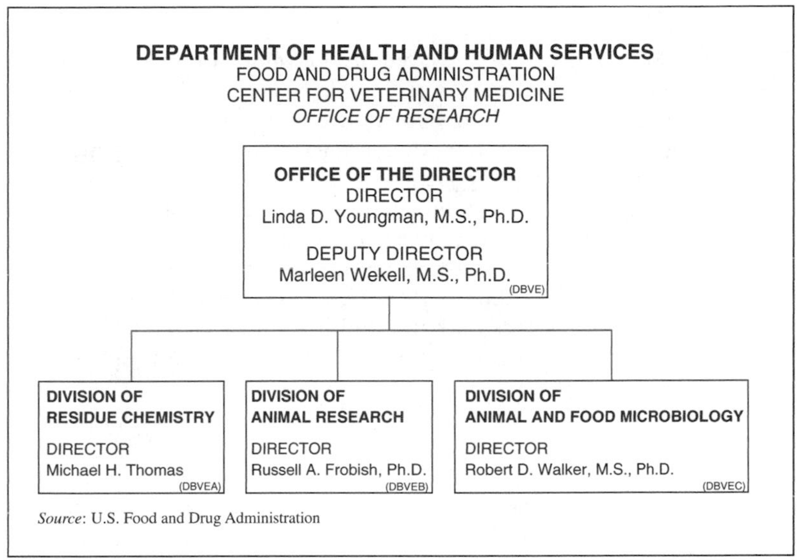

Source: U.S. Food and Drug Administration

FIGURE O–3. Organizational Chart

Advantages of Outlining

An outline provides structure to your writing by ensuring that it has a beginning (introduction), a middle (body), and an end (conclusion). Using an outline offers many other benefits.

- Larger and more complex subjects are easier to handle because an outline breaks them into manageable parts.
- Like a road map, an outline indicates a starting point and keeps you moving logically so that you do not lose your way before you arrive at your conclusion.
- Parts of an outline are easily moved around so that you can select the most effective arrangement of your ideas.
- Creating a good outline frees you from concerns of organization while you are writing a draft.
- An outline enables you to provide coherence and transition so that one part flows smoothly into the next without omitting important details.
- Logic errors are much easier to detect and correct in an outline than in a draft.
- An outline helps with collaborative writing because it enables a team to refine a project's scope, divide responsibilities, and maintain focus.

Types of Outlines

Two types of outlines are most common: short topic outlines and lengthy sentence outlines. A *topic outline* consists of short phrases arranged to reflect your primary method of development. A topic outline is especially useful for short documents such as letters, e-mails, or memos. See also correspondence.

For a large writing project, create a topic outline first, and then use it as a basis for creating a sentence outline. A *sentence outline* summarizes each idea in a complete sentence that may become the topic sentence for a paragraph in the rough draft. If most of your notes can be shaped into topic sentences for paragraphs in your rough draft, you can be relatively sure that your document will be well organized. See also note-taking and research.

Creating an Outline

When you are outlining large and complex subjects with many pieces of information, the first step is to group related notes into categories. Sort the notes by major and minor division headings. Use an appropriate method of development to arrange items and label them with Roman numerals. For example, the major divisions for this discussion of outlining could be as follows:

I. Advantages of outlining
II. Types of outlines
III. Creating an outline

O

The second step is to establish your minor points by deciding on the minor divisions within each major division. Arrange your minor points using a method of development under their major division and label them with capital letters.

II. Types of outlines
  A. Topic outlines
  B. Sentence outlines
  [Division and Classification]
III. Creating an outline
  A. Establish major and minor divisions.
  B. Sort notes by major and minor divisions.
  C. Complete the sentence outline.
  [Sequential]

You will often need more than two levels of headings. If your subject is complicated, you may need three or four levels of headings to better organize all of your ideas in proper relationship to one another. In that event, use the following numbering scheme:

I. First-level heading
    A. Second-level heading
        1. Third-level heading
            a. Fourth-level heading

The third step is to mark each of your notes with the appropriate Roman numeral and capital letter. Arrange the notes logically within each minor heading, and mark each with the appropriate, sequential Arabic number. As you do, make sure your organization is logical and your headings have parallel structure. For example, all the second-level headings under "III. Creating an outline" are complete sentences in the active voice.

Treat visuals as an integral part of your outline, and plan approximately where each should appear. Either include a rough sketch of the visual or write "illustration of . . ." at each place. As with other information in an outline, freely move, delete, or add visuals as needed.

The outline samples shown earlier use a combination of numbers and letters to differentiate the various levels of information. You could also use a decimal numbering system, such as the following, for your outline.

1. FIRST-LEVEL HEADING
    1.1 Second-level heading
    1.2 Second-level heading
        1.2.1 Third-level heading
        1.2.2 Third-level heading
            1.2.2.1 Fourth-level heading
            1.2.2.2 Fourth-level heading
    1.3 Second-level heading
2. FIRST-LEVEL HEADING

This system should not go beyond the fourth level because the numbers get too cumbersome beyond that point. In many documents, such as manuals, the decimal numbering system is carried over from the outline to the final version of the document for ease of cross-referencing sections.

Create your draft by converting your notes into complete sentences and paragraphs. If you have a sentence outline, the most difficult part

O

DIGITAL TIPS CREATING AN OUTLINE

Using the outline feature of your word-processing software permits you to format your outline automatically—fill in, rearrange, and update your outline, as well as create an alphanumeric or a decimal numbering outlining style. For step-by-step instructions, see <*bedfordstmartins.com/alredtech*> and select *Digital Tips*, "Creating an Outline."

of the writing job is over. However, whether you have a topic or a sentence outline, remember that an outline is not set in stone; it may need to change as you write the draft, but it should always be your point of departure and return.

outside [of]

In the phrase *outside of*, the word *of* is redundant.

- Place the rack outside ~~of~~ the incubator.

Do not use *outside of* to mean "aside from" or "except for."

- *Except for* ~~Outside of~~ his frequent absences, Jim has a good work record.

over [with]

In the expression *over with*, the word *with* is redundant; such words as *completed* or *finished* often better express the thought.

- You may enter the test chamber when the experiment is over. ~~with.~~
- You may enter the test chamber when the experiment is *completed.* ~~over with.~~

P

pace

Pace is the speed at which you present ideas to the reader. Your goal should be to achieve a pace that fits your readers, purpose, and subject. The more knowledgeable the reader is about the subject, the faster your pace can be. Be careful, though, not to lose control of the pace. In the first version of the following passage, facts are piled on top of each other at a rapid pace. In the second version, even though its length is no greater, the same facts are presented in two more easily assimilated sentences. In addition, the second version achieves a different and more desirable emphasis.

RAPID The hospital's generator produces 110 volts at 60 hertz and is powered by a 90-horsepower engine. It is designed to operate under normal conditions of temperature and humidity, for use under emergency conditions, and may be phased with other units of the same type to produce additional power when needed.

CONTROLLED The hospital's generator, which is powered by a 90-horsepower engine, produces 110 volts at 60 hertz under normal conditions of temperature and humidity. Designed especially for use under emergency conditions, this generator may be phased with other units of the same type to produce additional power when needed.

paragraphs

A paragraph performs three functions: (1) It develops the unit of thought stated in the topic sentence; (2) it provides a logical break in the material; and (3) it creates a visual break on the page, which signals a new topic.

Topic Sentence

A topic sentence states the paragraph's main idea; the rest of the paragraph supports and develops that statement with related details. The topic sentence is often the first sentence because it tells the reader what the paragraph is about.

- *The arithmetic of searching for oil is stark.* For all the scientific methods of detection, the only way the oil driller can actually know for sure that there is oil in the ground is to drill a well. The average cost of drilling an oil well is over $300,000, and drilling a single well may cost over $8,000,000. And once the well is drilled, the odds against its containing any oil at all are 8 to 1!

The topic sentence is usually most effective early in the paragraph, but a paragraph can lead up to the topic sentence, which is sometimes done to achieve emphasis.

- Energy does far more than simply make our daily lives more comfortable and convenient. Suppose you wanted to stop—and reverse—the economic progress of this nation. What would be the surest and quickest way to do it? Find a way to cut off the nation's oil resources! . . . The economy would plummet into the abyss of national economic ruin. *Our economy, in short, is energy-based.*
  —*The Baker World* (Los Angeles: Baker Oil Tools)

P

On rare occasions, the topic sentence may logically fall in the middle of a paragraph.

- . . . [It] is time to insist that science does not progress by carefully designed steps called "experiments," each of which has a well-defined beginning and end. *Science is a continuous and often a disorderly and accidental process.* We shall not do the young psychologist any favor if we agree to reconstruct our practices to fit the pattern demanded by current scientific methodology.
  —B. F. Skinner, "A Case History in Scientific Method"

Paragraph Length

A paragraph should be just long enough to deal adequately with the subject of its topic sentence. A new paragraph should begin whenever the subject changes significantly. A series of short, undeveloped paragraphs can indicate poor organization and sacrifice unity by breaking a single idea into several pieces. A series of long paragraphs, however, can fail to provide the reader with manageable subdivisions of thought. Paragraph length should aid the reader's understanding of ideas.

Occasionally, a one-sentence paragraph is acceptable if it is used as a transition between longer paragraphs or as a one-sentence introduction or conclusion in correspondence.

Writing Paragraphs

Careful paragraphing reflects the writer's logical organization and helps the reader follow the writer's thoughts. A good working outline makes it easy to group ideas into appropriate paragraphs. (See also outlining.) Although paragraphs can vary from their outlines, the following partial topic outline plots the course of the subsequent paragraphs:

TOPIC OUTLINE (PARTIAL)

I. Advantages of Chicago as location for new facility
    A. Transport infrastructure
        1. Rail
        2. Air
        3. Truck
        4. Sea (except in winter)
    B. Labor supply
        1. Engineering and scientific personnel
            a. Similar companies in area
            b. Major universities
        2. Technical and manufacturing personnel
            a. Community college programs
            b. Custom programs

RESULTING PARAGRAPHS

Probably the greatest advantage of Chicago as a location for our new facility is its excellent transport facilities. The city is served by three major railroads. Both domestic and international air cargo service is available at O'Hare International Airport; Midway Airport's convenient location adds flexibility for domestic air cargo service. Chicago is a major hub of the trucking industry, and most of the nation's large freight carriers have terminals there. Finally, except in the winter months when the Great Lakes are frozen, Chicago is a seaport, accessible through the St. Lawrence Seaway.

Chicago's second advantage is its abundant labor force. An ample supply of engineering and scientific staff is assured not only by the presence of many companies engaged in activities similar to ours but also by the presence of several major universities in the metropolitan area. Similarly, technicians and manufacturing personnel are in abundant supply. The colleges in the Chicago City

College system, as well as half a dozen other two-year colleges in the outlying areas, produce graduates with associate's degrees in a wide variety of technical specialties appropriate to our needs. Moreover, three of the outlying colleges have expressed an interest in developing off-campus courses attuned specifically to our requirements.

Paragraph Unity and Coherence

A good paragraph has unity and coherence, as well as adequate development. *Unity* is singleness of purpose, based on a topic sentence that states the core idea of the paragraph. When every sentence in the paragraph develops the core idea, the paragraph has unity.

Coherence is holding to one point of view, one attitude, one tense; it is the joining of sentences into a logical pattern. A careful choice of transitional words ties ideas together and thus contributes to coherence in a paragraph. Notice how the boldfaced, italicized words tie together the ideas in the following paragraph.

TOPIC SENTENCE *Over the past several months, I have heard complaints about the Merit Award Program.* ***Specifically***, many employees feel that this program should be linked to annual ***salary increases***. They believe that ***salary increases*** would provide a much better incentive than the current $500 to $700 cash awards for exceptional service. ***In addition***, these ***employees believe*** that their supervisors consider the cash awards a satisfactory alternative to salary increases. Although I don't think this practice is widespread, the fact that the ***employees believe*** that it is justifies a reevaluation of the Merit Award Program.

P

Simple enumeration (*first*, *second*, *then*, *next*, and so on) also provides effective transition within paragraphs. Notice how the boldfaced, italicized words and phrases give coherence to the following paragraph.

- Most adjustable office chairs have nylon tubes that hold metal spindle rods. To keep the chair operational, lubricate the spindle rods occasionally. ***First***, loosen the set screw in the adjustable bell. ***Then*** lift the chair from the base. ***Next***, apply the lubricant to the spindle rod and the nylon washer. ***When you have finished***, replace the chair and tighten the set screw.

parallel structure

Parallel sentence structure requires that sentence elements that are alike in function be alike in grammatical form as well. This structure achieves an economy of words, clarifies meaning, expresses the equality of the ideas, and achieves emphasis. Parallel structure assists readers because it allows them to anticipate the meaning of a sentence element on the basis of its construction.

Parallel structure can be achieved with words, phrases, or clauses.

- If you want to earn a satisfactory grade in the training program, you must be *punctual*, *courteous*, and *conscientious*. [parallel words]
- If you want to earn a satisfactory grade in the training program, you must recognize the importance *of punctuality*, *of courtesy*, and *of conscientiousness*. [parallel phrases]
- If you want to earn a satisfactory grade in the training program, *you must arrive punctually*, *you must behave courteously*, and *you must study conscientiously*. [parallel clauses]

Correlative conjunctions (*either . . . or*, *neither . . . nor*, *not only . . . but also*) should always use parallel structure. Both parts of the pairs should be followed immediately by the same grammatical form: two similar words, two similar phrases, or two similar clauses.

- Viruses carry either *DNA* or *RNA*, never both. [parallel words]
- Clearly, neither *serological tests* nor *virus isolation studies* alone would have been adequate. [parallel phrases]
- Either *we must increase our production efficiency* or *we must decrease our production goals.* [parallel clauses]

To make a parallel construction clear and effective, it is often best to repeat an article, a pronoun, a helping verb, a preposition, a subordinating conjunction, or the mark of an infinitive (*to*).

- The association has *a* mission statement and *a* code of ethics.
- The driver *must* check the gauge regularly and *must* act quickly when the indicator falls below the red line.

Parallel structure is especially important in creating lists, outlines, tables of contents, and headings, because it lets readers know the relative value of each item in a table of contents and each heading in the body of a document. See also outlining.

Faulty Parallelism

Faulty parallelism results when joined elements are intended to serve equal grammatical functions but do not have equal grammatical form. Faulty parallelism sometimes occurs because a writer tries to compare items that are not comparable.

NOT PARALLEL The company offers special college training to help nonexempt employees move into professional careers like engineering management, software development, service technicians, and sales trainees. [Notice faulty comparison of occupations—*engineering management* and *software development*—to people—*service technicians* and *sales trainees.*]

To avoid faulty parallelism, make certain that each element in a series is similar in form and structure to all others in the same series.

PARALLEL The company offers special college training to help nonexempt employees move into professional careers like *engineering management*, *software development*, *technical services*, and *sales*.

paraphrasing

Paraphrasing is restating or rewriting in your own words the essential ideas of another writer. The following example is an original passage explaining the concept of object blur. The paraphrased version restates the essential information of the passage in a form appropriate for a report.

P

ORIGINAL One of the major visual cues used by pilots in maintaining precision ground reference during low-level flight is that of object blur. We are acquainted with the object-blur phenomenon experienced when driving an automobile. Objects in the foreground appear to be rushing toward us, while objects in the background appear to recede slightly.

—Wesley E. Woodson and Donald W. Conover, *Human Engineering Guide for Equipment Designers*

PARAPHRASED Object blur refers to the phenomenon by which observers in a moving vehicle report that foreground objects appear to rush at them, while background objects appear to recede slightly (Woodson & Conover, 1964).

✦ ETHICS NOTE Because the paraphrase does not quote the source word for word, quotation marks are not used. However, paraphrased material should be credited because the *ideas* are taken from someone else. See also ethics in writing, note-taking, plagiarism, and quotations. ✦

parentheses

Parentheses are used to enclose explanatory or digressive words, phrases, or sentences. Material in parentheses often clarifies or defines the preceding text without altering its meaning.

- She severely bruised her tibia (or shinbone) in the accident.

Parenthetical information may not be essential to a sentence—in fact, parentheses deemphasize the enclosed material—but it may be helpful to some readers.

Parenthetical material does not affect the punctuation of a sentence, and any punctuation (such as a comma or period) should appear following the closing parenthesis.

- She could not fully extend her knee because of a torn meniscus (or cartilage), and she suffered pain from a severely bruised tibia (or shinbone).

When a complete sentence within parentheses stands independently, the ending punctuation goes inside the final parenthesis.

- The project director listed the problems her staff faced. (This was the third time she had complained to the board.)

For some constructions, however, you should consider using subordination rather than parentheses.

- The scans showed little damage *, which pleased the attending physician,* ~~(the attending physician was pleased),~~ but later tests revealed extensive bruising.

Parentheses also are used to enclose numerals or letters that indicate sequence.

- The following sections deal with (1) preparation, (2) research, (3) organization, (4) writing, and (5) revision.

Do not follow spelled-out numbers with numerals in parentheses representing the same numbers.

- Send five ~~(5)~~ copies of the report.

Use brackets to set off a parenthetical item that is already within parentheses.

- We should be sure to give Emanuel Foose (and his brother Emilio [1812–1882]) credit for his part in founding the institute.

See also documenting sources and quotations.

parts of speech

The term *parts of speech* describes the class of words to which a particular word belongs according to its function in a sentence (naming, asserting, describing, joining, acting, modifying, exclaiming). Many words can function as more than one part of speech. See also functional shift.

PART OF SPEECH	FUNCTION
noun, pronoun	naming/referring
verb	asserting
adjective, adverb	describing/modifying
conjunction, preposition	joining/linking
interjection	exclaiming

party

In legal language, *party* refers to an individual, a group, or an organization. (The injured *party* sued my client.) The term is inappropriate in all but legal writing; when you are referring to a person, use the word *person.*

P

- The ~~party~~ *person* whose file you requested is here now.

Party is, of course, appropriate when it refers to a group. (Jim arranged a tour of the facility for the members of our *party*.)

passive voice (*see* voice)

per

When *per* is used to mean "for each," "by means of," "through," or "on account of," it is appropriate (*per* annum, *per* capita, *per* diem, *per* head). When used to mean "according to" (*per* your request, *per* your order), the expression is jargon and should be avoided.

- As ~~per our discussion,~~ *we discussed,* I will send revised instructions.

percent / percentage

Percent is normally used instead of the symbol % (only 15 *percent*), except in tables, where space is at a premium. *Percentage*, which is never used with numbers, indicates a general size (only a small *percentage*).

periods

A period usually indicates the end of a declarative or an imperative sentence. Periods also link when used as leaders (as in rows of periods in a table of contents) and indicate omissions when used as ellipses. Periods are also used to end questions that are actually polite requests or instructions to which an affirmative response is assumed. (Will you call me as soon as he arrives.) See also sentence construction.

Periods in Quotations

Use a comma, not a period, after a declarative sentence that is quoted in the context of another sentence.

- "There is every chance of success," she stated.

A period is conventionally placed inside quotation marks. See also quotations.

- He stated clearly, "My vote is yes."

Periods with Parentheses

Place a period outside the final parenthesis when a parenthetical element ends a sentence.

- The institute was founded by Harry Denman (1902–1972).

Place a period inside the final parenthesis when a complete sentence stands independently within parentheses.

- The project director listed the problems her staff faced. (This was the third time she had complained to the board.)

Other Uses of Periods

Use periods after initials in names (Wilma T. Grant, J. P. Morgan). Use periods as decimal points with numbers (27.3 degrees Celsius, $540.26, 6.9 percent). Use periods to indicate certain abbreviations (Ms., Dr., Inc.). When a sentence ends with an abbreviation that ends with a period, do not add another period. (Please meet me at 3:30 p.m.) Use periods following the numerals in a numbered list and following lists with complete sentences.

- 1. Enter your name and PIN.
  2. Enter your address with zip code.
  3. Enter your home telephone number.

Period Faults

The incorrect use of a period is sometimes referred to as a *period fault.* When a period is inserted prematurely, the result is a sentence fragment.

FRAGMENT	After a long day at the office during which we finished the quarterly report. We left hurriedly for home.
SENTENCE	After a long day at the office, during which we finished the quarterly report, we left hurriedly for home.

When two independent clauses are joined without any punctuation, the result is a *fused,* or *run-on, sentence.* Adding a period between the clauses is one way to correct a run-on sentence.

RUN-ON	Bill was late for ten days in a row Ms. Sturgess had to fire him.
CORRECT	Bill was late for ten days in a row. Ms. Sturgess had to fire him.

P

Other options are to add a comma and a coordinating conjunction (*and, but, for, or, nor, so, yet*) between the clauses, to add a semicolon, or to add a semicolon with a conjunctive adverb, such as *therefore* or *however.*

person

Person refers to the form of a personal pronoun that indicates whether the pronoun represents the speaker, the person spoken to, or the person or thing spoken about. A pronoun representing the speaker is in the *first* person. (*I* could not find the answer in the manual.) A pronoun that represents the person or people spoken to is in the *second* person. (*You* will be a good manager.) A pronoun that represents the person or people spoken about is in the *third* person. (*They* received the news quietly.) The following list shows first-, second-, and third-person pronouns. See also case, number, and one.

PERSON	SINGULAR	PLURAL
First	I, me, my, mine	we, us, our, ours
Second	you, your, yours	you, your, yours
Third	he, him, his, she, her, hers, it, its	they, them, their, theirs

persuasion

Persuasive writing attempts to convince the reader to adopt the writer's point of view or take a particular action. Much technical writing uses persuasion to reinforce ideas that readers already have, to convince readers to change their current ideas, or to lobby for a particular suggestion or policy (as in Figure P–1 on page 376). You may find yourself pleading for safer working conditions, justifying the expense of a new program, or writing a proposal for a large purchase. See also audience and purpose.

In persuasive writing, the way you present your ideas is as important as the ideas themselves. You must support your appeal with logic and a sound presentation of facts, statistics, and examples. See also logic errors.

ETHICS NOTE Avoid ambiguity: Do not wander from your main point, and above all never make false claims. You should also acknowledge any real or potentially conflicting opinions; doing so allows you to anticipate and overcome objections and builds your credibility. See also ethics in writing. ✦

A writer also gains credibility, and thus persuasiveness, through the readers' impressions of the document's appearance. For this reason, consideration of layout and design is important, especially in documents such as résumés.

The memo shown in Figure P–1 on page 376 was written to persuade the engineering staff to accept and participate in a change to a new computer system. Notice that not everything in this memo is presented in a positive light. Change brings disruption, and the writer acknowledges that fact. See also "you viewpoint."

P

Interoffice Memo

TO: Engineering Staff
FROM: Bernadine Kovak, MIS Administrator *BK*
DATE: April 21, 2006
SUBJECT: Plans for the Changeover to the New Computer System

As you all know, the merger with Datacom has resulted in dramatic growth in our workload—a 30 percent increase in our technical support services during the last several months. To cope with this expansion, we will soon install the NRT/R4 server and QCS Enterprise software with Web-based applications.

Let me briefly describe the benefits of this system and the ways we plan to help you cope with the changeover.

The QCS system will help us access up-to-date technical information when we need it. This system will speed processing dramatically and give us access to all relevant company-wide databases. Because we anticipate that our workload will increase another 30 percent in the next several months, we need to get the QCS system online and working smoothly as soon as possible.

The changeover to this system, unfortunately, will cause some disruption at first. We will need to transfer many of our legacy programs and software applications to the new system. In addition, all of us will need to learn to navigate in the R4 and QCS environments. Once we have made these adjustments, however, I believe we will welcome the changes.

To help everyone cope with the changeover, we will offer training sessions that will begin next week. I have attached a sign-up form with specific class times. We will also provide a technical support hotline at extension 4040, which will be available during business hours, e-mail support at qcs-support@conco.com, and online help documentation.

I would like to urge you to help us make a smooth transition to the QCS system. Please e-mail me with your suggestions or questions about the impact of the changeover on your department. I look forward to working with you to make this system a success.

Enclosure: Training Session Schedule

FIGURE P–1. Persuasive Memo

P

phenomenon / phenomena

A *phenomenon* is an observable thing, fact, or occurrence (a natural *phenomenon*). Its plural form is *phenomena*.

photographs

Photographs are often the best way to show the appearance of an object, record an event, or demonstrate the development of a phenomenon over a period of time. Photographs, however, cannot depict the internal workings of a mechanism or below-the-surface details of objects or structures. Such details are better represented in drawings. Photographs are also effective in catching the readers' attention and adding personal relevance to such documents as brochures and newsletters.

An effective photograph shows important details and indicates relative size of the subject by including a familiar object—such as a ruler or a person—near the subject being photographed. Figure P–2 shows a photograph from an interactive Web presentation for buyers of corporate aircraft. This photograph is one in a series that simulates a pilot's "walk-around"—a procedure in which pilots visually examine an aircraft in a 360-degree safety inspection prior to takeoff. In this photo, the stair steps are lowered to help show the relative size of the aircraft.

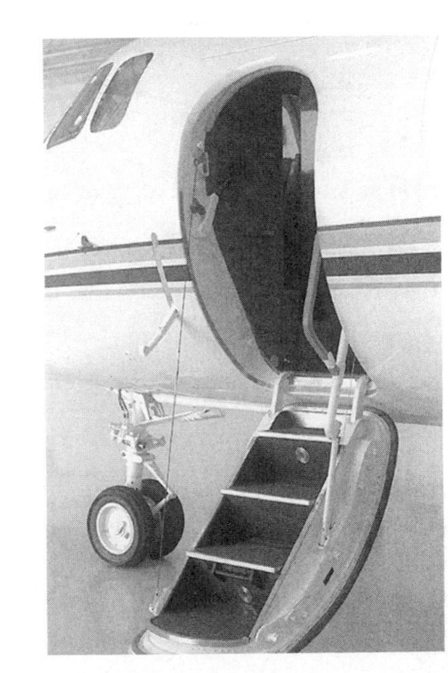

FIGURE P–2. Photo (of Aircraft Door). Photo courtesy of Ken Cook Company.

For reports, treat photographs as you do other visuals. Give the photograph a figure number, call-outs (labels) to identify key features, and a caption, if needed. Position the figure number and caption so that the reader can view them and the photograph from the same orientation.

ETHICS NOTE Be careful to avoid plagiarism by appropriately documenting sources for photographs and to obtain copyright permission if you plan to publish photographs that you do not take yourself. ✦

phrases

DIRECTORY

Phrases are groups of words within sentences that are based on nouns, nonfinite verb forms, or verb combinations without subjects. See also clauses and sentence construction.

- She encouraged her staff *by her calm confidence.* [phrase]

A phrase may function as an adjective, an adverb, a noun, or a verb.

- The subjects *on the agenda* were all discussed. [adjective]
- We discussed the project *with great enthusiasm.* [adverb]
- *Working hard* is her way of life. [noun]
- The chief engineer *should have been notified.* [verb]

Even though phrases function as adjectives, adverbs, nouns, or verbs, they are normally named for the kind of word around which they are constructed—preposition, participle, infinitive, gerund, verb, or noun. A phrase that begins with a preposition is a *prepositional phrase*, a phrase that begins with a participle is a *participial phrase*, and so on. For typical verb phrases and prepositional phrases that give speakers of English as a second language trouble, see idioms.

Prepositional Phrases

A preposition is a word that shows relationship and combines with a noun or pronoun (its object) to form a modifying phrase. A prepositional phrase, then, consists of a preposition plus its object and the object's modifiers.

- *After the meeting*, the district managers adjourned *to the executive dining room.*

Prepositional phrases, because they normally modify nouns or verbs, usually function as adverbs or adjectives. A prepositional phrase may function as an adverb of motion (Turn the dial four degrees *to the left*) or an adverb of manner (Answer customers' questions *in a courteous fashion*). A prepositional phrase may function as an adverb of place and may appear in different places in the sentence.

- *In home and office computer systems*, security is essential.
- Security is *essential in home and office computer systems.*

Prepositional phrases may function as adjectives; when they do, they follow the nouns they modify.

- Garbage *with a high protein content* can be processed into animal food.

Be careful when you use prepositional phrases because separating a prepositional phrase from the noun it modifies can cause ambiguity.

AMBIGUOUS	*The woman* standing by the security guard *in the gray suit* is our division manager.
CLEAR	*The woman in the gray suit* who is standing by the security guard is our division manager.

Watch as well for the overuse of prepositional phrases where modifiers would be more economical.

OVERUSED	The man *with gray hair in the blue suit with pinstripes* is the former president *of the company.*
ECONOMICAL	The *gray-haired* man in the *blue pin-striped* suit is the former *company* president.

P

Participial Phrases

A participle is any form of a verb that is used as an adjective. A participial phrase consists of a participle plus its object and its modifiers.

- The division *having the largest number of patents* will work with NASA.

The relationship between a participial phrase and the rest of the sentence must be clear to the reader. For that reason, every sentence containing a participial phrase must have a noun or pronoun that the participial phrase modifies; if it does not, the result is a dangling participial phrase.

Dangling Participial Phrases. A dangling participial phrase occurs when the noun or pronoun that the participial phrase is meant to modify is not stated but only implied in the sentence. See also <u>dangling modifiers</u>.

DANGLING	*Being unhappy with the job*, his efficiency suffered. [His efficiency was not unhappy with the job; what the participial phrase really modifies—*he*—is not stated but merely implied.]
CORRECT	*Being unhappy with the job*, he grew less efficient. [Now what that participial phrase modifies—*he*—is explicitly stated.]

Misplaced Participial Phrases. A participial phrase is misplaced when it is too far from the noun or pronoun it is meant to modify and so appears to modify something else. Such an error can make the writer look ridiculous.

MISPLACED	We saw a large warehouse *driving down the highway*.
CORRECT	*Driving down the highway*, we saw a large warehouse.

Infinitive Phrases

An infinitive is the bare form of a verb (go, run, talk) without the restrictions imposed by <u>person</u> and <u>number</u>. An infinitive is generally preceded by the word *to* (which is usually a preposition but in this use is called the *sign*, or *mark*, of the infinitive). An infinitive phrase consists of the word *to* plus an infinitive and any objects or modifiers.

- *To succeed in this field*, you must be willing *to assume responsibility*.

Do not confuse a prepositional phrase beginning with *to* with an infinitive phrase. In an infinitive phrase, *to* is followed by a verb; in a prepositional phrase, *to* is followed by a noun or a pronoun.

PREPOSITIONAL PHRASE	We went *to the building site*.
INFINITIVE PHRASE	Our firm tries *to provide a comprehensive training program*.

The implied subject of an introductory infinitive phrase should be the same as the subject of the sentence. If it is not, the phrase is a dangling modifier. In the following example, the implied subject of the infinitive is *you* or *one*, not *practice*.

- To learn a new language, ~~practice is needed.~~ *you must practice.*

Gerund Phrases

A gerund is a verbal ending in *-ing* that is used as a noun. A gerund phrase consists of a gerund plus any objects or modifiers and always functions as a noun.

SUBJECT	*Writing a technical manual* is a difficult task.
DIRECT OBJECT	She liked *designing the system.*

Verb Phrases

A verb phrase consists of a main verb and its helping verb.

- He *is* [helping verb] *working* [main verb] hard this summer.

Words can appear between the helping verb and the main verb of a verb phrase. (He *is* ***always*** *working.*) The main verb is always the last verb in a verb phrase.

Questions often begin with a verb phrase. (*Will* he *verify* the results?) The adverb *not* may be appended to a helping verb in a verb phrase. (He *did not work* today.)

Noun Phrases

A noun phrase consists of a noun and its modifiers. (Have *the two new employees* fill out *these forms.*)

plagiarism

P

Plagiarism is the use of someone else's unique ideas without acknowledgment, or the use of someone else's exact words without quotation marks and appropriate credit. Plagiarism is considered to be the theft of someone else's creative and intellectual property and is not accepted in business, science, journalism, academia, and other fields. See also ethics in writing.

Quoting a passage—including cutting and pasting a passage from the Internet into your work—is permissible only if you enclose the passage in quotation marks and properly cite the source. For detailed guidance on quoting correctly, see quotations. If you intend to publish, reproduce, or distribute material that includes quotations from published works, including Web sites, you may need to obtain written permission from the copyright holders of those works.

Likewise, you may paraphrase the words and ideas of another *if you document your source.* (See also paraphrasing and documenting sources.) Although you do not enclose paraphrased ideas or materials in quotation marks, you must document their sources. Paraphrasing a passage without citing the source is permissible only when the information paraphrased is common knowledge.

Common knowledge generally refers to information that is widely known and readily available in handbooks, manuals, atlases, and other references. For example, that "conservation of energy" is the first law of thermodynamics is common knowledge and is found in virtually every physics textbook. Common knowledge also refers to information within a specific field that is generally known and understood by most others in that field—even though it is not widely known by those outside the field. An indication that something is common knowledge is whether it is repeated in multiple sources without citation. The best advice is, when in doubt, document.

ETHICS NOTE In the workplace, employees often borrow from in-house manuals, reports, and other company documents. Using such boilerplate material is neither plagiarism nor a violation of copyright. ✦

WEB LINK **AVOIDING PLAGIARISM**

For a tutorial on using sources correctly, see <*bedfordstmartins.com/alredtech*> and select *Tutorials*, "The St. Martin's Tutorial on Avoiding Plagiarism." For links to other helpful resources, select *Links for Handbook Entries*.

point of view

Point of view is the writer's relation to the information presented, as reflected in the use of grammatical person. The writer usually expresses point of view in first-, second-, or third-person personal pronouns. Use of first person indicates that the writer is a participant or an observer. Use of second or third person indicates that the writer is giving directions, instructions, or advice, or writing about other people or something impersonal.

FIRST PERSON	*I* scrolled down to find the settings option.
SECOND PERSON	*You* must scroll down to find the settings option. Scroll down to find the settings option. [*You* is understood.]
THIRD PERSON	*He* scrolled down to find the settings option.

Consider the following sentence, revised from an impersonal to a more personal point of view. Although the meaning of the sentence does not change, the revision indicates that people are involved in the communication.

- ~~It is regrettable~~ *I regret* that *we cannot accept* the equipment shipped on the 12th ~~is unacceptable.~~.

Many people think they should avoid the pronoun *I* in technical writing. Such practice, however, often leads to awkward sentences, with people referring to themselves in the third person as *one* or as *the writer* instead of as *I*.

- ~~One~~ *I* can only conclude that the absorption rate is too fast.

However, do not use the personal point of view when an impersonal point of view would be more appropriate or more effective because you need to emphasize the subject matter over the writer or the reader. In the following example, it does not help to personalize the situation; in fact, the impersonal version may be more tactful.

PERSONAL	I received objections to my proposal from several of your managers.
IMPERSONAL	Several managers have raised objections to the proposal.

TIPS FOR STATING AN OPINION

In some cultures, stating an opinion in writing is considered impolite or unnecessary, but in the United States, readers expect to see a writer's opinion stated clearly and explicitly. The opinion should be followed by specific examples to help the reader understand the writer's point of view.

Whether you adopt a personal or an impersonal point of view depends on the purpose and the readers of the document. For example, in an informal e-mail to an associate, you would most likely adopt a personal point of view. However, in a report to a large group, you would probably emphasize the subject by using an impersonal point of view.

ETHICS NOTE In correspondence on company stationery, use of the pronoun *we* may be interpreted as reflecting company policy, whereas *I* clearly reflects personal opinion. Which pronoun to use should be decided according to whether the matter discussed in the letter is an individual (*I*) or a corporate (*we*) concern.

- *I* understand your frustration with the price increase, but *we* must now add the import tax to the sales price. ✦

positive writing

Presenting positive information as though it were negative is confusing to readers.

NEGATIVE If the error does *not* involve data transmission, the backup function will *not* be used.

In this sentence, the reader must reverse two negatives to understand the exception that is being stated. (See also double negatives.) The following sentence presents the exception in a positive and straightforward manner.

POSITIVE The backup function is used only when the error involves data transmission.

ETHICS NOTE Negative facts or conclusions, however, should be stated negatively; stating a negative fact or conclusion positively is deceptive because it can mislead the reader.

DECEPTIVE In the first quarter of this year, employee exposure to airborne lead was within 10 percent of acceptable state health standards.

ACCURATE In the first quarter of this year, employee exposure to airborne lead was 10 percent below acceptable state health standards.

See also ethics in writing. ✦

Even if what you are saying is negative, do not use more negative words than necessary.

NEGATIVE We are withholding your shipment because we have not received your payment.

POSITIVE We will forward your shipment as soon as we receive your payment.

See also correspondence and "you" viewpoint.

possessive case

A noun or pronoun is in the possessive case when it represents a person, place, or thing that possesses something. Possession is generally expressed with an apostrophe and an *s* (the *report's* title), with a prepositional phrase using *of* (the title *of the report*), or with the possessive form of a pronoun (*our* report).

Practices vary for some possessive forms, but the following guidelines are widely used. Above all, be consistent.

Singular Nouns

Most singular nouns show the possessive case with *'s*.

- a *manager's* office

  an *employee's* job satisfaction

  the *computer's* hard drive

  the *witness's* testimony

  the *bus's* schedule

When pronunciation with *'s* is difficult or when a multisyllable noun ends in a *z* sound, you may use only an apostrophe.

- *New Orleans'* convention hotels

Plural Nouns

Plural nouns that end in *-s* or *-es* show the possessive case with only an apostrophe.

- the *managers'* reports

  the *employees'* paychecks

  the *companies'* joint project

  the *witnesses'* reports

  the *buses'* schedules

Plural nouns that do not end in *-s* show the possessive with *'s*.

- *children's* clothing, *women's* resources, *men's* room

Apostrophes are not used in official names (*Consumers* Union) or for words that may appear to be possessive nouns but function as adjectives (a *computer peripherals* supplier).

Compound Nouns

Compound words form the possessive with *'s* following the final letter.

- the *vice president's* proposal, the *editor in chief's* desk

Plurals of some compound expressions are often best expressed with a prepositional phrase (presentations *of the editors in chief*).

Coordinate Nouns

Coordinate nouns show joint possession with *'s* following the last noun.

- *Fischer and Goulet's* partnership was the foundation of their business.

Coordinate nouns show individual possession with *'s* following each noun.

- The difference between *Barker's* and *Washburne's* test results was statistically insignificant.

Possessive Pronouns

The possessive pronouns (*its, whose, his, her, our, your, their*) are also used to show possession and do not require apostrophes. (Even good systems have *their* flaws.) Only the possessive form of a pronoun should be used with a gerund (a noun formed from an *-ing* verb).

- The safety officer insisted on *our* wearing protective clothing. [*Wearing* is the gerund.]

P

Possessive pronouns are also used to replace nouns. (The responsibility was *theirs*.) See also its/it's.

Indefinite Pronouns

Some indefinite pronouns (*all, any, each, few, most, none, some*) form the possessive case with the preposition *of*.

- Both desks were stored in the warehouse, but water ruined the surface *of each*.

Other indefinite pronouns (*everyone, someone, anyone, no one*), however, use *'s*.

- *Everyone's* contribution is welcome.

prefixes

A prefix is a letter or group of letters placed in front of a root word that changes the meaning of the root word. When a prefix ends with a vowel and the root word begins with a vowel, the prefix is often separated from the root word with a hyphen (*re-enter*, *pro-active*, *anti-inflammatory*). Some words with the double vowel are written without a hyphen (cooperate) and others with or without a hyphen (*re-elect* or *reelect*).

Prefixes, such as *neo-* (derived from a Greek word meaning "new"), are often hyphenated when used with a proper noun (*neo-Darwinism*). Such prefixes are not normally hyphenated when used with common nouns, unless the base word begins with the same vowel (neonatal, neo-orthodoxy).

A hyphen may be necessary to clarify the meaning of a prefix; for example, *reform* means "correct" or "improve," and *re-form* means "change the shape of." When in doubt, check a current dictionary.

preparation

The preparation stage of the writing process is essential. By determining the readers' needs, your primary purpose, the context, and the scope of coverage, you understand the information you will need to gather during research. See also "Five Steps to Successful Writing."

Writer's Checklist: Preparing to Write

- ☑ Determine who your readers are and learn certain key facts about them—their knowledge, attitudes, and needs relative to your subject.
- ☑ Determine the document's primary purpose: What exactly do you want your readers to know, believe, or be able to do when they have finished reading your document?
- ☑ Consider the context of your message and how it will affect your writing.
- ☑ Establish the scope of your document—the type and amount of detail you must include—not only by understanding your readers' needs and purpose but also by considering any external constraints, such as word limits for trade journal articles or the space limitations of Web pages. See also writing for the Web.
- ☑ Select the medium appropriate to your readers and purpose. See also selecting the medium.

prepositions

A preposition is a word that links a noun or pronoun (the preposition's object) to another sentence element by expressing such relationships as direction (*to, into, across, toward*), location (*at, in, on, under, over, beside, among, by, between, through*), time (*before, after, during, until, since*), or position (*for, against, with*). Together, the preposition, its object, and the object's modifiers form a prepositional phrase that acts as a modifier.

- Answer customers' questions *in a courteous manner.*
  [The prepositional phrase *in a courteous manner* modifies the verb *answer.*]

The object of a preposition (the word or phrase following the preposition) is always in the objective case. When the object is a compound, both nouns and pronouns should be in the objective case. For example, the phrase "between you and *me*" is frequently and incorrectly written as "between you and *I*." *Me* is the objective form of the pronoun, and *I* is the subjective form.

Many words that function as prepositions also function as adverbs. If the word takes an object and functions as a connective, it is a preposition; if it has no object and functions as a modifier, it is an adverb.

PREPOSITIONS	The manager sat *behind* the desk *in* her office.
ADVERBS	The customer lagged *behind*; then he came *in* and sat down.

Certain verbs, adverbs, and adjectives are used with certain prepositions (interested *in*, aware *of*, equated *with*, adhere *to*, capable *of*, object *to*, infer *from*). See also idioms.

Prepositions at the End of a Sentence

A preposition at the end of a sentence can be an indication that the sentence is awkwardly constructed.

She was at the .
- ~~The~~ branch office ~~is where she was at.~~

However, if a preposition falls naturally at the end of a sentence, leave it there. (I don't remember which file name I saved it *under*.)

Prepositions in Titles

Capitalize prepositions in titles when they are the first or last words, or when they contain more than four letters (unless you are following a style that recommends otherwise). See also capitalization.

- The newspaper column "*In* My Opinion" included a review of the article "New Concerns *About* Distance Education."

Preposition Errors

Do not use redundant prepositions, such as "off *of*," "in back *of*," "inside *of*," and "at *about*."

EXACT	The client arrived at ~~about~~ four o'clock.
APPROXIMATE	The client arrived ~~at~~ about four o'clock.

Avoid unnecessarily adding the preposition *up* to verbs.

- Call ~~up and~~ *to* see if he is in his office.

Do not omit necessary prepositions.

- He was oblivious *to* and not distracted by the view from his office window.

See also conciseness and English as a second language.

presentations

P

DIRECTORY

The steps required to prepare an effective presentation parallel the steps you follow to write a document. As with writing a document, determine your purpose and analyze your audience. Then gather the facts that will support your point of view or proposal and logically organize that information. Presentations do, however, differ from written documents in a number of important ways. They are intended for listeners, not readers. Because you are speaking, your manner of delivery, the way you organize the material, and your supporting visuals require as much attention as your content.

Determining Your Purpose

Every presentation is given for a purpose, even if it is only to share information. To determine the primary purpose of your presentation, use the following question as a guide: What do I want the audience to know, to believe, or to do when I have finished the presentation? Based on the answer to that question, write a purpose statement that answers the *what?* and *why?* questions.

- The purpose of my presentation is to convince my company's chief information officer of the need to improve the appearance, content, and customer use of our company's Web site [*what*] so that she will be persuaded to allocate additional funds for site-development work in the next fiscal year [*why*].

Analyzing Your Audience

Once you have determined the desired end result of the presentation, you need to analyze your audience so that you can tailor your presentation to their needs. (See readers.) Ask yourself these questions about your audience:

- What is your audience's level of experience or knowledge about your topic?
- What is the general educational level and age of your audience?
- What is your audience's attitude toward the topic you are speaking about, and—based on that attitude—what concerns, fears, or objections might your audience have?
- Do any subgroups in the audience have different concerns or needs?
- What questions might your audience ask about this topic?

P

Gathering Information

Once you have focused the presentation, you need to find the facts and arguments that support your point of view or the action you propose. As you gather information, keep in mind that you should give the audience only what will accomplish your goals; too much detail will overwhelm them, and too little will not adequately inform your listeners or support your recommendations. For detailed guidance about gathering information, see research.

Structuring the Presentation

When structuring the presentation, focus on your audience. Listeners are freshest at the outset and refocus their attention near the end. Take advantage of that pattern. Give your audience a brief overview of your presentation at the beginning, use the body to develop your ideas, and end with a summary of what you covered and, if appropriate, a call to action. See also methods of development.

The Introduction. Include in the introduction an opening that focuses your audience's attention, such as in the following examples.

- You have to write an important report, but you'd like to incorporate lengthy handwritten notes from several meetings you attended. Your scanner will not read these notes, and you will have to type many pages. You groan because that seems an incredible waste of time. Have I got a solution for you! [*Definition of a problem*]
- As many as 50 million Americans have high blood pressure. [*An attention-getting statement*]
- Would you be interested in a full-sized computer keyboard that is waterproof and noiseless, and can be rolled up like a rubber mat? [*A rhetorical question*]
- As I sat at my computer one morning, deleting my eighth spam message of the day, I decided that it was time to take action to eliminate this time-waster. [*A personal experience*]
- According to researchers at the Massachusetts Institute of Technology, "Garlic and its cousin, the onion, confer major health benefits—including fighting cancer, infections, and heart disease." [*An appropriate quotation*]

P

Following your opening, use the introduction to set the stage for your audience by providing an overview of the presentation, which can include general or background information that will be needed to understand any more detailed information in the body of your presentation. It can also show how you have organized the material.

- This presentation analyzes three high-volume, on-demand printers for us to consider purchasing. Based on a comparison of all three, I will recommend the one I believe best meets our needs. To do so, I'll discuss the following five points:

  1. Why we need a high-volume printer [*the problem*]
  2. The basics of on-demand technology [*general information*]

3. The criteria I used to compare the three printer models [*comparison*]
4. The printer models I compared and why [*possible solutions*]
5. The printer I propose we buy [*proposed solution*]

The Body. If applicable, present the evidence that will persuade the audience to agree with your conclusions and act on them. (See persuasion.) If there is a problem, demonstrate that it exists and offer a solution or range of possible solutions. For example, if your introduction stated that the problem is low profits, high costs, outdated technology, or high employee absenteeism, you could use the following approach.

1. Prove your point.
   a. Marshal the facts and data you need.
   b. Present the information using easy-to-understand visuals.
2. Offer solutions.
   a. "Increase profits by lowering production costs."
   b. "Cut overhead to reduce costs or abolish specific programs or product lines."
   c. "Replace outdated technology or upgrade existing technology."
   d. "Offer employees more flexibility in their work schedules or other incentives."
3. Anticipate questions ("How much will it cost?") and objections ("We're too busy now—when would we have time to learn the new software?") and incorporate the answers into your presentation.

The Closing. Fulfill the goals of your presentation in the closing. If your purpose is to motivate the listeners to take action, ask them to do what you want them to do; if your purpose is to get your audience to think about something, summarize what you want them to think about. Many presenters make the mistake of not actually closing—they simply quit talking, shuffle papers, and then walk away.

Because your closing is what your audience is most likely to remember, use that time to be strong and persuasive. Consider the following possible closing.

P

- Based on all the data, I believe that the Worthington TechLine 5510 Production Printer best suits our needs. It produces 40 pages per minute more than its closest competitor and provides modular systems that can be upgraded to support new applications. The Worthington is also compatible with our current computer network, and staff training at our site is included with our purchase. Although the initial cost is higher than that for the other two models, the additional capabilities, compatibility with most standard environments, and lower maintenance costs make it a better value.

I recommend we allocate the funds necessary for this printer by the 15th of this month in order to be well prepared for the production of next quarter's customer publications.

The closing brings the presentation full circle and asks the audience to fulfill the purpose of the presentation—exactly what a conclusion should do.

Transitions. Planned transitions should appear between the introduction and the body, between points in the body, and between the body and the closing. Transitions are simply a sentence or two to let the audience know that you are moving from one topic to the next. They also prevent a choppy presentation and provide the audience with assurance that you know where you are going and how to get there.

- Before getting into the specifics of each printer I compared, I'd like to present the benefits of networked, on-demand printers in general. That information will provide you with the background you'll need to compare the differences among the printers and their capabilities discussed in this presentation.

It is also a good idea to pause for a moment after you have delivered a transition between topics to let your listeners shift gears with you. Remember, they do not know your plan.

Using Visuals

Well-planned visuals not only add interest and emphasis to your presentation, they also clarify and simplify your message because they communicate clearly, quickly, and vividly. Charts, graphs, and illustrations greatly increase audience understanding and retention of information, especially for complex issues and technical information that could otherwise be misunderstood or overlooked.

You can create and present the visual components of your presentation by using a variety of media—flip charts, whiteboard or chalkboard, overhead transparencies, slides, or computer presentation software. See also layout and design.

P

Flip Charts. Flip charts are ideal for smaller groups in a conference room or classroom and are also ideal for brainstorming with your audience.

Whiteboard or Chalkboard. The whiteboard or chalkboard common to classrooms is convenient for creating sketches and for jotting notes during your presentation. If your presentation requires extensive notes or complex drawings, create them before the presentation to save time and to minimize audience restlessness.

Overhead Transparencies. With transparencies you can create a series of overlays to explain a complex device or system, adding (or removing) the overlays one at a time. You can also lay a sheet of paper over a list of items on a transparency, uncovering one item at a time as you discuss it, to focus audience attention on each point in the sequence.

Presentation Software. Presentation software, such as PowerPoint, Corel Presentations, and Freelance Graphics, lets you create your presentation on your computer. You can develop charts and graphs with data from spreadsheet software or locate visuals on the Web, and then import those files into your presentation. This software also offers standard templates and other features that help you design effective visuals and integrated text. Enhancements include a selection of typefaces, highlighting devices, background textures and colors, and clip-art images. Images can also be printed out for use as overhead transparencies or handouts. Avoid using too many enhancements, however, which may distract viewers from your message. Figure P–3 shows slides for a brief presentation.

Rehearse your presentation using your electronic slides, and practice your transitions from slide to slide. Also practice loading your presentation and anticipate any technical difficulties that might arise. Should you encounter a technical snag during the presentation, stay calm and give yourself time to solve the problem. If you cannot solve the problem, move on without the technology. As a backup, carry a printout of your electronic presentation as well as an extra electronic copy on a storage medium.

WEB LINK **PREPARING PRESENTATION SLIDES**

For a helpful tutorial on creating effective slides, see <*bedfordstmartins.com/alredtech*> and select *Tutorials*, "Preparing Presentation Slides." For links to additional information and tutorials for using presentation software, select *Links for Handbook Entries*.

TRIDENT CONSULTING

Trident Software

Jamie Gurasev
Trident Technology, Inc.
Technical Services Group
May 22, 2006

1

Why Trident?

- We handle any size account.
- Our designs compare favorably to those of our competitors.
- We support your specific needs.

2

Design Information

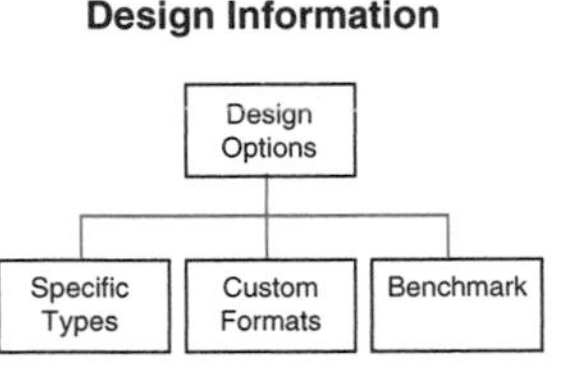

3

Relative Cost

HAZET	WAKO	TRIDENT
$69,000	$62,000	$57,000

4

Additional Resources

- Exceptional Backup
- On-site Support
- Reliability Evaluations

5

Thank You for Coming

For more information, contact:

Jamie Gurasev
Technical Services Group
jgurasev@trident.com

TRIDENT CONSULTING

6

FIGURE P–3. Slides for a Brief Presentation

Writer's Checklist: Using Visuals in a Presentation

- ☑ Use text sparingly in visuals. Use bulleted or numbered lists, keeping them in parallel structure and with balanced content. Use numbers if the sequence is important and bullets if it is not.
- ☑ Limit the number of bulleted or numbered items to 5 or 6 per visual. Each visual should contain no more than 40 to 45 words. Any more will clutter the visual and force you to use a smaller font that could impair the audience's ability to read it.
- ☑ Make your visuals consistent in type style, size, and spacing.
- ☑ Use a type size visible to members of the audience in the back of the room. Type should be boldface and no smaller than 30 points. For headings, 45- or 50-point type works even better.
- ☑ Use graphs and charts to show data trends. Use only one or two illustrations per visual to avoid clutter and confusion.
- ☑ Make the contrast between your text and the background sharp. Use light backgrounds with dark lettering and avoid textured or decorated backgrounds.
- ☑ Use no more than 12 visuals per presentation. Any more will tax the audience's concentration.
- ☑ Match your delivery of the content to your visuals. Do not put one set of words or images on the screen and talk about the previous visual or, even worse, the next one.
- ☑ Do not read the text on your visual word for word. Your audience can read the visuals; they look to you to explain the key points in detail.

Delivering a Presentation

Once you have outlined and drafted your presentation and prepared your visuals, you are ready to practice your presentation and delivery techniques.

Practice. Familiarize yourself with the sequence of the material—major topics, notes, and visuals—in your outline. Once you feel comfortable with the content, you are ready to practice the presentation itself.

PRACTICE ON YOUR FEET AND OUT LOUD. Try to practice in the room where you will give the presentation. Practicing on-site helps you get the feel of the room: the acoustics, the lighting, the arrangement of the chairs, the position of electrical outlets and switches, and so forth. Practice out loud to make clear exactly how long your presentation will take, highlight problems such as awkward transitions, and help eliminate verbal tics, such as "um," "you know," and "like."

PRACTICE WITH YOUR VISUALS AND TEXT. Integrate your visuals into your practice sessions to help your presentation go more smoothly. Operate the equipment (computer, slide projector, or overhead projector) until you are comfortable with it. Decide if you want to use a remote control or to have someone else advance your slides. Even if things go wrong, being prepared and practiced will give you the confidence and poise to continue.

Delivery Techniques That Work. Your delivery is both audible and visual. In addition to your words and message, your nonverbal communication affects your audience. Be animated—your words will make an impression and have more staying power when they are delivered with physical and vocal animation. If you want listeners to share your point of view, show enthusiasm for your topic. The most common delivery techniques include making eye contact; using movement and gestures; and varying voice inflection, projection, and pace.

EYE CONTACT. The best way to establish rapport with your audience is through eye contact. In a large audience, directly address those people who seem most responsive to you in different parts of the room. Doing that helps you establish rapport with your listeners by holding their attention and gives you important visual cues that let you know how your message is being received. Are people engaged and actively listening, or are they looking around or staring at the floor? Such cues tell you that you may need to speed up or slow down the pace of your presentation.

MOVEMENT. Animate the presentation with physical movement. Take a step or two to one side after you have been talking for a minute or so. That type of movement is most effective at transitional points in your presentation between major topics or after pauses or emphases. Too much movement, however, can be distracting, so try not to pace.

Another way to integrate movement into your presentation is to walk to the screen and point to the visual as you discuss it. Touch the screen with the pointer and then turn back to the audience before beginning to speak (remember the three *t*'s: touch, turn, and talk).

GESTURES. Gestures both animate your presentation and help communicate your message. Most people gesture naturally when they talk; nervousness, however, can inhibit gesturing during a presentation. Keep one hand free and use that hand to gesture.

VOICE. Your voice can be an effective tool in communicating your sincerity, enthusiasm, and command of your topic. Use it to your advantage to project your credibility. *Vocal inflection* is the rise and fall of your voice at different times, such as the way your voice naturally rises at the end of a question ("You want it *when*?"). A conversational delivery and eye contact promote the feeling among members of the audience that you are addressing them directly. Use vocal inflection to highlight differences between key and subordinate points in your presentation.

PROJECTION. Most speakers think they are projecting more loudly than they are. Remember that your presentation is ineffective for anyone in the audience who cannot hear you. If listeners must strain to hear you, they may give up trying to listen. Correct projection problems by practicing out loud with someone listening from the back of the room.

PACE. Be aware of the speed at which you deliver your presentation. If you speak too fast, your words will run together, making it difficult for your audience to follow. If you speak too slowly, your listeners will become impatient and distracted.

Presentation Anxiety. Everyone experiences nervousness before a presentation. Survey after survey reveals that for most people dread of public speaking ranks among their top five fears. Instead of letting fear inhibit you, focus on channeling your nervous energy into a helpful stimulant. The best way to master anxiety is to know your topic thoroughly—knowing what you are going to say and how you are going to say it will help you gain confidence and reduce anxiety as you become immersed in your subject.

Writer's Checklist: Preparing for and Delivering a Presentation

- ☑ Practice your presentation with visuals; practice in front of listeners, if possible.
- ☑ Visit the location of the presentation ahead of time to familiarize yourself with the surroundings.
- ☑ Prepare a set of notes that will trigger your memory during the presentation.
- ☑ Make as much eye contact as possible with your audience to establish rapport and maximize opportunities for audience feedback.
- ☑ Animate your delivery by integrating movement, gestures, and vocal inflection into your presentation. However, keep your movements and speech patterns natural.
- ☑ Speak loudly and slowly enough to be heard and understood.
- ☑ Do not read the text on your visuals word for word; explain the key points in detail.

For information and tips on communicating with cross-cultural audiences, see global communication, global graphics, and international correspondence.

press releases

Companies and organizations write press releases (or *news releases*) to announce new products and services, new policies, special events (such as branch openings, company anniversaries, mergers, and grand openings), management changes, and sponsorship of social-action and cultural programs. The purpose of a press release is both to inform the public about the company and its products and services and to promote a favorable image. Even the announcement of what might be regarded as an unfavorable event, such as the closing of a division, can put a company in a good light by demonstrating candor, the ability to act effectively in a crisis, or the potential long-term benefits of a change. Large corporations and institutions usually have their own public relations staffs or use outside agencies. However, if you work for a small company without public relations resources, you may be called on to write a press release.

The press release should be clear, concise, and written with particular attention to the five *w*'s: *who*, *what*, *where*, *when*, and *why*. Begin the first paragraph with the place and date of the announcement, as shown in Figure P–4 on page 400. Write the release using the decreasing order-of-importance method of development. Put all critical information in the first paragraph, information of the next level of importance in the second paragraph, and so on. Editors may need to make your release fit the space they have available; if they must cut your release, they will delete the last paragraph first, and then the next to last, and so on. Make sure your facts are accurate and be careful to define any unfamiliar terms.

Releases are usually sent to local newspapers, television and radio stations, and other special groups, such as trade publications or professional associations. Send the release to a specific person whenever possible. Otherwise, try to address the news release to a particular editor, such as to the business editor or the technology editor. See also newsletter articles.

Writer's Checklist: Preparing Press Releases

☑ Print the release on company stationery with a minimum of one-inch margins on each side of the page.

☑ Provide contact information for the person who can supply further details.

☑ Use boldface type for the headlines and double-space text for easy reading and editing.

Writer's Checklist: Preparing Press Releases (continued)

- ☑ Use *-more-* centered at the bottom of the page when you need to indicate that another page follows.
- ☑ Allow one blank line and center # # # or *-30-* or *-End-* to indicate where the press release ends.

WEB LINK WRITING PRESS RELEASES

For links to Web sites that offer tips, samples, and resources for writing press releases, see <*bedfordstmartins.com/alredtech*> and select *Links for Handbook Entries.*

News Release

BENTLEY PLASTICS
3535 Michigan Avenue ■ Chicago, IL 60653

Contact: Marjorie Kohls
E-mail: mkohls@bentley.com
Phone: (312) 712-1946
Fax: (312) 712-1950

FOR IMMEDIATE RELEASE

Mark Williams Joins Bentley Plastics as Vice President

Chicago, Illinois, May 22, 2006: Marketing expert and author Mark Williams has been appointed Vice President for Marketing at Bentley Plastics, a manufacturer of polymer tubing and coils. He will direct the Illinois and Indiana district sales offices and coordinate overseas distribution through the company's Singapore office. Williams will travel extensively throughout Southeast Asia while developing marketing channels for Bentley.

Formerly, Williams was director of services with International Marketing Associates, a consulting group in New York. While at IMA, he developed a computer-based marketing center that linked textile firms in the United States, Finland, and Great Britain. A graduate of the Columbia University Graduate School of Business, Williams is the author of *Marketing Dynamics*, an informal examination of psychological appeals to the buying public. The book has been used in marketing classrooms at several universities.

Bentley Plastics, headquartered in Chicago, has main plants in Skokie, Illinois, and Gary, Indiana. Since 1978, Bentley has also been producing tubing for French distribution through the firm of Jourdan and Sons, Paris.

# # #

FIGURE P–4. Press Release

principal / principle

Principal, meaning "an amount of money on which interest is earned or paid" or "a chief official in a school or court proceeding," is sometimes confused with *principle,* which means "a basic truth or belief."

- The bank will pay 6.5 percent on the *principal.*
- He sent a letter to the *principal* of the high school.
- She is a person of unwavering *principles.*

Principal is also an adjective, meaning "main" or "primary." (My *principal* objection is that it will be too expensive.)

process explanation

A process explanation may describe the steps in a process, an operation, or a procedure, such as the steps necessary to design and manufacture a product. The introduction often presents a brief overview of the process or lets readers know why it is important for them to become familiar with the process you are explaining. Be sure to define terms that readers might not understand and provide visuals to clarify the process. See also defining terms and instructions.

In describing a process, use transitional words and phrases to create unity within paragraphs, and select headings to mark the transition from one step to the next. Visuals, like the one used in Figure P–5 on page 402, can help convey your message.

P

progress and activity reports

Progress reports keep readers informed about major workplace projects, whereas *activity reports* focus on the ongoing work of individual employees. Both are sometimes called *status reports.*

Progress Reports

A progress report provides information about a project—its status, whether it is on schedule and within budget, and so on. Progress reports are often submitted by a contracting company to a client company, as shown in Figure P–6 on page 403. They are used mainly for projects that involve many steps over a period of time and are issued at

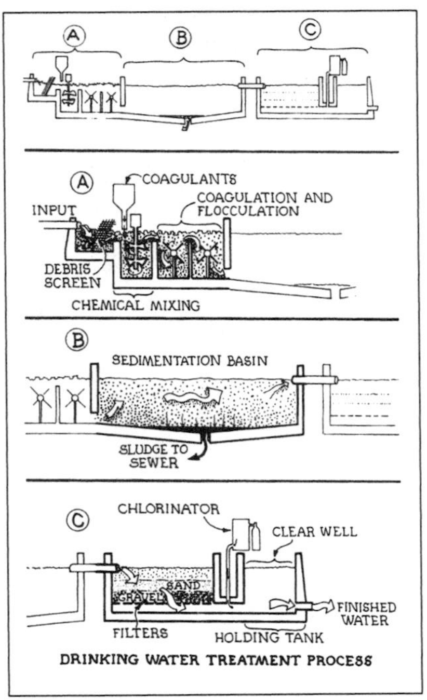

DRINKING WATER TREATMENT PROCESS

Drinking Water Treatment Process

After surface water has been transported from its source to a local water system, most of it must be processed in a treatment plant before it can be used. Some groundwater, on the other hand, is considered chemically and biologically pure enough to pass directly from a well into the distribution system that carries it to the home.

Although there are innumerable variations, surface water is usually treated as follows: First, it enters a storage lagoon where a chemical, usually copper sulfate, is added to control algae growth. From there, water passes through one or more screens that remove large debris. Next, a coagulant, such as alum, is mixed into the water to encourage the settling of suspended particles. The water flows slowly through one or more sedimentation basins so that larger particles settle to the bottom and the sludge can be removed. Water then passes through a filtration basin partially filled with sand and gravel where yet more suspended particles are removed.

At that point in the process, the Safe Drinking Water Act mandates an additional step for communities using surface water. Water must be filtered through activated carbon to remove any remaining microscopic organic material and chemicals. The chlorinator is used to treat. . . .

FIGURE P–5. Process Explanation (with Illustration)

Hobard Construction Company
9032 Salem Avenue
Lubbock, TX 79409

www.hobardcc.com
(808) 769-0832
Fax: (808) 769-5327

August 15, 2006

Walter M. Wazuski
County Administrator
109 Grand Avenue
Manchester, NH 03103

Dear Mr. Wazuski:

Subject: Progress Report 8 for July 1–July 29, 2006

The renovation of the County Courthouse is progressing on schedule and within budget. Although the cost of certain materials is higher than our original bid indicated, we expect to complete the project without exceeding the estimated costs because the speed with which the project is being completed will reduce overall labor expenses.

Costs
Materials used to date have cost $78,600, and labor costs have been $193,000 (including some subcontracted plumbing). Our estimate for the remainder of the materials is $59,000; remaining labor costs should not exceed $64,000.

Work Completed
As of July 29, we had finished the installation of the circuit-breaker panels and meters, the level-one service outlets, and all the subfloor wiring. The upgrading of the courtroom, the upgrading of the records-storage room, and the replacement of the air-conditioning units are in the preliminary stages.

Work Schedule
We have scheduled the upgrading of the courtroom to take place from August 29 to October 7, the upgrading of the records-storage room from October 11 to November 18, and the replacement of the air-conditioning units from November 21 to December 16. We see no difficulty in having the job finished by the scheduled date of December 23.

Sincerely yours,

Tran Nuguélen

Tran Nuguélen
ntran@hobardcc.com

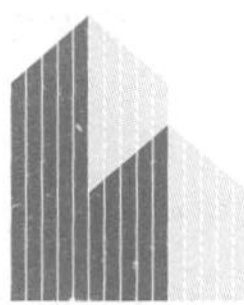

FIGURE P–6. Progress Report

P

regular intervals to state what has been done and what remains to be done. Progress reports help keep projects running smoothly by helping managers assign work, adjust schedules, allocate budgets, and order supplies and equipment. All progress reports for a particular project should have the same format.

The introduction to the first progress report should identify the project, any materials needed, and the project's completion date. Subsequent reports summarize the progress to date; include the status of schedules and costs; list the steps that remain to be taken; and conclude with recommendations about changes in the schedule, materials, and so on.

Activity Reports

Within an organization, employees often submit activity reports on the progress of ongoing projects. Managers may combine the activity reports of several individuals or teams into larger activity reports and, in turn, submit those larger reports to their own managers. The activity report shown in Figure P–7 was submitted by a manager (Wayne Tribinski) who supervises 11 employees; the reader of the report (Kathryn Hunter) is Tribinski's manager.

Because the activity report is issued periodically (usually monthly) and contains material familiar to its readers, it normally needs no introduction or conclusion, although it may need a brief opening to provide context. Although the format varies from company to company, the following sections are typical: Current Projects, Current Problems, Plans for the Next Period, and Current Staffing Level (for managers).

pronoun reference

A pronoun should refer clearly to a specific antecedent. Avoid vague and uncertain references.

, which was a big one,
- We got the account ^ after we wrote the proposal. ~~It was a big one.~~

For coherence, place pronouns as close as possible to their antecedents—distance increases the likelihood of ambiguity.

, praised for its architectural design, is .
- The office building ^ next to City Hall ^ ~~is praised for its architectural design.~~

A general (or broad) reference or one that has no real antecedent is a problem that often occurs when the word *this* is used by itself.

INTEROFFICE MEMO

Date: June 6, 2006
To: Kathryn Hunter, Director of IT
From: Wayne Tribinski, Manager, Applications Programs *WT*
Subject: Activity Report for May 2006

We are dealing with the following projects and problems, as of May 31.

Projects

1. For the *Software Training Mailing Campaign*, we anticipate producing a set of labels for mailing software training information to customers by June 10.
2. The *Search Project* is on hold until the PL/I training has been completed, probably by the end of June.
3. The project to provide a database for the *Information Management System* has been expanded in scope to provide a database for all training activities. We are rescheduling the project to take the new scope into account.

Problems

The *Information Management System* has been delayed. The original schedule was based on the assumption that a systems analyst who was familiar with the system would work on this project. Instead, the project was assigned to a newly hired systems analyst who was inexperienced and required much more learning time than expected.

Bill Michaels, whose activity report is attached, is correcting a problem in the *CNG Software*. This correction may take a week.

Plans for Next Month

- Complete the *Software Training Mailing Campaign.*
- Resume the *Search Project.*
- Restart the project to provide a database on information management with a schedule that reflects its new scope.
- Write a report to justify the addition of two software developers to my department.
- Congratulate publicly the recipients of Meritorious Achievement Awards: Bill Thomasson and Nancy O'Rourke.

Current Staffing Level

Current staff: 11
Open requisitions: 0

Attachment

FIGURE P–7. Activity Report

- He deals with personnel problems in his work. This *experience* helps him in his personal life.

Another problem is a hidden reference, which has only an implied antecedent.

- A high-lipid, low-carbohydrate diet is "ketogenic" because it favors ~~their~~ *the* formation *of ketone bodies*.

Do not repeat an antecedent in parentheses following the pronoun. If you feel you must identify the pronoun's antecedent in that way, rewrite the sentence.

AWKWARD	The senior partner first met Bob Evans when he (Evans) was a trainee.
IMPROVED	Bob Evans was a trainee when the senior partner first met him.

For advice on avoiding pronoun-reference problems with gender, see biased language.

pronouns

DIRECTORY

P

A pronoun is a word that is used as a substitute for a noun (the noun for which a pronoun substitutes is called the *antecedent*). Using pronouns in place of nouns relieves the monotony of repeating the same noun over and over. See also pronoun reference.

Personal pronouns refer to the person or people speaking (*I, me, my, mine; we, us, our, ours*); the person or people spoken to (*you, your, yours*); or the person, people, or thing(s) spoken of (*he, him, his; she, her, hers; it, its; they, them, their, theirs*). See also person and point of view.

- If *their* figures are correct, *ours* must be in error.

Demonstrative pronouns (*this, these, that, those*) indicate or point out the thing being referred to.

- *This* is my desk. *These* are my coworkers. *That* will be a difficult job. *Those* are incorrect figures.

Relative pronouns (*who, whom, which, that*) perform a dual function: (1) They take the place of nouns, and (2) they connect and establish the relationship between a dependent **clause** and its main clause.

- The department manager decided *who* would be hired.

Interrogative pronouns (*who, whom, what, which*) are used to ask questions.

- *What* is the trouble?

Indefinite pronouns specify a class or group of persons or things rather than a particular person or thing (*all, another, any, anyone, anything, both, each, either, everybody, few, many, most, much, neither, nobody, none, several, some, such*).

- Not *everyone* liked the new procedures; *some* even refused to follow them.

A *reflexive pronoun*, which always ends with the suffix *-self* or *-selves*, indicates that the subject of the sentence acts upon itself. See also **sentence construction**.

- The electrician accidentally shocked *herself*.

The reflexive pronouns are *myself, yourself, himself, herself, itself, oneself, ourselves, yourselves*, and *themselves*. *Myself* is not a substitute for *I* or *me* as a personal pronoun.

- Victor and ~~myself~~ I completed the report on time.
- The assignment was given to Ingrid and ~~myself.~~ me.

Intensive pronouns are identical in form to the reflexive pronouns, but they perform a different function: Intensive pronouns emphasize their antecedents.

- I *myself* asked the same question.

Reciprocal pronouns (*one another, each other*) indicate the relationship of one item to another. *Each other* is commonly used when referring to two persons or things and *one another* when referring to more than two.

- Lashell and Kara work well with *each other*.
- The crew members work well with *one another*.

Case

Pronouns have forms to show the subjective, objective, and possessive cases, as shown in the entry case.

A pronoun that is used as the subject of a clause or sentence is in the subjective case (*I, we, he, she, it, you, they, who*). The subjective case is also used when the pronoun follows a linking verb.

- *She* is my boss.
- My boss is *she*.

A pronoun that is used as the object of a verb or preposition is in the objective case (*me, us, him, her, it, you, them, whom*).

- Ms. Davis hired Tom and *me*. [object of verb]
- Between *you* and *me*, she's wrong. [object of preposition]

A pronoun that is used to express ownership is in the possessive case (*my, mine, our, ours, his, her, hers, its, your, yours, their, theirs, whose*).

- He took *his* notes with him on the business trip.
- We took *our* notes with us on the business trip.

ESL TIPS FOR USING POSSESSIVE PRONOUNS

In many languages, possessive pronouns agree in number and gender with the nouns they modify. In English, however, possessive pronouns agree in number and gender with their antecedents. Check your writing carefully for agreement between a possessive pronoun and the word, phrase, or clause to which it refers.

- The *woman* brought *her* brother a cup of soup.
- *Robert* sent *his* mother flowers on Mother's Day.

A pronoun appositive takes the case of its antecedent.

- Two systems analysts, Joe and *I*, were selected to represent the company.
  [*Joe and I* is in apposition to the subject, *systems analysts*, and must therefore be in the subjective case.]
- The manager selected *two representatives*—Joe and *me*.
  [*Joe and me* is in apposition to two representatives, which is the object of the verb, *selected*, and therefore must be in the objective case.]

If you have difficulty determining the case of a compound pronoun, try using the pronoun singly.

- In his letter, Eldon mentioned *him* and *me*.

  In his letter, Eldon mentioned *him*.

  In his letter, Eldon mentioned *me*.

- *They* and *we* must discuss the terms of the merger.

  They must discuss the terms of the merger.

  We must discuss the terms of the merger.

When a pronoun modifies a noun, try it without the noun to determine its case.

- [*We*/*Us*] pilots fly our own planes.

  We fly our own planes.
  [You would not write, "*Us* fly our own planes."]

- He addressed his remarks directly to [*we*/*us*] technicians.

  He addressed his remarks directly to *us*.
  [You would not write, "He addressed his remarks directly to *we*."]

Gender

A pronoun must agree in gender with its antecedent. A problem sometimes occurs because the masculine pronoun has traditionally been used to refer to both sexes. To avoid the sexual bias implied in such usage, use *he or she* or the plural form of the pronoun, *they*.

- ~~Each~~ ^All may stay or go as ~~he chooses.~~ ^they choose.

As in this example, when the singular pronoun (*he*) changes to the plural (*they*), the singular indefinite pronoun (*each*) must also change to its plural form (*all*). See also biased language.

Number

Number is a frequent problem with only a few indefinite pronouns (*each*, *either*, *neither*, and those ending with *-body* or *-one*, such as *anybody*, *anyone*, *everybody*, *everyone*, *nobody*, *no one*, *somebody*, *someone*) that are normally singular and so require singular verbs and are referred to by singular pronouns.

- As *each member arrives* for the meeting, please hand *him or her* a copy of the confidential report. *Everyone* must return the copy before *he or she* leaves. *Everybody* on the committee *understands* that *neither* of our major competitors *is* aware of the new process we have developed.

Person

Third-person personal pronouns usually have antecedents.

- Gina presented the report to the members of the board of directors. *She* [Gina] first summarized *it* [the report] for *them* [the directors] and then asked for questions.

First- and second-person personal pronouns do not normally require antecedents.

- *I* like my job.
- *You* were there at the time.
- *We* all worked hard on the project.

proofreaders' marks

Publishers have established symbols called *proofreaders' marks* that writers and editors use to communicate in the production of publications. Familiarity with those symbols makes it easy for you to communicate your changes to others. Figure P–8 lists standard proofreaders' marks.

MARK/SYMBOL	MEANING	EXAMPLE	CORRECTED TYPE
ℓ	Delete	the ~~manager's~~ report	the report
^	Insert	the ^ report (manager's)	the manager's report
dots (stet)	Let stand	the ~~manager's~~ report	the manager's report
≡ (cap)	Capitalize	the monday meeting	the Monday meeting
/ (lc)	Lowercase	the Monday Meeting	the Monday meeting
∾ (tr)	Transpose	the cover lettre	the cover letter
⁀	Close space	a loud speaker	a loudspeaker
#	Insert space	a loudspeaker	a loud speaker
¶	Paragraph	...report. ¶ The meeting...	...report. The meeting...

FIGURE P–8. Proofreaders' Marks

MARK/SYMBOL	MEANING	EXAMPLE	CORRECTED TYPE
(run-in mark)	Run in with previous line or paragraph	...report. The meeting...	...report. The meeting...
— (ital)	Italicize	the New York Times	the *New York Times*
~~ (bf)	Boldface	Use boldface sparingly.	Use **boldface** sparingly.
⊙	Insert period	I wrote the e-mail	I wrote the e-mail.
^,	Insert comma	However we cannot...	However, we cannot...
=	Insert hyphen	clear cut decision	clear-cut decision
1/M	Insert em dash	Our goal productivity	Our goal—productivity
^:v or :/	Insert colon	We need the following	We need the following:
^;v or ;/	Insert semicolon	we finished we achieved	we finished; we achieved
" "	Insert quotation marks	He said, I agree.	He said, "I agree."
'	Insert apostrophe	the managers report	the manager's report

FIGURE P–8. Proofreaders' Marks (*continued*)

proofreading

Computer grammar checkers and spell checkers, while a help to proofreading, can make writers overconfident. If a typographical error results in a legitimate English word (for example, *coarse* instead of *course*), the spell checker will not flag the misspelling. Therefore, you still must proofread your work carefully—both on-screen and on paper. You may find some of the tactics discussed in revision useful when proofreading; in fact, you may find passages during proofreading that will require further revision.

Whether the material you proofread is your own writing or that of someone else, consider proofreading in several stages. Although you need to tailor the stages to the specific document and to your own problem areas, the following *Writer's Checklist* should provide a useful starting point for proofreading.

P

Writer's Checklist: Proofreading in Stages

FIRST-STAGE REVIEW

- ☑ Appropriate format, as for reports or correspondence
- ☑ Typographical consistency (headings, spacing, fonts)
- ☑ Correct numbering of figures and tables

Writer's Checklist: Proofreading in Stages (continued)

SECOND-STAGE REVIEW

- ☑ Specific grammar and usage problems
- ☑ Appropriate punctuation
- ☑ Correct abbreviations and capitalization
- ☑ Correct spelling (especially names and places)
- ☑ Complete Web or e-mail addresses
- ☑ Accurate data in tables and lists
- ☑ Cut-and-paste errors; for example, a result of moved or deleted text and numbers

FINAL-STAGE REVIEW

- ☑ Survey of your overall goals: readers' needs and purpose
- ☑ Appearance of the document (see layout and design)
- ☑ Review by a trusted colleague, especially for crucial documents (see collaborative writing)

Consider using standard proofreaders' marks for proofreading someone else's document.

DIGITAL TIPS **PROOFREADING FOR FORMAT CONSISTENCY**

Viewing whole pages on-screen is an effective way to check formatting, spacing, and typographical consistency, as well as the general appearance of documents. For comparing layouts, view multiple pages or "tile" separate documents side by side. For instructions, see *<bedfordstmartins.com/alredtech>* and select *Digital Tips*, "Proofreading for Format Consistency."

proposals

DIRECTORY

A proposal is a document written to persuade readers to follow a plan or course of action that you believe will solve a problem or fulfill a need. (See also persuasion.) You may send a proposal to others within your organization (an internal proposal) or to potential customers or clients outside the organization (an external or a sales proposal). A proposal may be written in response to a request for proposals (RFP) that describes a particular need. Often, you will collaborate with others in preparing a proposal. See collaborative writing.

Strategies

Regardless of the type of proposal you write, begin by considering its audience and purpose, the project management, and the proposal structure.

Audience and Purpose. A proposal offers a plan to fill a need, and readers will evaluate your plan based on how well you answer their questions about what you are proposing to do, how and when you plan to do it, and how much it will cost. Because proposals often require more than one level of approval, take all your readers into account as you answer their questions. Consider especially their levels of technical knowledge of the subject. For example, if your primary reader is an expert on your subject but a supervisor who must also approve the proposal is not, provide an executive summary written in nontechnical language. You might also include a glossary of terms used in the body of the proposal or an appendix that explains highly detailed information in nontechnical language. If your primary reader is not an expert but a supervisor is, write the proposal with the nonexpert in mind and include an appendix that contains the technical details.

Writing a persuasive and even complex proposal can be simplified by composing a concise statement of exactly the problem or opportunity that your proposal is designed to address. Stating the problem or opportunity concisely helps you and your readers understand the value, scope, and limitations of your proposed solution.

Project Management. Proposal writers are often faced with writing high-quality, persuasive proposals under tight organizational deadlines. Breaking the task down into manageable parts is the key to accomplishing your goal. For example, you might set your own deadlines for completing various sections of the proposal or for scheduling the stages of the writing process, as described in "Five Steps to Successful Writing."

Proposal Structure

Proposals vary widely in length, formality, and structure.

Short Proposals. A short or medium-length proposal typically consists of an introduction, a body, and a conclusion.

INTRODUCTION. The introduction should state the purpose and scope of your proposal, as well as the problem you propose to solve and your solution to it. It should also indicate the dates on which you propose to begin and complete work, any special benefits of your proposed approach, and the cost of the project. For a sales (external) proposal, refer to any previous positive associations your organization has had with the potential client or customer.

BODY. The body should offer the details of your solution to the problem and explain (1) the product or service you are offering, (2) how the job will be done, (3) how you will perform the work and any special materials you may use, (4) a schedule when each phase of the project will be completed, and (5) a breakdown of project costs.

CONCLUSION. The conclusion should persuasively resell your proposal by emphasizing the benefits of your solution, product, or service over any competing ideas. Also include details about the time period during which the proposal is valid. Effective conclusions show confidence in your solution, your appreciation for the opportunity to submit the proposal, and your willingness to provide further information, as well as encouraging your reader to act on your proposal.

Long Proposals. The sections that follow are found in longer, more formal proposals. The number of sections in any particular proposal depends on the audience, the purpose, and the scope of the proposal, as well as on the standard practice within an organization. Typically, however, the sections can be grouped into *front matter*, *body* (which includes the introduction and conclusion), and *back matter*. For more information on many of the following components, see formal reports.

P

FRONT MATTER

- *Cover Letter or Letter of Transmittal.* In the cover letter, express appreciation for the opportunity to submit your proposal, any assistance from the customer, and any previous positive associations. Then summarize the proposal's recommendations and express confidence that they will satisfy the customer's needs.
- *Title Page.* Include the title of the proposal, the date, the organization to which it is being submitted, and your company name.
- *Table of Contents.* Include a table of contents in longer proposals to guide readers to important headings, which should be listed according to beginning page numbers.
- *List of Figures.* If your proposal has six or more figures, include a list of figures with captions as well as figure and page numbers. See visuals.

BODY

- *Executive Summary.* Briefly summarize the proposal's highlights in persuasive, nontechnical language for decision-makers.
- *Introduction.* Explain the reasons for the proposal; emphasize reader benefits; and, when appropriate, discuss your understanding of the problem.
- *Background or Problem.* Describe the problem or opportunity your proposal addresses. To make your proposal more persuasive, your problem statement should illustrate how your proposal will benefit your client's organization.
- *Product Description.* If your proposal offers products as well as services, include a general description of the products and any technical specifications.
- *Detailed Solutions (Rationale).* Explain in a detailed section that will be read by technical specialists exactly how you plan to do what you are proposing.
- *Cost Analysis.* Itemize the estimated costs of all the products and services you are offering.
- *Delivery Schedule.* Outline how you will accomplish the work and show a timetable for each phase of the project.
- *Staffing.* Summarize the expertise (education, experience, and certifications) of key personnel who will work on the project. Include their <u>résumés</u> in an appendix.
- *Site Preparation.* If your recommendations include modifying the customer's physical facilities, include a site-preparation description that details the required modifications.
- *Training Requirements.* If the products and services you are proposing require training the customer's employees, specify the required training and its cost.
- *Statement of Responsibilities.* To prevent misunderstandings about what you and your customer's responsibilities will be, state those responsibilities.
- *Organizational Sales Pitch.* Describe your company, its history, and its present position in the industry. An organizational sales pitch is designed to sell your company and its general capability in the field. It promotes the company and concludes the proposal on an upbeat, persuasive note.
- *Authorization Request and Deadline.* Close with a request for approval and a deadline that explains how long the proposed prices are valid.
- *Conclusion.* Include a persuasive conclusion that summarizes the proposal's key points and stresses your company's strong points.

BACK MATTER

- *Appendixes.* Provide résumés of key personnel or material of interest to some readers, such as statistical analyses, organizational charts, and workflow diagrams.
- *Bibliography.* List all sources consulted to prepare the proposal. See bibliographies and documenting sources.
- *Glossary.* If your proposal contains terms that will be unfamiliar to your intended audience, list and define them in the glossary.

Internal Proposals

Two common types of internal proposals are often distinguished from each other by the frequency with which they are written and by the degree of change proposed.

Routine Internal Proposals. Routine internal proposals are most frequent and typically include small spending requests, requests for permission to hire new employees or increase salaries, and requests to attend conferences or purchase new equipment. In writing routine proposals, follow the introduction-body-conclusion format for short proposals on page 414, and highlight any benefits to be realized.

Formal Internal Proposals. Formal internal proposals are usually proposals to commit relatively large sums of money. They have various names, but the most common designation is a *capital appropriations request* or a *capital appropriations proposal.*

The introduction to a formal internal proposal should persuasively establish that a problem exists that needs a solution and provide any background information your reader needs to help him or her make the decision in question. The introduction should also briefly describe any supporting evidence you are including, such as a feasibility study you may have conducted. See also feasibility reports.

P

The body of a formal internal proposal should offer a practical solution to the problem; compare possible alternatives; respond to possible objections; and describe all the equipment, property, or services you are proposing to purchase and their key benefits to the organization. You should include any justifications for each expenditure: the calculated return on investment or the internal rate of return; the volume of use over, for example, the next two years; the product life expectancy, technical support, or warranties; and the most economical purchasing option.

The conclusion should emphasize the benefits of what you are proposing and express your willingness to provide any further information that may be required. Figure P–9 shows a formal internal proposal. The writer of the proposal responded to management's concern about the issue of reducing company health-care costs. The writer, Leslie Galusha, concluded that helping employees improve their fitness would be a major step in solving the problem of escalating health-care costs. She examined government data and researched through personal visits and interviews the approaches of other companies.

External Proposals

External proposals to be submitted to other companies or organizations could be either solicited or unsolicited.

Solicited Proposals. Solicited proposals are prepared in response to a request for goods or services. Procuring organizations that would like competing companies to bid for a job commonly issue a request for proposals (RFP) or an invitation for bids (IFB).

ABO, Inc.
Interoffice Memo

To: Joan Marlow, Director, Human Resources Division
From: Leslie Galusha, Chief LG
Employee Benefits Department
Date: June 13, 2006
Subject: Employee Fitness and Health-Care Costs

Health-care and workers'-compensation insurance costs at ABO, Inc., have risen 100 percent over the last five years. In 2001, costs were $5,675 per employee per year; in 2006, they have reached $11,560 per employee per year. This doubling of costs mirrors a national trend, with health-care costs anticipated to continue to rise at the same rate for the next ten years. Controlling these escalating expenses will be essential. They are eating into ABO's profit margin because the company currently pays 70 percent of the costs for employee coverage.

Healthy employees bring direct financial benefits to companies in the form of lower employee insurance costs, lower absenteeism rates, and reduced turnover. Regular physical exercise promotes fit, healthy people by reducing the risk of coronary heart disease, diabetes, osteoporosis, hypertension, and stress-related problems. I propose that to promote regular, vigorous physical exercise for our employees, ABO implement a health-care program that focuses on employee fitness. . . .

FIGURE P–9. Formal Internal Proposal (Introduction)

P

Joan Marlow 2 June 13, 2006

Problem of Health-Care Costs

The U.S. Department of Health and Human Services recently estimated that health-care costs in the United States will triple by the year 2015. Corporate expenses for health care are rising at such a fast rate that, if unchecked, in eight years they will significantly erode corporate profits.

Researchers agree that people who do not participate in a regular and vigorous exercise program incur double the health-care costs and are hospitalized 30 percent more days than people who exercise regularly. Nonexercisers are also 41 percent more likely to submit medical claims over $10,000 at some point during their careers than are those who exercise regularly.

My study of Tenneco, Inc., found that the average health-care claim for unfit men was $2,006 per illness compared with an average claim of $862 for those who exercised regularly. For women, the average claim for those who were unfit was $2,535, more than double the average claim of $1,039 for women who exercised. Additionally, Control Data Corporation found that each nonexerciser cost the company an extra $515 a year in health-care expenses.

These figures are further supported by data from independent studies. A model created by the National Institutes of Health (NIH) estimates that the average white-collar company could save $596,000 annually in medical costs (per 1,000 employees) just by promoting wellness. NIH researchers estimated that for every $1 a firm invests in a health-care program, it saves up to $3.75 in health-care costs. Another NIH study of 667 insurance-company employees showed savings of $2.65 million over a five-year period. The same study also showed a 400 percent drop in absentee rates after the company implemented a company-wide fitness program.

Possible Solutions for ABO

The benefits of regular, vigorous physical activity for employees and companies are compelling. To achieve these benefits at ABO, I propose that we choose from one of two possible options: Build in-house fitness centers at our warehouse facilities, or offer employees several options for membership at a national fitness club. The following analysis compares . . .

FIGURE P–9. Formal Internal Proposal (*continued*) (Body)

Joan Marlow 3 June 13, 2006

Conclusion and Recommendation

I recommend that ABO, Inc., participate in the corporate membership program at AeroFitness Clubs, Inc., by subsidizing employee memberships. By subsidizing memberships, ABO shows its commitment to the importance of a fit workforce. Club membership allows employees at all five ABO warehouses to participate in the program. The more employees who participate, the greater the long-term savings in ABO's health-care costs. Building and equipping fitness centers at all five warehouse sites would require an initial investment of nearly $2.5 million. These facilities would also occupy valuable floor space—on average, 4,000 square feet at each warehouse. Therefore, this option would be very costly.

Enrolling employees in the corporate program at AeroFitness would allow them to receive a one-month free trial membership. Those interested in continuing could then join the club and pay half of the one-time membership fee of $900 and receive a 30 percent discount on the $600 yearly fee. The other half of the membership fee ($450) would be paid for by ABO. If em-ployees leave the company, they would have the option of purchasing ABO's share of the membership to continue at AeroFitness or selling their half of the membership to another ABO employee wishing to join AeroFitness.

Implementing this program will help ABO, Inc., reduce its health-care costs while building stronger employee relations by offering employees a desirable benefit. If this proposal is adopted, I have some additional thoughts about publicizing the program to encourage employee participation. I look forward to discussing the details of this proposal with you and answering any questions you may have.

FIGURE P–9. Formal Internal Proposal (*continued*) (Conclusion)

An IFB is commonly issued by government agencies to solicit bids on clearly defined products or services. An IFB is restrictive, binding the bidder to produce an item that meets the exact requirements of the agency. The goods or services to be procured are defined in the IFB by references to performance standards stated in specifications. Bidders must be prepared to prove that their product will meet all requirements of the specifications.

In contrast to an IFB, an RFP is flexible. Often an RFP will define a problem and allow those who respond to suggest possible solutions. Sometimes, RFPs are presented in several stages: (1) development of a concept, (2) construction of a prototype or mock-up, and (3) production of the device or plan selected.

The procuring organization generally publishes its RFP or IFB in trade journals, on its Web site, or in a specialized venue such as Federal Business Opportunities at <www.fedbizopps.gov/>.

Unsolicited Proposals. Unsolicited proposals are submitted to a company without a prior request for a proposal. Companies often operate for years with a problem they have never recognized (unnecessarily high maintenance costs, for example, or poor inventory-control methods). You might prepare an unsolicited proposal if you were convinced that the potential customer could realize substantial benefits by adopting your solution to a problem. Of course, you would need to convince the customer of the need for what you are proposing and that your solution would be the best one. Many unsolicited proposals are preceded by an inquiry to determine potential interest. If you receive a positive response, you would conduct a detailed study of the prospective customer's needs to determine whether you can be of help and, if so, exactly how. You would then prepare your proposal on the basis of your study.

Sales Proposals. A persuasive sales proposal must demonstrate above all that the prospective customer's purchase of the seller's products or services will solve a problem, improve operations, or offer other benefits. Sales proposals vary greatly in size and sophistication—from several pages written by one person, to dozens of pages written collaboratively by several people, to hundreds of pages written by a team of professional proposal writers. A short sales proposal might bid for the construction of a single home, a moderate-length proposal might bid for the installation of a computer network, and a large proposal might bid for the construction of a multimillion-dollar water purification system.

ETHICS NOTE Always keep in mind that, once submitted, a sales proposal is a legally binding document that promises to offer goods or services within a specified time and for a specified price. ✦

Your first task in writing a sales proposal is to find out exactly what your prospective customer needs. To do that, survey your potential customer's business, and determine whether your organization can satisfy the customer's needs. Before preparing a sales proposal, try to find out who your principal competitors are. Then compare your company's strengths with those of the competing firms, determine your advantages over your competitors, and emphasize those advantages in your proposal. Figure P–10 on pages 422–32 shows an example of a major sales proposal, which uses the format recommended for long proposals (but not all elements are shown).

WEB LINK **SAMPLE SALES PROPOSALS**

For a complete and annotated version of the proposal shown in Figure P–10, as well as additional sample proposals and the RFP to which Figure P–10 responded, see <*bedfordstmartins.com/alredtech*> and select *Model Documents Gallery*.

Grant and Research Proposals. Grant and research proposals are written to request the approval of, and usually funding for, particular projects. For example, a professor of education might submit a research or grant proposal to the Department of Education to request funding for research on the relationship of class size to educational performance. Many government and private agencies solicit research and grant proposals. The granting agencies usually have their own requirements for the format and content of proposals submitted to them, but the proposals must always be persuasive. Tailor your grant or research proposal to your audience carefully by explaining the project's goals, your plan for achieving those goals, and your qualifications to perform the project.

Writer's Checklist: Writing Persuasive Proposals

- ☑ Analyze your audience carefully to determine how to best meet your readers' needs or requirements.
- ☑ Write a concise purpose statement to clarify your proposal's goals.
- ☑ Emphasize the proposal's benefits to readers and anticipate their questions or objections.
- ☑ Incorporate evidence to support the claims of your proposal.
- ☑ Divide the writing task into manageable segments and develop a work schedule.
- ☑ Review the descriptions of proposal sections and their uses in this entry.
- ☑ Select an appropriate, visually appealing format. (See layout and design.)
- ☑ Use a confident, upbeat tone throughout the proposal.

P

The Waters Corporation
17 North Waterloo Blvd.
Tampa, Florida 33607
Phone: (813) 919-1213 Fax: (813) 919-4411
www.waters.comp.com

September 2, 2006

Mr. John Yeung, General Manager
Cookson's Retail Stores, Inc.
101 Longuer Street
Savannah, Georgia 31499

Dear Mr. Yeung:

The Waters Corporation appreciates the opportunity to respond to Cookson's Request for Proposals dated July 25, 2006. We would like to thank Mr. Becklight, Director of your Management Information Systems Department, for his invaluable contributions to the study of your operations that we conducted before preparing our proposal.

It has been Waters's privilege to provide Cookson's with retail systems and equipment since your first store opened many years ago. Therefore, we have become very familiar with your requirements as they have evolved during the expansion you have experienced since that time. Waters's close working relationship with Cookson's has resulted in a clear understanding of Cookson's philosophy and needs.

Our proposal describes a Waters Interactive Terminal/Retail Processor System designed to meet Cookson's network and processing needs. It will provide all of your required capabilities, from the point-of-sale operational requirements at the store terminals to the host processor. The system uses the proven Retail III modular software, with its point-of-sale applications, and the superior Interactive Terminal, with its advanced capabilities and design. This system is easily installed without extensive customer reprogramming.

FIGURE P–10. Sales Proposal (Cover Letter)

Mr. J. Yeung
Page 2
September 2, 2006

The Waters Interactive Terminal/Retail Processor System, which is compatible with much of Cookson's present equipment, not only will answer your present requirements but will provide the flexibility to add new features and products in the future. The system's unique hardware modularity, efficient microprocessor design, and flexible programming capability greatly reduce the risk of obsolescence.

Thank you for the opportunity to present this proposal. You may be sure that we will use all the resources available to the Waters Corporation to ensure the successful implementation of the new system.

Sincerely yours,

Janet A. Curtain

Janet A. Curtain
Executive Account Manager
General Merchandise Systems
(JCurtain@netcom.TF.com)

Enclosure: Proposal

FIGURE P–10. Sales Proposal (*continued*) (Cover Letter)

P

The Waters Proposal September 2, 2006

EXECUTIVE SUMMARY

The Waters 319 Interactive Terminal/615 Retail Processor System will provide your management with the tools necessary to manage people and equipment more profitably with procedures that will yield more cost-effective business controls for Cookson's.

The equipment and applications proposed for Cookson's were selected through the combined effort of Waters and Cookson's Management Information Systems Director, Mr. Becklight. The architecture of the system will respond to your current requirements and allow for future expansion.

The features and hardware in the system were determined from data acquired through the comprehensive survey we conducted at your stores in February of this year. The total of 71 Interactive Terminals proposed to service your four store locations is based on the number of terminals currently in use and on the average number of transactions processed during normal and peak periods. The planned remodeling of all four stores was also considered, and the suggested terminal placement has been incorporated into the working floor plan. The proposed equipment configuration and software applications have been simulated to determine system performance based on the volumes and anticipated growth rates of the Cookson's stores.

The information from the survey was also used in the cost justification, which was checked and verified by your controller, Mr. Deitering. The cost-effectiveness of the Waters Interactive Terminal/Retail Processor System is apparent. Expected savings, such as the projected 46 percent reduction in sales audit expenses, are realistic projections based on Waters's experience with other installations of this type.

– 1 –

FIGURE P–10. Sales Proposal (*continued*) (Executive Summary)

The Waters Proposal September 2, 2006

GENERAL SYSTEM DESCRIPTION

The point-of-sale system that Waters is proposing for Cookson's includes two primary Waters products. These are the 319 Interactive Terminal and the 615 Retail Processor.

Waters 319 Interactive Terminal

The primary component in the proposed retail system is the Interactive Terminal. It contains a full microprocessor, which gives it the flexibility that Cookson's has been looking for.

The 319 Interactive Terminal provides you with freedom in sequencing a transaction. You are not limited to a preset list of available steps or transactions. The terminal program can be adapted to provide unique transaction sets, each designed with a logical sequence of entry and processing to accomplish required tasks. In addition to sales transactions recorded on the selling floor, specialized transactions such as theater-ticket sales and payments can be designed for your customer-service area.

The 319 Interactive Terminal also functions as a credit authorization device, either by using its own floor limits or by transmitting a credit inquiry to the 615 Retail Processor for authorization.

Data-collection formats have been simplified so that transaction editing and formatting are much more easily accomplished. The IS manager has already been provided with documentation on these formats and has outlined all data-processing efforts that will be necessary to transmit the data to your current systems. These projections have been considered in the cost justification.

Waters 615 Retail Processor

The Waters 615 Retail Processor is a minicomputer system designed to support the Waters family of retail terminals. The . . .

[*The proposal next describes the Waters 615 Retail Processor before moving on to the detailed solution section.*]

FIGURE P–10. Sales Proposal (*continued*) (General Description of Products)

P

PAYROLL APPLICATION

Current Procedure
Your current system of reporting time requires each hourly employee to sign a time sheet; the time sheet is reviewed by the department manager and sent to the Payroll Department on Friday evening. Because the week ends on Saturday, the employee must show the scheduled hours for Saturday and not the actual hours; therefore, the department manager must adjust the reported hours on the time sheet for employees who do not report on the scheduled Saturday or who do not work the number of hours scheduled.

The Payroll Department employs a supervisor and three full-time clerks. To meet deadlines caused by an unbalanced workflow, an additional part-time clerk is used for 20 to 30 hours per week. The average wage for this clerk is $9.00 per hour.

Advantage of Waters's System
The 319 Interactive Terminal can be programmed for entry of payroll data for each employee on Monday morning by department managers, with the data reflecting actual hours worked. This system would eliminate the need for manual batching, controlling, and data input. The Payroll Department estimates conservatively that this work consumes 40 hours per week.

Hours per week	40
Average wage (part-time clerk)	× 9.00
Weekly payroll cost	$360.00
Annual savings	$18,720

Elimination of the manual tasks of tabulating, batching, and controlling can save 0.25 hourly unit. Improved workflow resulting from timely data in the system without data-input processing will allow more efficient use of clerical hours. This would reduce payroll by the 0.50 hourly unit currently required to meet weekly check disbursement.

Eliminate manual tasks	0.25
Improve workflow	0.75
40-hour unit reduction	1.00
Hours per week	40
Average wage (full-time clerk)	11.00
Savings per week	$440.00
Annual savings	$22,880

TOTAL SAVINGS: $41,600

FIGURE P–10. Sales Proposal (*continued*) (Detailed Solution)

The Waters Proposal September 2, 2006

COST ANALYSIS

This section of our proposal provides detailed cost information for the Waters 319 Interactive Terminal and the Waters 615 Retail Processor. It then multiplies these major elements by the quantities required at each of your four locations.

319 Interactive Terminal

Equipment	Price	Maint. (1 yr.)
Terminal	$2,895	$167
Journal Printer	425	38
Receipt Printer	425	38
Forms Printer	525	38
Software	220	–
TOTALS	$4,490	$281

[*The cost section goes on to describe other costs to install the system before summarizing the costs.*]

The following table summarizes all costs.

Location	Hardware	Maint. (1 yr.)	Software
Store No. 1	$72,190	$4,975	$3,520
Store No. 2	89,190	6,099	4,400
Store No. 3	76,380	5,256	3,740
Store No. 4	80,650	5,537	3,960
Data Center	63,360	6,679	12,480
Subtotals	$381,770	$28,546	$28,100

TOTAL $438,416

DELIVERY SCHEDULE

Waters is normally able to deliver 319 Interactive Terminals and 615 Retail Processors within 30 days of the date of the contract. This can vary depending on the rate and size of incoming orders.

All the software recommended in this proposal is available for immediate delivery. We do not anticipate any difficulty in meeting your tentative delivery schedule.

–8–

P

FIGURE P–10. Sales Proposal (*continued*) (Cost Analysis and Delivery Schedule)

The Waters Proposal September 2, 2006

SITE PREPARATION

Waters will work closely with Cookson's to ensure that each site is properly prepared prior to system installation. You will receive a copy of Waters's installation and wiring procedures manual, which lists the physical dimensions, service clearance, and weight of the system components in addition to the power, logic, communications-cable, and environmental requirements. Cookson's is responsible for all building alterations and electrical facility changes, including the purchase and installation of communications cables, connecting blocks, and receptacles.

Wiring
For the purpose of future site considerations, Waters's in-house wiring specifications for the system call for two twisted-pair wires and twenty-two shielded gauges. The length of communications wires must not exceed 2,500 feet.

As a guide for the power supply, we suggest that Cookson's consider the following.

1. The branch circuit (limited to 20 amps) should service no equipment other than 319 Interactive Terminals.
2. Each 20-amp branch circuit should support a maximum of three Interactive Terminals.
3. Each branch circuit must have three equal-size conductors—one hot leg, one neutral, and one insulated isolated ground.
4. Hubbell IG 5362 duplex outlets or the equivalent should be used to supply power to each terminal.
5. Computer-room wiring will have to be upgraded to support the 615 Retail Processor.

–9–

P

FIGURE P–10. Sales Proposal (*continued*) (Site-Preparation Section)

The Waters Proposal September 2, 2006

TRAINING

To ensure a successful installation, Waters offers the following training course for your operators.

Interactive Terminal/Retail Processor Operations
Course number: 8256
Length: three days
Tuition: $500.00

This course provides the student with the skills, knowledge, and practice required to operate an Interactive Terminal/Retail Processor System. Online, clustered, and stand-alone environments are covered.

We recommend that students have a department-store background and that they have some knowledge of the system configuration with which they will be working.

– 10 –

FIGURE P–10. Sales Proposal (*continued*) (Training Section)

The Waters Proposal September 2, 2006

RESPONSIBILITIES

On the basis of its years of experience in installing information-processing systems, Waters believes that a successful installation requires a clear understanding of certain responsibilities.

Waters's Responsibilities
Generally, it is Waters's responsibility to provide its users with needed assistance during the installation so that live processing can begin as soon thereafter as is practical. The following items describe our specific responsibilities.

- Provide operations documentation for each application that you acquire from Waters.
- Provide forms and other supplies as ordered.
- Provide specifications and technical guidance for proper site planning and installation.
- Provide adviser assistance in the conversion from your present system to the new system.

Cookson's Responsibilities
Cookson's will be responsible for the suggested improvements described earlier, as well as the following.

- Identify an installation coordinator and system operator.
- Provide supervisors and clerical personnel to perform conversion to the system.
- Establish reasonable time schedules for implementation.
- Ensure that the physical site requirements are met.
- Provide personnel to be trained as operators and ensure that other employees are trained as necessary.
- Assume the responsibility for implementing and operating the system.

– 11 –

FIGURE P–10. Sales Proposal (*continued*) (Statement of Responsibilities)

DESCRIPTION OF VENDOR

The Waters Corporation develops, manufactures, markets, installs, and services total business information-processing systems for selected markets. These markets are primarily in the retail, financial, commercial, industrial, health-care, education, and government sectors.

The Waters total system concept encompasses one of the broadest hardware and software product lines in the industry. Waters computers range from small business systems to powerful general-purpose processors. Waters computers are supported by a complete spectrum of terminals, peripherals, and data-communication networks, as well as an extensive library of software products. Supplemental services and products include data centers, field service, systems engineering, and educational centers.

The Waters Corporation was founded in 1934 and presently has approximately 26,500 employees. The Waters headquarters is located at 17 North Waterloo Boulevard, Tampa, Florida, with district offices throughout the United States and Canada.

WHY WATERS?

Corporate Commitment to the Retail Industry
Waters's commitment to the retail industry is stronger than ever. We are continually striving to provide leadership in the design and implementation of new retail systems and applications that will ensure our users of a logical growth pattern.

Research and Development
Over the years, Waters has spent increasingly large sums on research-and-development efforts to ensure the availability of products and systems for the future. In 2005, our research-and-development expenditure for advanced systems design and technological innovations reached the $70-million level.

Leading Point-of-Sale Vendor
Waters is a leading point-of-sale vendor, having installed over 150,000 units. The knowledge and experience that Waters has gained over the years from these installations ensure well-coordinated and effective systems implementations.

– 12 –

FIGURE P–10. Sales Proposal (*continued*) (Vendor Description and Organizational Sales Pitch)

The Waters Proposal September 2, 2006

CONCLUSION

Waters welcomes the opportunity to submit this proposal to Cookson's. The Waters Corporation is confident that we have offered the right solution at a competitive price. Based on the hands-on analysis we conducted, our proposal takes into account your current and projected workloads and your plans to expand your facilities and operations. Our proposal will also, we believe, afford Cookson's future cost-avoidance measures in employee time and in enhanced accounting features.

Waters has a proven track record of success in the manufacture, installation, and servicing of retail business information systems stretching over many decades. We also have a demonstrated record of success in our past business associations with Cookson's. We believe that the system we propose will extend and strengthen this partnership.

Should you require additional information about any facet of this proposal, please contact Janet A. Curtain, who will personally arrange to meet with you or arrange for Waters's technical staff to meet with you or send you the information you need.

We look forward to your decision and to continued success in our working relationship with Cookson's.

– 13 –

FIGURE P–10. Sales Proposal (*continued*) (Conclusion)

pseudo- / quasi-

As a prefix, *pseudo-*, meaning "false or counterfeit," is joined to the root word without a hyphen unless the root word begins with a capital letter (*pseudo*science, *pseudo*-Newtonian). *Pseudo-* is sometimes confused with *quasi-*, meaning "somewhat" or "partial." Unlike *semi-*, *quasi-* means "resembling something" rather than "half." *Quasi-* is usually hyphenated in combinations (*quasi*-scientific theories). See also bi-/semi-.

punctuation

Punctuation helps readers understand the meaning and relationships of words, phrases, clauses, and sentences. Marks of punctuation link, separate, enclose, indicate omissions, terminate, and classify. Most punctuation marks can perform more than one function. See also sentence construction.

The use of punctuation is determined by grammatical conventions and the writer's intention. Understanding punctuation is essential for writers because it enables them to communicate with clarity and precision. See also grammar.

Detailed information on each mark of punctuation is given in its own entry. The following are the 13 marks of punctuation.

apostrophe	'	parentheses	( )
brackets	[ ]	period	.
colon	:	question mark	?
comma	,	quotation marks	" "
dash	—	semicolon	;
exclamation mark	!	slash	/
hyphen	-		

See also abbreviations, capitalization, contractions, dates, ellipses, italics, and numbers.

WEB LINK PRACTICING PUNCTUATION

For online exercises that provide practice in using commas, semicolons, apostrophes, quotation marks, and more, see <*bedfordstmartins.com/alredtech*> and select *Exercise Central*.

purpose

What do you want your readers to know, to believe, or to do when they have read your document? When you answer that question, you have determined the primary purpose, or objective, of your document. Be careful not to state a purpose too broadly. A purpose such as "to explain a fax machine" is too general to be helpful to you as you write. In contrast, "to instruct the reader how to use the IT-770 fax machine to send and retrieve faxes" is a specific purpose that will help you focus on what you need your document to accomplish. Often the context will help you focus your purpose.

However, the writer's primary purpose is often more complex than simply "to explain" something. To fully understand this complexity, you need to ask yourself not only *why* you are writing the document but *what* you want to influence your reader to believe or to do after reading it. Suppose a writer for a newsletter has been assigned to write an article about cardiopulmonary resuscitation (CPR). In answer to the question *what?* the writer could state the purpose as "to emphasize the importance of CPR." To the question *why?* the writer might respond, "to encourage employees to sign up for evening CPR classes." Putting the answers to the two questions together, the writer's purpose might be stated as, "To write a document that will emphasize the importance of CPR and encourage employees to sign up for evening CPR classes." Note that the primary purpose of the document on CPR was to persuade the readers of the importance of CPR, and the secondary goal was to motivate them to register for a class. Secondary goals often involve such abstract notions as to motivate, to persuade, to reassure, or to inspire your reader. See also persuasion.

If you answer the questions *what?* and *why?* and put the answers into writing as a stated purpose that includes both primary and secondary goals, you will simplify your writing task and more likely achieve your purpose. For a collaborative writing project, it is especially important to collectively write a statement of your purpose to ensure that the document achieves its goals. Do not lose sight of that purpose as you become engrossed in the other steps of the writing process. See also "Five Steps to Successful Writing."

Q

question marks

The question mark (?) has several uses, but it most often ends a sentence that is a direct question or request.

- Where did you put the specifications? [direct question]
- Will you e-mail me if your shipment does not arrive by June 10? [request]

Use a question mark to end a statement that has an interrogative meaning—a statement that is declarative in form but asks a question.

- The lab report is finished? [question in declarative form]

Question marks may follow a series of separate items within an interrogative sentence.

- Do you remember the date of the contract? Its terms? Whether you signed it?

Use a question mark to end an interrogative clause within a declarative sentence.

- It was not until July (or was it August?) that we submitted the report.

Retain the question mark in a title that is being cited, even though the sentence in which it appears has not ended.

- *Can Quality Be Controlled?* is the title of her book.

Never use a question mark to end a sentence that is an indirect question.

- He asked me where I put the specifications?

When a question is a polite request or instruction to which an affirmative response is assumed, a question mark is not necessary.

- Will you call me as soon as he arrives. [polite request]

When used with quotations, the placement of the question mark is important. When the writer is asking a question, the question mark belongs outside the quotation marks.

- Did she say, "I don't think the project should continue"?

If the quotation itself is a question, the question mark goes inside the quotation marks.

- She asked, "Do we have enough funding?"

If both cases apply—the writer is asking a question and the quotation itself is a question—use a single question mark inside the quotation marks.

- Did she ask, "Do we have enough funding?"

questionnaires

A questionnaire—a series of questions on a particular topic sent out to a number of people—serves the same purpose as an interview but does so on paper, as an e-mail attachment, or online. The sample cover memo and questionnaire in Figure Q–1 were sent to employees in a large organization who had participated in a six-month program of flexible working hours.

Questionnaires have several advantages over the personal interview as well as several disadvantages.

ADVANTAGES

- A questionnaire allows you to gather information from more people more quickly than you could by conducting personal interviews.
- It enables you to obtain responses from people who are difficult to reach or who are in various geographical locations.
- Those responding to a questionnaire have more time to think through their answers than when faced with the pressure of composing thoughtful and complete answers to an interviewer.
- The questionnaire may yield more objective data because it reduces the possibility that the interviewer's tone of voice or facial expressions might influence an answer.
- The cost of distributing and tabulating a questionnaire is lower than the cost of conducting numerous personal interviews.

Luxwear Products Corporation
MEMO

To: All Company Employees
From: Nelson Barrett, Director *NB*
Date: October 17, 2006
Subject: Review of Flexible Working Hours Program

Please complete and return the questionnaire enclosed regarding Luxwear's trial program of flexible working hours. Your answers will help us decide whether we should make the program permanent.

Return the completed questionnaire to Ken Rose, Mail Code 12B, by October 28. Your signature on the questionnaire is not necessary. All responses will be confidential and given serious consideration. Feel free to raise additional issues pertaining to the program.

If you want to discuss any item in the questionnaire, call Pam Peters in the Human Resources Department at extension 8812 or e-mail at pp1@lpc.com.

Enclosure: Questionnaire

FIGURE Q–1. Questionnaire (Cover Memo)

Flexible Working Hours Program
Questionnaire

1. What kind of position do you occupy?

☐ Supervisory
☐ Nonsupervisory

2. Indicate to the nearest quarter of an hour your starting time under flextime.

☐ 7:00 a.m.	☐ 8:15 a.m.
☐ 7:15 a.m.	☐ 8:30 a.m.
☐ 7:30 a.m.	☐ 8:45 a.m.
☐ 7:45 a.m.	☐ 9:00 a.m.
☐ 8:00 a.m.	☐ Other (specify) ________________

3. Where do you live?

☐ Talbot County	☐ Greene County
☐ Montgomery County	☐ Other (specify) ________________

4. How do you usually travel to work?

☐ Drive alone	☐ Walk
☐ Bus	☐ Car pool
☐ Train	☐ Motorcycle
☐ Bicycle	☐ Other (specify) ________________

5. Has flextime affected your commuting time?

☐ Increase: Approximate number of minutes ________
☐ Decrease: Approximate number of minutes ________
☐ No change

6. If you drive alone or in a car pool, has flextime increased or decreased the amount of time it takes you to find a parking space?

☐ Increased ☐ Decreased ☐ No change

7. Has flextime had an effect on your productivity?

a. Quality of work
☐ Increased ☐ Decreased ☐ No change

b. Accuracy of work
☐ Increased ☐ Decreased ☐ No change

c. Quiet time for uninterrupted work
☐ Increased ☐ Decreased ☐ No change

Q

FIGURE Q–1. Questionnaire (*continued*)

8. Have you had difficulty getting in touch with coworkers who are on different work schedules from yours?

   ☐ Yes ☐ No

9. Have you had trouble scheduling meetings within flexible starting and quitting times?

   ☐ Yes ☐ No

10. Has flextime affected the way you feel about your job?

    ☐ Yes ☐ No

    If yes, please answer (a) or (b):

    a. Feel better about job
       ☐ Slightly ☐ Considerably

    b. Feel worse about job
       ☐ Slightly ☐ Considerably

11. How important is it for you to have flexibility in your working hours?

    ☐ Very ☐ Not very ☐ Somewhat ☐ Not at all

12. Has flextime allowed you more time to be with your family?

    ☐ Yes ☐ No

13. If you are responsible for the care of a young child or children, has flextime made it easier or more difficult for you to obtain babysitting or day-care services?

    ☐ Easier ☐ More difficult ☐ No change

14. Do you recommend that the flextime program be made permanent?

    ☐ Yes ☐ No

15. Please describe below or attach any major changes you recommend for the program.

Thank you for your assistance.

FIGURE Q–1. Questionnaire (*continued*)

DISADVANTAGES

- The results of a questionnaire may be slanted in favor of those people who have strong opinions on a subject because they are more likely to respond than those with only moderate views.
- The questionnaire does not allow specific follow-up to answers; at best, a questionnaire can be designed to let one question lead logically to another.
- Distributing questionnaires and waiting for replies may take considerably longer than conducting a personal interview.

See also interviewing for information and research.

Selecting the Recipients

Selecting the proper recipients for your questionnaire is crucial if you are to gather representative and usable data. If you wanted to survey the opinions of large groups in the general population—for example, all medical technologists working in private laboratories or all independent garage owners—your task would not be easy. Because you cannot include everybody in your survey, you need to choose a representative cross section. For example, you would want to include enough people from around the country, respondents of both genders, and people with an assortment of educational training. Only then could you make a generalized statement based on your findings from the sample. (The best sources of information on sampling techniques are market-research and statistics texts.)

WEB LINK **ONLINE SURVEYS**

Surveymonkey.com offers help for designing online surveys as well as collecting and analyzing the results. For links to this site and more, see <*bedfordstmartins.com/alredtech*> and select *Links for Handbook Entries*.

Q

Preparing the Questions

A key goal in designing the questionnaire is to keep it as brief as possible. The longer a questionnaire is, the less likely the recipient will be to complete and return it. Also, the questions should be easy to understand. A confusing question will yield confusing results, whereas a carefully worded question will be easy to answer. Ideally, recipients should be able to answer most questions with a "yes" or "no" or by checking or circling a choice among several options. Such answers are easy to tabulate and require minimum effort on the part of the respondent, thus increasing your chances of obtaining a response. See also forms design.

- Do you recommend that the flextime program be made permanent?

  ☐ Yes ☐ No ☐ No opinion

If you need more information than such questions produce, provide an appropriate range of answers, as in the following example.

- How many hours of overtime would you be willing to work each week?

  ☐ 4 hours ☐ 8 hours ☐ Over 10 hours

  ☐ 6 hours ☐ 10 hours ☐ No overtime

Questions should be neutral; they should not be worded in such a way as to lead respondents to give a particular answer, which can result in inaccurate or skewed data.

SLANTED Would you prefer the freedom of a four-day workweek?

NEUTRAL Would you choose to work a four-day workweek, ten hours a day, with every Friday off?

Writer's Checklist: Designing a Questionnaire

☑ Prepare a cover letter (memo or e-mail) explaining who you are, the questionnaire's purpose, the date by which you need a response, and how and where to send the completed questionnaire.

☑ Include a stamped, self-addressed envelope if you are using regular mail.

☑ Construct as many questions as possible for which the recipient does not have to compose an answer.

☑ Include a section on the questionnaire for additional comments, where the recipient may clarify his or her overall attitude toward the subject.

☑ State whether the information provided as well as the recipient's identity will be kept confidential.

☑ Include questions about the respondent's age, gender, education, occupation, and so on, only if such information will be of value in interpreting the answers.

☑ Include your contact information (mailing address, phone number, and e-mail address).

☑ Consider offering some tangible appreciation to those who answer the questionnaire by a specific date, such as a copy of the results or, for a product questionnaire, a gift certificate.

quid pro quo

Quid pro quo, which is Latin for "one thing for another," suggests mutual cooperation or "tit for tat" in a relationship between two groups or individuals. The term may be appropriate to business and legal contexts if you are sure your readers understand its meaning. (Before approving the plan, we insisted on a fair *quid pro quo*.) See also foreign words in English.

quotation marks

Quotation marks (" ") are used to enclose a direct quotation of spoken or written words. Quotation marks have other special uses, but they should not be used for emphasis.

Direct Quotations

Enclose in quotation marks anything that is quoted word for word (a direct quotation) from speech or written material.

- She said clearly, "I want the progress report by three o'clock."

Do not enclose indirect quotations—usually introduced by the word *that*—in quotation marks. Indirect quotations are paraphrases of a writer's or speaker's words or ideas. See also paraphrasing.

- She said that she wanted the progress report by three o'clock.

ETHICS NOTE When you use quotation marks to indicate that you are quoting, do not make any changes or omissions in the quoted material unless you clearly indicate what you have done. For further information on incorporating quoted material and inserting comments, see plagiarism and quotations. ✦

Use single quotation marks (' ') to enclose a quotation that appears within a quotation.

- John said, "Jane told me that she was going to 'stay with the project if it takes all year.' "

Words and Phrases

Use quotation marks to set off special words or terms only to point out that the term is used in context for a unique or special purpose (that is, in the sense of the term *so-called*).

- A remarkable chain of events caused the sinking of the "unsinkable" *Titanic* on its maiden voyage.

Slang, colloquial expressions, and attempts at humor, although infrequent in workplace writing, should seldom be set off by quotation marks.

- Our first six months amounted to a ~~"shakedown cruise."~~ shakedown cruise.

Titles of Works

Use quotation marks to enclose titles of reports, short stories, articles, essays, single episodes of radio and television programs, and short musical works (including songs). However, do not use quotation marks for titles of books and periodicals, which should appear in italics.

- His report, "Effects of Government Regulations on Motorcycle Safety," cited the article "No-Fault Insurance and Motorcycles," published in *American Motorcyclist* magazine.

Use quotation marks for parts of publications, such as chapters of books and articles or sections within periodicals.

- "Bad Writing" was an article by Barbara Wallraff in the "On Language" column of the *New York Times*.

Some titles, by convention, are not set off by quotation marks, underlining, or italics, although they are capitalized.

- Professional Writing [college course title], the Bible, the Constitution, Lincoln's Gettysburg Address, the Lands' End Catalog

Punctuation

Commas and periods always go inside closing quotation marks.

- "Reading *Computer World* gives me the insider's view," he says, adding, "It's like a conversation with the top experts."

Semicolons and colons always go outside closing quotation marks.

- He said, "I will pay the full amount"; this statement surprised us.

All other punctuation follows the logic of the context: If the punctuation is part of the material quoted, it goes inside the quotation marks; if the punctuation is not part of the material quoted, it goes outside the quotation marks.

quotations

DIRECTORY

Using direct and indirect quotations is an effective way to make or support a point. However, avoid the temptation to overquote during the note-taking phase of your research; concentrate on summarizing what you read.

ETHICS NOTE When you do use a quotation (or an idea of another writer), cite your source properly. If you do not, you will be guilty of plagiarism. For specific details on citation systems, see documenting sources. ✦

Direct Quotations

A direct quotation is a word-for-word copy of the text of an original source. Choose direct quotations (which can be of a word, a phrase, a sentence, or, occasionally, a paragraph) carefully and use them sparingly. Enclose direct quotations in quotation marks and separate them from the rest of the sentence by a comma or colon. Use the initial capital letter of a quotation if the quoted material originally began with a capital letter.

- The Dean of the Medical College stated, "We must attract more students to careers in medicine in order to meet the health-care needs of older Americans."

Q

When dividing a quotation, set off the material that interrupts the quotation with commas, and use quotation marks around each part of the quotation.

- "We must attract more students to careers in medicine," he said in a recent interview, "in order to meet the health-care needs of older Americans."

Indirect Quotations

An indirect quotation is a paraphrased version of an original text. It is usually introduced by the word *that* and is not set off from the rest of the sentence by punctuation marks. See also paraphrasing.

- In a recent interview he said that recruiting students to medical schools is essential to meeting the demands of older Americans.

Deletions or Omissions

Deletions or omissions from quoted material are indicated by three ellipsis dots (. . .) within a sentence and a period plus three ellipsis points (. . . .) at the end of a sentence.

- "If monopolies could be made to respond . . . we would be able to enjoy the benefits of . . . large-scale efficiency. . . ."

When a quoted passage begins in the middle of a sentence rather than at the beginning, ellipsis dots are not necessary; the fact that the first letter of the quoted material is not capitalized tells the reader that the quotation begins in midsentence.

- Rivero goes on to conclude that "the Research and Development Center is essential to keeping high-tech companies in the state."

Inserting Material into Quotations

When it is necessary to insert a clarifying comment within quoted material, use brackets.

- "The industry is an integrated system that serves an extensive [geographic] area, with divisions existing as islands within the larger system's sphere of influence."

When quoted material contains an obvious error or might be questioned in some other way, insert the expression *sic* (Latin for "thus"), in italic type and enclosed in brackets following the questionable material to indicate that the writer has quoted the material *exactly as it appeared in the original.*

- The company considers the Baker Foundation to be a "guilt-edged [*sic*] investment."

Incorporating Quotations into Text

Quote word for word only when your source concisely sums up a great deal of information or reinforces a point you are making. Quotations must also match logically, grammatically, and syntactically the rest of the sentence and surrounding text. Notice in Figure Q–2 that the quotation blends with the content of the surrounding text, which uses transition to introduce and comment on the quotation.

Depending on the citation system, the style of incorporating quotations varies. For examples of three different styles, see documenting

After reviewing a large number of works on technical communication, Alred sees differences in the focus of practitioners and educators:

> Some works, for example, are essential to the immediate needs of professional technical communicators who face specific and demanding workplace tasks under tight deadlines. Other works are crucial to educators who must be concerned with the long-range implications of research or theoretical insights as they prepare students for professional careers in technical communication. (ix–x)

The aims of practitioners and educators, however, are not as distinct as this statement by itself might suggest. For example, many practicing technical communicators, as they design Web pages or produce documentation, see themselves as educators. Specifically, in preparing material for users of systems, practitioners must . . .

FIGURE Q–2. Long Quotation (MLA Style)

sources. Figure Q–2 shows MLA style for a long quotation. At the end of the document, the following entry would appear in the MLA-style list of works cited as the source of the quotation in Figure Q–2.

- Alred, Gerald J. "Essential Works on Technical Communication." Technical Communication 50.4 (November 2003): 585–616.

Do not rely too heavily on the use of quotations in the final version of your document. Generally, avoid quoting anything that is longer than one paragraph.

R

raise / rise

Both *raise* and *rise* mean "move to a higher position." However, *raise* is a transitive verb and always takes an object (*raise* crops), whereas *rise* is an intransitive verb and never takes an object (heat *rises*).

re

Re (and its variant form, *in re*) is business and legal jargon meaning "in reference to" or "in the case of." Although *re* is sometimes used in memos and e-mails, *subject* is a preferable term.

readers

The first rule of effective writing is to *help your readers*. If you overlook this commitment, your writing will not achieve its purpose, either for you or for your business or organization.

Determining Your Readers' Needs

To help your readers, first determine their needs relative to your purpose and goals by asking key questions during preparation.

- Who specifically is your reader? Do you have multiple readers? Who needs to see or use the document?
- What do your readers already know about your subject? What are your readers' attitudes about the subject? (Skeptical? Supportive? Anxious? Bored?)

- What does the context suggest about the readers' expectations for content or layout and design?
- Do you need to adapt your message for international readers? If so, see global communication, global graphics, and international correspondence.

In the workplace, your readers are usually less familiar with the subject than you are. You have to be careful, therefore, when writing on a topic that is unique to your area of specialization. Be sensitive to the needs of those whose training or experience lies in other areas; provide definitions of nonstandard terms and explanations of principles that you, as a specialist, take for granted. Note that even if you write a trade journal article for others in your field, you should explain new or special uses of standard terms and principles. See also defining terms.

Writing to Individual Readers

When you write to an individual reader, empathize with that reader by considering the "you" viewpoint and writing in an appropriate tone. You may find it useful to visualize that person sitting across from you as you write. Likewise, when you write to a group of readers who are relatively homogeneous, you might create an image of a composite reader and write for *that* reader. You could also list that reader's characteristics (experience, training, attitudes, and work habits, for example) to help you write at the appropriate level. See also technical writing style.

Writing for Diverse Audiences

R

For documents aimed at multiple readers with different needs, consider segmenting the document for different groups of readers: an executive summary for top managers, an appendix with detailed data for technical specialists, and the body for those readers who need to make decisions based on the details. See also formal reports and proposals.

When you have multiple readers with various needs but cannot segment your document, first determine your primary or most important readers—such as those who will make decisions based on the document—and be sure to meet their needs. Then, meet the needs of secondary readers, such as those who need only some of the document's contents, as long as you do not sacrifice the needs of your primary readers. See also correspondence, e-mail, persuasion, and "Five Steps to Successful Writing."

really

Really is an adverb meaning "actually" or "in fact." Although both *really* and *actually* are often used as intensifiers for emphasis or sarcasm in speech, avoid such use in writing.

- Did he ~~really~~ finish the report on time?

reason is [because]

Replace the redundant phrase *the reason is because* with *the reason is that* or simply *because*. See also conciseness.

reference letters

Writing a reference letter (or letter of recommendation) can range from completing an admission form for a prospective student to composing a detailed description of professional accomplishments and personal characteristics for someone seeking employment. In Figure R–1, a former employer has written a letter for someone who is seeking an advanced position as a researcher.

To write an effective letter of recommendation, you must be familiar enough with the applicant's abilities and performance to offer an evaluation, and you must keep in mind the following.

- Communicate truthfully and without embellishment.
- Address specifically the applicant's skills, abilities, knowledge, and personal characteristics.
- Respond directly to the inquiry, carefully addressing the specific questions asked.
- Identify yourself by name, title or position, employer, and address.

You could begin, as in Figure R–1 on page 450, by stating the circumstances of your acquaintance and how long you have known the person for whom you are writing the letter. You should mention, with as much substantiation as possible, one or two outstanding characteristics of the applicant. Organize the details in your letter using the decreasing order-of-importance method of development. Conclude with a brief summary of the applicant's qualifications and a clear statement of recommendation.

ETHICS NOTE When you are asked to serve as a reference or to supply a letter of reference, be aware that applicants have a legal right

City of Springfield
Legislative Reference Bureau
200 East Main Street
Springfield, AK 99501
(414) 224-5555

January 17, 2006

Mr. Phillip Lester
Human Resources Director
Thompson Enterprises
201 State Street
St. Louis, MO 63102

Dear Mr. Lester:

How long writer has known applicant and the circumstances

As Kerry Hawkins's former employer, I am happy to have the opportunity to recommend her. I've known Kerry for five years, first as an intern in our office and for two years as a full-time research associate. This past year she has remained in contact as she completed graduate school.

Outstanding characteristics of applicant

Our office is the official research arm for the Springfield City Council, so we approved Kerry as an intern not only because of her outstanding grades but also because the university internship coordinator reported that she possessed excellent research skills. Kerry proved her worth in our office as an intern and was offered a full-time position as a research associate to work on projects under my supervision. I found Kerry not only to be a careful researcher but also to be able to complete her assignments on schedule within tight deadlines. The material provided in her well-written reports unfailingly met the requirements for my work and more. During her time in our office, Kerry also proved herself to be a most valued colleague.

Recommendation and summary of qualifications

I strongly recommend Kerry for her ability to work independently, to organize her time efficiently, and to write clearly and articulately. I regret that we have no position to offer her at this time. Please do not hesitate to let me know if I can provide further information.

Sincerely yours,

Michelle Paul

Michelle Paul, Manager
mpaul@springfield.ar.gov

FIGURE R–1. Reference Letter

to examine what you have written about them, unless they sign a waiver. See correspondence for letter format and general advice. ✦

refusal letters

A refusal letter delivers a negative message (or bad news) in the form of a letter, a memo, or an e-mail message. The ideal refusal letter says "no" in such a way that you not only avoid antagonizing your reader but also maintain goodwill.

When the stakes are high, you must convince your reader *before* you present the bad news that your reasons for refusing are logical or understandable. (See also correspondence.) Stating a negative message in your opening may cause your reader to react too quickly and dismiss your explanation. The following pattern, used in the message shown in Figure R-2 on page 452, is an effective way to handle this problem.

1. In the opening, provide a context (often called a "buffer").
2. Review the facts leading to the refusal or bad news.
3. Give the negative message based on the facts.
4. In the closing, establish or reestablish a positive relationship.

Your opening can establish a positive and professional tone, for example, by expressing appreciation for your reader's time, effort, or interest.

- The Screening Procedures Committee appreciates the time and effort you spent on your proposal for a new security-clearance procedure.

Next, review the circumstances of the situation sympathetically by placing yourself in the reader's position. Clearly establish the reasons you cannot do what the reader wants—even though you have not yet said you cannot do it. A good explanation, as shown in the following example, should detail the reasons for your refusal so thoroughly that the reader will accept the negative message as a logical conclusion.

R

- We reviewed the potential effects of implementing your proposed security-clearance procedure company-wide. We asked the Security Systems Department to review the data, surveyed industry practices, sought the views of senior management, and submitted the idea to our legal staff. As a result of this process, we have reached the following conclusions:
    - The cost savings you project are correct only if the procedure could be required universally.

From: Ted Richardson <tedr@crb.com>
To: Javier Lopez <lopez@tnco.com>
Sent: Friday, February 24, 2006 10:47 AM
Subject: Regional Meeting Speech

Context

Dear Mr. Lopez:

I am honored to have been invited to address your regional meeting in St. Louis on May 17. That you would consider me as a potential contributor to such a gathering of experts is indeed flattering.

Review of facts and refusal

On checking my schedule, I find that I will be attending the annual meeting of our parent corporation's Board of Directors on that date. Therefore, as much as I would enjoy addressing your members, I must decline.

Goodwill close

I have been very favorably impressed over the years with your organization's contributions to the engineering profession, and I would welcome the opportunity to participate in a future meeting.

Sincerely,

Ted Richardson

FIGURE R–2. Refusal Letter (Sent as E-mail)

- The components of your procedure are legal, but most are not widely accepted by our industry.
- Based on our survey, some components could alienate employees who would perceive them as violating an individual's rights.
- Enforcing company-wide use would prove costly and impractical.

R

Do not belabor the negative message—state your refusal quickly, clearly, and as positively as possible.

- For those reasons, the committee recommends that divisions continue their current security-screening procedures.

Close your message in a way that reestablishes goodwill—do not repeat the bad news (avoid writing "Again, we are sorry we cannot use your idea"). You might provide an option, offer a friendly remark, assure the reader of your high opinion of his or her product or service, or merely wish the reader success.

- Because some components of your procedure may apply in certain circumstances, we would like to feature your ideas in the next issue of *The Guardian*. I have asked the editor to contact you next week. On behalf of the committee, thank you for the thoughtful proposal.

For responding to a complaint letter, see adjustment letters. For refusing a job offer, see acceptance/refusal letters.

regarding / with regard to

In regards to and *with regards to* are incorrect idioms for *in regard to* and *with regard to*. Both *as regards* and *regarding* are acceptable variants.

- In ~~regards~~ *regard* to your last e-mail, I think a meeting is a good idea.
- ~~With regards to~~ *Regarding* your last e-mail, I think a meeting is a good idea.

regardless

Always use *regardless* instead of *irregardless*, which expresses a double negative and is nonstandard. The prefix *ir-* renders the base word negative, but *regardless* is already negative, meaning "unmindful."

repetition

The deliberate use of repetition to build a sustained effect or to emphasize a feeling or an idea can be a powerful device. See also emphasis.

- Similarly, atoms *come and go* in a molecule, but the molecule *remains*; molecules *come and go* in a cell, but the cell *remains*; cells *come and go* in a body, but the body *remains*; persons *come and go* in an organization, but the organization *remains*.

  —Kenneth Boulding, *Beyond Economics*

Repeating keywords from a previous sentence or paragraph can also be used effectively to achieve transition.

- For many years, *oil* has been a major industrial energy source. However, *oil* supplies are limited, and other sources of energy must be developed.

R

Be consistent in the word or phrase you use to refer to something. In technical writing, it is generally better to repeat a word (so there will be no question in the reader's mind that you mean the same thing) than to use synonyms to avoid repetition.

SYNONYMS	Several recent *analyses* support our conclusion. These *studies* cast doubt on the feasibility of long-range forecasting. The *reports*, however, are strictly theoretical.
CONSISTENT TERMS	Several recent *studies* support our conclusion. These *studies* cast doubt on the feasibility of long-range forecasting. They are, however, strictly theoretical.

Purposeless repetition, however, makes a sentence awkward and hides its key ideas. See also conciseness.

- She said that the customer ~~said that he~~ was canceling the order.

See also affectation.

reports

A report is an organized presentation of factual information, often aimed at multiple audiences, that may present the results of an investigation, a trip, or a research project. For any report—whether formal or informal—assessing the readers' needs is essential. Following is a list of report entries in this book:

Formal reports often present the results of long-term projects or those that involve multiple participants. (See also collaborative writing.) Such projects may be done either for your own organization or as a contractual requirement for another organization. Formal reports generally follow a precise format and include such elements as abstracts and executive summaries. See also proposals.

Informal and short reports normally run from a few paragraphs to a few pages and ordinarily include only an introduction, a body, a conclusion, and (if necessary) recommendations. Because of their brevity, informal reports are customarily written as correspondence: letters (if sent outside your organization) and memos or e-mails (if internal).

The introduction announces the subject of the report, states its purpose, and gives any essential background information. It may also summarize the conclusions, findings, or recommendations made in the report. The body presents a clearly organized account of the report's subject—the results of a test carried out, the status of a project, and so on. The amount of detail to include depends on the complexity of the subject and on your readers' familiarity with it.

The conclusion summarizes your findings and tells readers what you think their significance may be. In some reports, a final, separate section gives recommendations; in others, the conclusions and the recommendations sections are combined into one section. This final section makes suggestions for a course of action based on the data you have presented. See also persuasion.

requests for proposals

When a company or another organization needs a task performed or a service provided outside its staff expertise or capabilities, it seeks help from vendors that can provide the expertise or capabilities needed. The company seeking help usually makes its requirements known to potential vendors by issuing a request for proposals (RFP). The RFP details the company's requirements so that vendors can evaluate whether to bid on the project.

The details provided in an RFP permit vendors to understand your needs and enable you to determine from their responses whether to select them to meet your requirements. Although RFPs lay out your requirements, they usually do not specify how these requirements are to be met.* Potential vendors prepare proposals to describe their solutions for meeting your requirements; you then choose from among them the one that best suits your needs, budget, and schedule.

RFP Structure

Like proposals, RFPs vary in length, formality, and structure. Some are a page or two, and others run to hundreds of pages. The scope of information included and even the terminology used to describe specific sections vary with the type of task or service being solicited. RFPs involving computer systems, for example, usually include a "Technical Requirements" section, specifying site preparation requirements and

*A related document is an invitation for bids (IFB). An IFB strictly defines the quantity, type, and specifications for an item that an organization, such as a government agency, intends to purchase.

describing the current technical infrastructure at the company issuing the RFP. The purpose and context of RFPs also vary. Each RFP involves unique legal, budgetary, confidentiality, and administrative considerations. The following sections describe typical components for RFPs.

Information About Your Company. One section of an RFP should describe your company's mission, goals, size, facility locations, position in the marketplace, and possibly a brief company history. It should also provide your company contact information. This section parallels the "Description of Vendor" section in proposals (see Figure P–10, page 431).

Project Description. Another section, sometimes called "Scope of Work" or "Workscope Description," describes the deliverable (the project or service) you need, with a detailed list of requirements. For example, an RFP prepared by a pharmaceutical company soliciting training for its sales representatives about a new drug divided the project description into three sections:

- Target Audience (members of sales staff and their technical background and computer experience)
- Content (specific curriculum and length of training)
- Deliverable (in-person training, CD-ROM, or Web page, using video, audio, and animation)

Vendors responding to this RFP would organize their proposals in line with these requirements. For some projects, requirements can be described in a summary statement, as shown in the following examples.

- The project deliverable will be a questionnaire to be sent by mail to approximately 400 adults regarding their consumer experiences with various health-care providers. . . .
- The deliverable will be a networked online forms package for approximately 1,200 employees, permitting information to be filled out online, stored and sorted in a database, routed via e-mail, and printed on paper. The package shall be integrated into the company's current network environment at its headquarters facility. . . .

The project description may also indicate whether the deliverable must be created new or can be obtained commercially and adapted to the project's needs. For example, if the project involves software development, the description may state whether the requesting company prefers commercially available software or expects the vendor to develop unique software for the project. The project-description section can also specify other details such as a project-management plan and any warranty or liability requirements.

Delivery Schedule. For complex projects lasting a month or longer, specify the time allotted for the project and a proposed schedule of tasks.

Proposal Description. Many RFPs provide how-to guidance with format requirements for proposals. The proposal-description section usually describes how the proposal should be organized; the number of copies expected; whether an electronic file is required; where, the date by which, and to whom it should be sent; and how it should be sent (registered mail, courier service, etc.). Some companies even provide an electronic proposal template for vendors to follow.

Vendor Qualifications. Most RFPs request information about the vendor as well—a summary of the vendor's experience, professional certifications, number of employees, years in business, quality-control procedures, awards and honors, and the like. An RFP may require references of recent customers as well as financial references. RFPs usually require that vendors submit the résumés of the principal employees assigned to the project.

Proposal-Evaluation Criteria. The proposal-evaluation section informs vendors of the criteria you will use to select the company to work for you. Will you decide solely on cost? a combination of cost and vendor past performance? technical expertise? Many companies organize the criteria by importance or weight while others grade proposals with a point system and include a table of criteria, each of which is given a point value on a scale of 100 or 1,000 total points, as in the following sample.

- Proposal-Evaluation Criteria

CRITERIA	POINTS
Administrative	50
Technical	700
Management	100
Price	100
Presentation and demonstration	50
Total	**1,000**

R

Appendixes. Some RFPs include one or more appendixes or "Attachments." These may include sample forms and questionnaires, a sample contract, technical information about a company's server and workstation configurations, workflow-analysis diagrams, dates and times when vendors can visit the work site before finalizing their proposals, and other essentials too detailed for the body of the RFP.

ETHICS NOTE If your RFP contains any company-confidential information, you could include a legally binding nondisclosure statement for vendors to sign before you send the full RFP. Many companies also include a guarantee not to open proposals before the due date and never to disclose information in one vendor's proposal to a competing vendor. ✦

WEB LINK **SAMPLE REQUEST FOR PROPOSALS**

To view the RFP that accompanies the sample proposal in Figure P–10, see <*bedfordstmartins.com/alredtech*> and select *Model Documents Gallery*.

research

DIRECTORY

R

Research is the process of investigation—the discovery of facts. To be focused, research must be preceded by preparation, especially consideration of your readers, purpose, and scope. See "Five Steps to Successful Writing."

In an academic setting, your preparatory resources include conversations with your peers, instructors, and especially a research librarian.

On the job, your main resources are your own knowledge and experience and that of your colleagues. In many workplaces, sources of information may include test reports, laboratory reports, and the like. In this setting, begin by brainstorming with colleagues about what sources will be most useful to your topic and how you can find them.

Primary Research

Primary research is the gathering of raw data compiled from interviews, direct observation, surveys and questionnaires, experiments, recordings, and the like. In fact, direct observation and interaction are the only ways to obtain certain kinds of information, such as human and animal behavior, certain natural phenomena, and the operation of systems and equipment. You can also conduct primary research on the Internet by participating in discussion groups and newsgroups and by using e-mail to request information from specific audiences.

ETHICS NOTE If you are planning research that involves observation, choose your sites and times carefully, and be sure to obtain permission in advance. During your observations, remain as unobtrusive as possible, and keep accurate, complete records that indicate date, time of day, duration of the observation, and so on. Save interpretations of your observations for future analysis. Be aware that observation can be valuable research, but it may also be time-consuming, complicated, and expensive, and you may inadvertently influence the subjects you are observing. ✦

Secondary Research

Secondary research is the gathering of information that has been analyzed, assessed, evaluated, compiled, or otherwise organized into accessible form. Sources include books, articles, reports, Web documents, e-mail discussions, correspondence, minutes of meetings, operating manuals, brochures, and so forth. The following sections, Library Research Strategies (page 460) and Internet Research Strategies (page 464), provide methods for finding secondary sources.

As you seek information, keep in mind that in most cases the more recent the information, the better. Recently published periodicals and newspapers, as well as academic (.edu), organizational (.org), and government (.gov) Web sites, can be good sources of current information and can include interviews, articles, papers, and conference proceedings. See especially the *Writer's Checklist: Evaluating Print and Online Sources* on page 469.

When a resource seems useful, read it carefully and take notes that include any additional questions about your topic. Some of your questions may eventually be answered in other sources; those that remain unanswered can guide you to further primary research. For example, you may discover that you need to talk with an expert. Not only can someone skilled in a field answer many of your questions, but he or she can also suggest further sources of information. See documenting sources, interviewing for information, listening, note-taking, paraphrasing, and plagiarism.

Library Research Strategies

The library is a gateway to all types of sources—including Internet sources—that can help you find the best, most reliable information. The first step in using library resources, either in an academic institution or in a workplace, is to develop a search strategy appropriate to the information needed for your topic. You may want to begin by meeting with a research librarian who can help you quickly find the best print or online resources for your topic—a brief conversation can focus your research and save you time. In addition, use your library's homepage for access to its catalogs, databases of articles, subject directories to the Web, and more. A sample of a library's homepage is shown in Figure R–3.

Your search strategy depends on the kind of information you are seeking. For example, if you need the latest data offered by government research, check the Web, as described later in this entry. Likewise, if you need a current article on a topic, search an online database—such as InfoTrac—subscribed to by your library. For an overview of a subject, you might turn to an encyclopedia; for historical background, your best resources are books, journals, and primary documents.

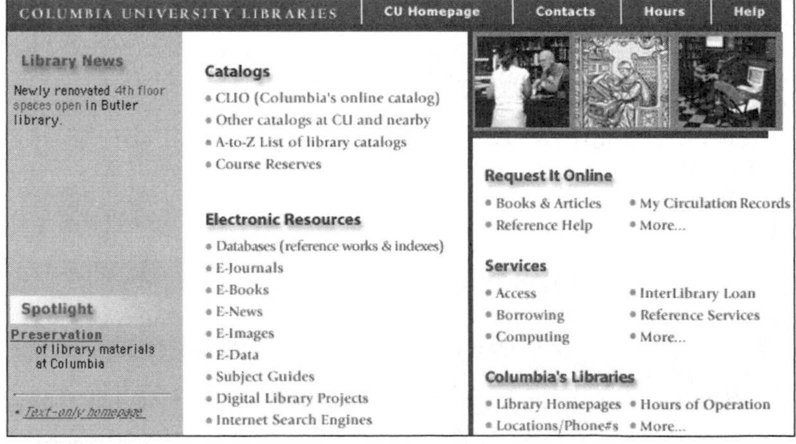

FIGURE R–3. Library Homepage

Online Catalogs (Locating Books). An online catalog—accessed through a library terminal and through the Internet—allows you to search a library's holdings, indicates an item's location and availability, and may allow you to arrange an interlibrary loan.

You can search a library's online catalog by author, title, keyword, or subject. The most typical ways of searching for a specific topic are by subject or by keyword. If your search turns up too many results, you can usually narrow it by using the "limit search" or "advanced search" option.

Online Databases and Indexes (Locating Articles). Most libraries subscribe to online databases, such as the following collections of online articles, many of which are available through a library's Web site.

- *InfoTrac*: a collection of databases of articles, many available in full text, which can also provide specialized databases in engineering, health, and other fields.
- *ProQuest*: a database of articles, many available in full text, which can also provide specialized databases for nursing, biology, and psychology.
- *EBSCOhost*: a database of articles, many available in full text, which can also provide specialized databases in a range of subjects.
- *FirstSearch*: a collection of specialized databases such as WorldCat (library collections) and ArticleFirst (articles, some in full text).
- *Lexis/Nexis Universe*: a collection of databases containing news, patent, legal, and congressional information (most in full text).

These databases, sometimes called *periodical indexes*, are excellent resources for articles published within the last ten to twenty years. Some include descriptive abstracts and full texts of articles. To find older articles, you may need to consult a print index, such as the *Readers' Guide to Periodical Literature* and the *New York Times Index*, or a reference librarian. (For more information on print indexes, see "General Guides" in the Reference Works section on page 463.)

R

To locate articles in a database, conduct a keyword search, as shown in Figure R–4 on page 462. If your search turns up too many results, refine your search by connecting two search terms with AND—"engineering AND employment"—or use other options offered by the database, such as a limited, a modified, or an advanced search.

Reference Works. In addition to articles, books, and Web sources, you may want to consult reference works such as encyclopedias, dictionaries, and atlases for a brief overview of your subject. Bibliographies,

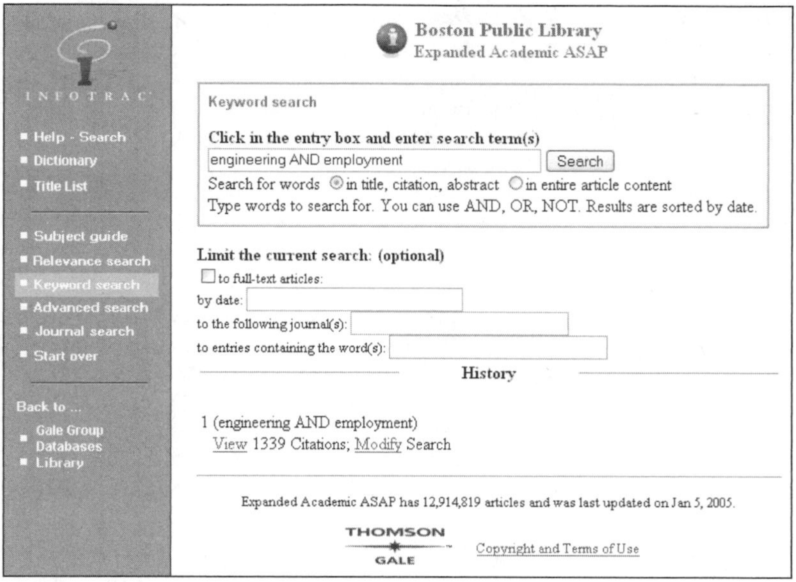

FIGURE R–4. InfoTrac Search Page

which are lists of works written about a topic, can direct you to more specialized sources. Ask your reference librarian to recommend reference works and bibliographies that are most relevant to your topic. Many are available in print, on CD-ROM, and online.

ENCYCLOPEDIAS. Encyclopedias are comprehensive, multivolume collections of articles arranged alphabetically. Some cover a wide range of subjects, while others, such as *The Encyclopedia of Careers and Vocational Guidance*, 12th ed., edited by Andrew Morkes (Chicago: Ferguson, 2002), focus on specific areas.

R

DICTIONARIES. General dictionaries can be compact or comprehensive, unabridged publications. Specialized dictionaries define terms used in a particular field, such as medicine, computers, architecture, or consumer affairs, and offer detailed definitions of field-specific terms, usually written in straightforward language.

HANDBOOKS AND MANUALS. Handbooks and manuals are typically one-volume compilations of frequently used information in a particular field. They offer brief definitions of terms or concepts, standards for presenting information, procedures for documenting sources, and visuals such as graphs and tables.

BIBLIOGRAPHIES. Bibliographies list books, periodicals, and other research materials published in areas such as engineering, medicine, the humanities, and the social sciences.

GENERAL GUIDES. The annotated *Guide to Reference Books*, 11th ed., by Robert Balay (Chicago: American Library Association, 1996), can help you locate reference books, indexes, and other research materials. The following are specialized indexes. Check your library's homepage or with your reference librarian to find out if your library subscribes to a particular index, and whether it is available online.

Applied Science and Technology Index, 1958–. Alphabetical subject listing; issued monthly.

Engineering Index, 1934–. Alphabetical subject listing containing brief abstracts; issued monthly.

Government Reports Announcements and Index, 1965–. Semimonthly index of reports, arranged by subject, author, and report number.

Index to the Times (London), 1790–. Monthly.

Monthly Catalog of U.S. Government Publications, 1895–. Unclassified publications of all federal agencies, listed by subject, author, and report number; issued monthly.

New York Times Index, 1851–. Alphabetical list of subjects covered in *New York Times* articles; issued bimonthly.

Readers' Guide to Periodical Literature, 1900–. Monthly index of about 200 general U.S. periodicals, arranged alphabetically by subject.

Safety Sciences Abstracts Journal, 1974–. Listing of literature on industrial and occupational safety, transportation, and environmental and medical safety; issued quarterly.

ATLASES. Atlases provide representations of the physical and political boundaries of countries, climate, population, or natural resources.

STATISTICAL SOURCES. Statistical sources are collections of numerical data. They are the best sources for such information as the U.S. gross domestic product, the consumer price index, or the demographic breakdown of the general population.

American Statistics Index. Washington, D.C.: Congressional Information Service, 1978–. Monthly, quarterly, and annual supplements.

United States Bureau of the Census. *Statistical Abstract of the United States*. Washington, D.C.: Government Printing Office, 1879–. Annual (<www.census.gov>).

Internet Research Strategies

The Internet provides access to a staggering amount of information. However, the information on the Web varies widely in its completeness and accuracy, and it can adversely affect the quality and accuracy of your completed project. You need to spend extra time and effort to critically evaluate Internet sources by following the advice in the *Writer's Checklist: Evaluating Print and Online Sources* on page 469.

As comprehensive as search engines and directories may seem, keep in mind that none is complete or objective. Most search sites are incomplete, carrying only a preselected range of content. Many, for example, do not index Adobe PDF files or Usenet newsgroups, and many cannot index databases and other non-HTML-based content. Further, search sites' ranking of the sites they believe will be relevant to you is based on a number of different strategies. Some sites base relevance on how high on the given page your search term appears, on the number of appearances of your term, or on the number of other sites that link to the page. Almost all major search sites now sell high rankings to the highest bidders, so your results may not highlight the pages most relevant to your search. Your best strategy is to research how your favorite search engines work; nearly all provide detailed methodologies on their help pages.

Search Engines. Search engines—like Google or AltaVista—are machine-generated databases created by "spiders" or "bots" that travel the Web looking for new or updated pages; search engines typically have databases of billions of pages. The following search engines are used widely on the Web:

AltaVista	<www.altavista.com>
Excite	<www.excite.com>
Google	<www.google.com>
Hotbot	<www.hotbot.lycos.com>
Lycos	<www.lycos.com>
WebCrawler	<www.webcrawler.com>
Yahoo!	<www.yahoo.com>

R

Search sites may use a hybrid approach. For example, some search engines, such as Northern Light, at <www.northernlight.com>, search not only the Web but also their own database of articles—content that is edited and compiled by staff librarians and not available elsewhere on the Web.

Many search engines allow advanced searches with options that provide more selective results. Figure R–5 shows an advanced search conducted on Google for graduate programs in technical writing—a search limited to results in English and to sites within the ".edu" do-

Google Advanced Search Advanced Search Tips | About Google

Find results	with **all** of the words	technical writing — 20 results
	with the **exact phrase**	graduate program — Google Search
	with **at least one** of the words	
	without the words	
Language	Return pages written in	English
File Format	Only — return results of the file format	any format
Date	Return web pages updated in the	past year
Numeric Range	Return web pages containing numbers between	and
Occurrences	Return results where my terms occur	in the title of the page
Domain	Only — return results from the site or domain	.edu *e.g. google.com, .org More info*
SafeSearch	No filtering — Filter using SafeSearch	

FIGURE R–5. Advanced Google Search

main. Although search engines vary in what and how they search, you can use some basic strategies, as in the *Writer's Checklist: Using Search Engines* that follows.

Writer's Checklist: Using Search Engines

- ☑ Enter words and phrases that are as specific to your topic as possible. For example, if you are looking for information about *nuclear power* and enter only the term *nuclear*, the search will also yield listings for *nuclear* family, *nuclear* medicine, and *nuclear* winter.
- ☑ Use Boolean operators (AND, OR, NOT) to narrow your search. For example, if you are searching for information on breast cancer and are finding references to nothing but prostate cancer, try "breast AND cancer NOT prostate."
- ☑ Consider conducting an advanced search (see Figure R–5).
- ☑ Check any search tips available at the engine you use. For example, some engines allow you to narrow your search by combining phrases with double quotation marks: "Usability testing" will return only pages that have the full compound phrase.
- ☑ Use a variety of search engines.
- ☑ Consider using a metasearch engine, such as Dogpile (<www.dogpile.com>) and Metacrawler (<www.metacrawler.com>), if you are interested in obtaining as many hits as possible. Be prepared to refine and narrow your results.

R

Web Directories. Directories are human generated, so they index fewer pages than search engines but offer subject directory trees to help organize their content, as shown in Figure R–6. In addition to the subject directories offered by many search engines, the following directories will help you to conduct selective, scholarly research on the Web:

Infomine	<http://infomine.ucr.edu>
The Internet Public Library	<www.ipl.org>
World Wide Web Virtual Library	<www.vlib.org>

The Web includes numerous directories and sites devoted to specific subject areas. Following are some suggested resources for researching a technology topic.

CIO's Resource Centers	<www.cio.com/research>
IEEE Spectrum Online	<www.spectrum.ieee.org>
National Science Foundation	<www.nsf.gov>
Yahoo's Science Directory	<http://dir.yahoo.com/Science/>
LSU Libraries Federal Agencies Directory	<www.lib.lsu.edu/gov/fedgov .html>
FedStats	<www.fedstats.gov>

Some sites combine search engines with directories; for example, Google operates both a standard search engine and directory and a special contributor-generated directory referred to as an "Open Directory" (<http://dmoz.org>).

The web organized by topic into categories.

Arts Movies, Music, Television,...	**Home** Consumers, Homeowners, Family,...	**Regional** Asia, Europe, North America,...
Business Industries, Finance, Jobs,...	**Kids and Teens** Computers, Entertainment, School,...	**Science** Biology, Psychology, Physics,...
Computers Hardware, Internet, Software,...	**News** Media, Newspapers, Current Events,...	**Shopping** Autos, Clothing, Gifts,...
Games Board, Roleplaying, Video,...	**Recreation** Food, Outdoors, Travel,...	**Society** Issues, People, Religion,...
Health Alternative, Fitness, Medicine,...	**Reference** Education, Libraries, Maps,...	**Sports** Basketball, Football, Soccer,...
World Deutsch, Español, Français, Italiano, Japanese, Korean, Nederlands, Polska, Svenska, ...		

FIGURE R–6. Google's Main Subject Directory

Evaluating Sources

The easiest way to ensure that information is valid is to obtain it from a reputable source. For Internet sources, be especially concerned about the validity of the information provided. Because anyone can publish on the Web, it is sometimes difficult to determine authorship of a document, and frequently a person's qualifications for speaking on a topic are absent or questionable. The Internet versions of established, reputable journals in medicine, management, engineering, computer software, and the like merit the same level of trust as the printed versions. However, as you move away from established, reputable sites, exercise more caution. Be especially wary of unmoderated discussion groups on Usenet and other public Web forums. Use the following domain abbreviations to help you determine an Internet site sponsor:

.aero	aerospace industry
.biz	business
.com	company or individual
.coop	business cooperative
.edu	college or university
.gov	federal government
.info	general use
.int	international
.mil	U.S. military
.museum	museum
.name	individual
.net	network provider
.org	nonprofit organization
.pro	professionals

In addition, keep in mind the following four criteria when evaluating Internet sources: authority, accuracy, bias, and currency.*

Authority. Because anyone can publish on the Web, it is sometimes difficult to determine authorship of a document, and frequently a person's qualifications for speaking on a topic are absent or questionable. If you do not recognize the author as well known and respected in the field, check the site's "About Us" page or mission statement if available. Following are some possible ways you can determine authority:

- The author's document was listed in or linked from a reliable source or document.

*From Leigh Ryan, *The Bedford Guide for Writing Tutors*, 3rd ed. (Boston: Bedford/St. Martin's, 2002).

- The document gives substantive biographical information about the author so that you can evaluate his or her credentials, or you can get this information by linking to another document.
- The author is referenced or mentioned positively by another author or organization whose authority you trust.

If the publisher or sponsor is an organization, you may generally assume that the document meets the standards and aims of the group. Consider also:

- The suitability of the organization to address this topic.
- Whether this organization or agency is recognized and respected in the field.
- The relationship of the author to the publisher or sponsor. (Does the document tell you something about the author's expertise or qualifications?)

Accuracy. Criteria for evaluating accuracy might include the following:

- Other sources that the document relies on are linked to or are included in a bibliography.
- Background information can be verified.
- Methodology is appropriate for the topic.
- With a research project, data that was gathered includes explanations of research methods and interpretations.
- The graphs and visuals are free of distortion.
- The site is modified or updated regularly.

Bias. To determine bias, consider how the context reveals the author's knowledge of the subject and his or her stance on the topic. Check the site for the following:

- The site identifies in some form the audience it targets.
- The site was developed by a recognized academic institution; government agency; or national, international, or commercial organization with an established reputation in the subject area.
- The author shows knowledge of theories, techniques, or schools of thought usually related to the topic.
- The author shows knowledge of related sources and attributes them properly.
- The author discusses the value and limitations of the approach, if it is new.

- The author acknowledges that the subject itself or his or her treatment of it is controversial, if you know that to be the case.

Currency. If currency is important, consider whether the document has a publication or "last updated" date or includes date of copyright, gives dates showing when information was gathered, or gives information about new material when appropriate.

Writer's Checklist: Evaluating Print and Online Sources

FOR ALL SOURCES

☑ Is the resource recent enough and relevant to your topic? Is it readily available?

☑ Who is the intended audience? The mainstream public? A small group of professionals?

☑ Who is the author(s)? Is the author(s) an authority on the subject?

☑ Does the author(s) provide enough supporting evidence and document sources so that you can verify the information's accuracy?

☑ Is the information presented in an objective, unbiased way? Are any biases made clear?

☑ Are opinions clearly labeled? Are viewpoints balanced, or are opposing opinions acknowledged?

☑ Are the language, tone, and style appropriate and compelling?

FOR A BOOK

☑ Does the preface or introduction indicate the author's or book's purpose?

☑ Does the table of contents relate to your topic? Does the index contain terms related to your topic?

☑ Are the chapters useful? Skim through one that seems related to your topic—notice especially the introduction, headings, and closing.

FOR AN ARTICLE

☑ Is the publisher of the magazine or other periodical well known?

☑ What is the article's purpose? For a journal article, read the abstract; for a newspaper article, read the headline and lead sentences.

☑ Does the article contain informative diagrams or other visuals that indicate its scope?

FOR A WEB SITE

☑ Does a reputable group or organization sponsor or maintain the site?

Writer's Checklist: Evaluating Print and Online Sources (continued)

- ☑ Are the purpose and scope of the site clearly stated? Check the "Mission Statement" or "About Us" pages. Are there any disclaimers?
- ☑ Is the site updated, thus current? Are the links functional and up to date?
- ☑ Is the documentation authoritative and credible? Check the links to other sources and cross-check facts at other reputable Web sites, such as academic ones.
- ☑ Is the site well designed? Is the material well written and error-free?

WEB LINK EVALUATING ONLINE SOURCES

For a tutorial on evaluating information online, see <*bedfordstmartins.com/alredtech*> and select *Tutorials*, "Evaluating Online Sources." For links to additional related resources, select *Links for Handbook Entries*.

resignation letters

Resignation letters (or memos) should be as positive as possible, regardless of the reason you are leaving a job. You usually write a resignation letter to your supervisor or to an appropriate person in the Human Resources Department. Use the following guidelines.

- Start on a positive note, regardless of the circumstances under which you are leaving.
- Consider pointing out how you have benefited from working for the company or say something complimentary about the company.
- Comment on something positive about the people with whom you have been associated.
- Explain why you are leaving in an objective, factual tone.
- Avoid angry recriminations, because your resignation will remain on file with the company and could haunt you in the future when you need references.

R

Your letter or memo should give enough notice to allow your employer time to find a replacement. It might be no more than two weeks, or it might be enough time to enable you to train your replacement. Some organizations may ask for a notice equivalent to the number of weeks of vacation you receive. Check the policy of your employer before you begin your letter.

The sample resignation letter in Figure R–7 is from an employee who is leaving to take a job offering greater opportunities. The memo of resignation in Figure R–8 on page 472 is written by an employee who is leaving because her position has been reclassified and her supervisor has not supported her advancement, although no personal conflict is mentioned. Notice that it opens and closes positively and that the reason for the resignation is stated without apparent anger or bitterness. For strategies concerning negative messages, see correspondence and refusal letters.

227 Kenwood Drive
Austin, TX 78719
January 6, 2006

R.W. Johnson, Director of Metallurgy
Hannibal Laboratories
1914 East 6th Street
Austin, TX 78702

Dear Mr. Johnson:

Positive opening

My three years at Hannibal Laboratories have been an invaluable period of learning and professional development. I arrived as a novice, and I believe that today I am a professional—primarily as a result of the personal attention and tutoring I have received from my superiors and the fine example set by both my superiors and my peers.

Reason for leaving

I believe, however, that the time has come for me to move on to a larger company that can give me an opportunity to continue my professional development. Therefore, I have accepted a position with Procter & Gamble, where I am scheduled to begin on January 23. Thus, my last day at Hannibal will be January 20. I will be happy to train my replacement during the next two weeks.

Positive closing

Many thanks for the experience I have gained and best wishes for the future.

Sincerely,

J. L. Washburne

J. L. Washburne

FIGURE R–7. Resignation Letter (to Accept a Better Position)

R

MEMORANDUM

To: T. W. Haney, Vice President, Administration
From: L. R. Rupp *LRR*
Date: February 13, 2006
Subject: Resignation from Winterhaven, effective March 3, 2006

Positive opening — My five-year stay with the Winterhaven Company has been a pleasant experience, and I believe that it has been mutually beneficial.

Reason for leaving — Because of the recent restructuring of my job, I have accepted a position with another company that I feel will offer me greater advancement opportunities. I am, therefore, submitting my resignation, to be effective on March 3, 2006.

Positive closing — I have enjoyed working with my coworkers at Winterhaven and wish the company success in the future.

FIGURE R–8. Resignation Memo (Under Negative Conditions)

respective / respectively

Respective is an adjective that means "pertaining to two or more things regarded individually." (The committee members returned to their *respective* offices.) *Respectively* is the adverb form of *respective*, meaning "singly, in the order designated."

- The first, second, and third engineering design award winners were Maria Juarez, Dan Wesp, and Simone Luce, *respectively*.

Respective and *respectively* are unnecessary if the meaning of individuality is clear.

- ~~The~~ *Each* committee ~~members~~ *member* prepared ~~their respective reports.~~ *a report.*

R

restrictive and nonrestrictive elements

Modifying phrases and clauses may be either restrictive or nonrestrictive. A *nonrestrictive phrase or clause* provides additional information about what it modifies, but it does not restrict the meaning of what it modifies. The nonrestrictive phrase or clause can be removed without changing the essential meaning of the sentence. It is a parenthetical element that is set off by commas to show its loose relationship with the rest of the sentence.

NONRESTRICTIVE	This instrument, *which is called a backscatter gauge*, fires beta particles at an object and counts the particles that bounce back.

A *restrictive phrase or clause* limits, or restricts, the meaning of what it modifies. If it were removed, the essential meaning of the sentence would change. Because a restrictive phrase or clause is essential to the meaning of the sentence, it is never set off by commas.

RESTRICTIVE	All employees *wishing to donate blood* may take Thursday afternoon off.

Writers need to distinguish between nonrestrictive and restrictive elements. The same sentence can take on two entirely different meanings, depending on whether a modifying element is set off by commas (because it is nonrestrictive) or not (because it is restrictive). A slip by the writer can not only mislead readers but also embarrass the writer.

MISLEADING	He gave a poor performance evaluation to the staff members who protested to the Human Resources Department. [This suggests that he gave the poor evaluation because the staff members had protested.]
ACCURATE	He gave a poor performance evaluation to the staff members, who protested to the Human Resources Department. [This suggests that the staff members protested because of the poor evaluation.]

Use *which* to introduce nonrestrictive clauses and *that* to introduce restrictive clauses.

NONRESTRICTIVE	After John left the restaurant, *which* is one of the finest in New York, he came directly to my office.
RESTRICTIVE	Companies *that* diversify usually succeed.

R

résumés

DIRECTORY

A résumé, the key tool of the job search, itemizes the qualifications that you summarize in your application letter (sometimes referred to as a *cover letter*). A résumé* should be limited to one page—or two pages if you have substantial experience. On the basis of the information in the résumé and application letter, prospective employers decide whether to ask you to come in for an interview. If you are invited to an interview, the interviewer can base specific questions on the contents of your résumé. See also interviewing for a job.

Because résumés affect a potential employer's first impression, make sure that yours is well organized, carefully designed, consistently formatted, easy to read, and free of errors. Consider first an organization that highlights your strengths and fits your goals, as suggested by the examples shown in this entry. Experiment to determine a layout and design that is attractive and uncluttered. Consistency is especially important on a résumé. Be sure to use, for example, the same date formats (5/2006 or May 2006), punctuation, and spacing throughout. Proofreading is essential. Verify the accuracy of the information and have someone else review it. Use a quality printer and high-grade paper.

R

Sample Résumés

The sample résumés in this entry are provided to stimulate your thinking about how to tailor your résumé to your own job search. Before you design and write your résumé, look at as many samples as possible, and

*A detailed résumé for someone in an academic and scientific area is often called a *curriculum vitae* (also *vita* or *c.v.*). It may include education, publications, projects, grants, and awards as well as a full work history. Outside the United States, the term *curriculum vitae* is often a synonym for the term *résumé*.

then organize and format your own to best suit your previous experience and your professional goals and to make the most persuasive case to your target employers. See also persuasion.

- Figure R–9 presents a conventional student résumé in which the student is seeking an entry-level position.
- Figure R–10 shows a résumé with a variation of the conventional headings to highlight professional credentials.
- Figure R–11 presents a student résumé with a format that is appropriately nonconventional because this student needs to demonstrate skills in graphic design for his potential audience.
- Figure R–12 shows a résumé that focuses on the applicant's management experience.
- Figure R–13 focuses on how the applicant advanced and was promoted within a single company.
- Figure R–14 illustrates how an applicant can organize a résumé by combining functional and chronological elements.
- Figure R–15 presents an electronic résumé in ASCII (American Standard Code for Information Interchange) format. Notice how this résumé emphasizes keywords so that potential employers searching online for applicants will be able to find it easily.

Analyzing Your Background

In preparing to write your résumé, determine what kind of job you are seeking. Then ask yourself what information about you and your background would be most important to a prospective employer. List the following:

- Schools you attended, degrees you hold, your major field of study, academic honors you were awarded, your grade point average, particular academic projects that reflect your best work
- Jobs you have held, your principal and secondary duties in each job, when and how long you held each job, promotions, skills you developed in your jobs that potential employers value and seek in ideal job candidates, projects or accomplishments that reflect your important contributions
- Other experiences and skills you have developed that would be of value in the kind of job you are seeking; extracurricular activities that have contributed to your learning experience; leadership, interpersonal, and communication skills you have developed; any collaborative work you have performed; computer skills you have acquired

Ana María López

School
148 University Drive
Bloomington, Indiana 47405
(812) 652-4781
aml@iu.edu

Home (after June 2006)
1436 West Schantz Avenue
Laurel, Pennsylvania 17322
(717) 399-2712
aml@yahoo.com

OBJECTIVE

Position as dental hygienist, with long-term goal of developing a community practice dental service.

EDUCATION

Bachelor of Science in Dental Hygiene, expected June 2006
Indiana University

Licensure: August 2004
Grade Point Average: 3.88 out of possible 4.0
Senior Honor Society
Minor: Management Information Systems

DENTAL EXPERIENCE

Northpoint Dental Associates, Bloomington, Indiana, 2004
Dental Assistant, Summer and Fall Quarters
Developed office and laboratory management system.

Rodriguez Dental Associates, Bloomington, Indiana, 2003
Dental Assistant Intern
Prepared patients for exams; processed X-rays; maintained patient treatment records.
Associate Editor, *Community Health Newsletter*, 2002–2003
Wrote articles on good dental health practices; researched community health needs for editor; edited submissions.

COMPUTER SKILLS

Software: Microsoft Word, Excel, PowerPoint, Lotus, MIS SX2
Hardware: Macintosh, IBM-PC, UNIX Host Systems
Medical: Magnus Patient Database System

REFERENCES

Available upon request.

FIGURE R–9. Student Résumé (for an Entry-Level Position)

CHRIS RENAULT, RN

3785 Raleigh Court, #46 • Phoenix, AZ 67903 •
(555) 467-1115 • chris@resumepower.com

Qualifications

- ➢ *Recent Honors Graduate of Approved Nursing Program*
- ➢ *Current Arizona Nursing Licensure and BLS Certification*
- ➢ *Presently Completing Clinical Nurse Internship Program*

Education & Licensure

ARIZONA STATE UNIVERSITY
Tempe, AZ
Bachelor of Science in Nursing (BSN), 2006
Graduated summa cum laude (GPA: 4.0)

MOHAVE COMMUNITY COLLEGE
Kingman, AZ
Associate Degree in Nursing (AN), 2004
Graduated cum laude (GPA 3.5)

Coursework Highlights: Family and Community Nursing, Health-Care Delivery Models, Health Assessment, Pathology, Microbiology, Nursing Research, Nursing of Older Adults, Health-Care Ethics

Arizona RN License, 2006
BLS Certification, 2006

Clinical Internship

CAMELBACK MEDICAL CENTER – Phoenix, AZ
Nurse Intern, 2005 to Present

- Accepted into new graduate RN training program and completing in-depth, eight-month rotation working under a trained preceptor.
- Gaining valuable clinical experience to assume the role of a professional nurse within an acute-care setting. Rotating through all medical center areas, including Post Surgical, Orthopedics, Pediatrics, Oncology, Emergency Department, Psychiatric Nursing, Cardiac Telemetry, and Critical Care.
- Developing speed and skill in the day-to-day functions of a staff nurse. Participating in patient assessment, treatment, medication disbursement, and surgical preparation as a member of the health-care team.
- Earned written commendations from preceptor for *". . . excellent ability to interact with patients and their families, showing a high degree of empathy, medical knowledge, and concern for quality and continuity of patient care."*

Community Involvement

Active Volunteer and Fundraising Coordinator, The American Cancer Society – Scottsdale, AZ, Chapter (2004 to Present)
Participant, Annual AIDS Walkathon (2001 to 2004) and "Find the Cure" Breast Cancer Awareness Marathon (2003, 2004)

FIGURE R–10. Résumé (Highlighting Professional Credentials). Prepared by Kim Isaacs, Advanced Career Systems, Inc.

Joshua S. Goodman
222 Morewood Avenue
Pittsburgh, PA 15212
Jgoodman@aol.com

OBJECTIVE

A position as a graphic designer with responsibilities in information design, packaging, and media presentations.

GRAPHIC DESIGN EXPERIENCE

Assistant Designer • Dyer/Khan, Los Angeles, California
Summer 2003, Summer 2004
Assistant Designer in a versatile design studio. Responsible for design, layout, comps, mechanicals, and project management.
Clients: Paramount Pictures, Mattel Electronics, and Motown Records.

Photo Editor • Paramount Pictures Corporation, Los Angeles, California
Summer 2002
Photo Editor for merchandising department. Established art files for movie and television properties. Edited images used in merchandising. Maintained archive and database.

Production Assistant • Grafis, Los Angeles, California
Summer 2001
Production assistant at fast-paced design firm. Assisted with comps, mechanicals, and miscellaneous studio work.
Clients: ABC Television, A&M Records, and Ortho Products Division.

EDUCATION

Carnegie Mellon University, Pittsburgh, Pennsylvania
BFA in Graphic Design — May 2005.

Graphic Design
Corporate Identity
Industrial Design
Graphic Imaging Processes
Color Theory
Computer Graphics
Typography
Serigraphy
Photography
Video Production

COMPUTER SKILLS

XML, HTML, JavaScript, Forms, Macromedia Dreamweaver 3, Macromedia Flash 4, Photoshop 5.5, Image Ready (Animated GIFs), CorelDRAW, DeepPaint, iGrafx Designer, MapEdit (Image Mapping), Scanning, Microsoft Access/Excel, QuarkXPress.

ACTIVITIES

Member, Pittsburgh Graphic Design Society; Member, The Design Group.

FIGURE R–11. Student Résumé (for a Graphic Design Job)

R

ROBERT MANDILLO
7761 Shalamar Drive
Dayton, Ohio 45424
(513) 255-4137
mand@juno.com

OBJECTIVE

A management position in the aerospace industry with responsibility for developing new designs and products.

MANAGEMENT EXPERIENCE

MANAGER, ENGINEERING DRAFTING DEPARTMENT — May 1998–Present
Wright-Patterson Air Force Base, Dayton, Ohio

Supervise 17 drafting mechanics in support of the engineering design staff. Develop, evaluate, and improve materials and equipment for the design and construction of exhibits. Write specifications, negotiate with vendors, and initiate procurement activities for exhibit design support.

SUPERVISOR, GRAPHICS ILLUSTRATORS—June 1985–April 1998
Henderson Advertising Agency, Cincinnati, Ohio

Supervised five illustrators and four drafting mechanics after promotion from Graphics Technician. Analyzed and approved work-order requirements. Selected appropriate media and techniques for orders. Rendered illustrations in pencil and ink. Converted department to CAD system.

EDUCATION

BACHELOR OF SCIENCE IN MECHANICAL ENGINEERING TECHNOLOGY, 1985
Edison State College, Wooster, Ohio

ASSOCIATE'S DEGREE IN MECHANICAL DRAFTING, 1983
Wooster Community College, Wooster, Ohio

PROFESSIONAL AFFILIATION

National Association of Mechanical Engineers and Drafting Mechanics

REFERENCES

References, letters of recommendation, and a portfolio of original designs and drawings available online at <www.juno.com/mand>.

FIGURE R–12. Résumé (Applicant with Management Experience)

CAROL ANN WALKER
1436 West Schantz Avenue
Laurel, Pennsylvania 17322
(717) 399-2712
caw@yahoo.com

FINANCIAL EXPERIENCE

KERFHEIMER CORPORATION, Philadelphia, Pennsylvania

Senior Financial Analyst, June 2001–Present
Report to Senior Vice President for Corporate Financial Planning. Develop manufacturing cost estimates totaling $30 million annually for mining and construction equipment with Department of Defense.

Financial Analyst, November 1998–June 2001
Developed $50-million funding estimates for major Department of Defense contracts for troop carriers and digging and earth-moving machines. Researched funding options, resulting in savings of $1.2 million.

FIRST BANK, INC., Bloomington, Indiana

Planning Analyst, September 1993–November 1998
Developed successful computer models for short- and long-range planning.

EDUCATION

Ph.D. in Finance: expected, June 2006
The Wharton School of the University of Pennsylvania

M.S. in Business Administration, 1997
University of Wisconsin–Milwaukee
"Executive Curriculum" for employees identified as promising by their employers.

B.S. in Business Administration (*magna cum laude*), 1993
Indiana University
Emphasis: Finance Minor: Professional Writing

PUBLISHING AND MEMBERSHIP

Published "Developing Computer Models for Financial Planning," *Midwest Finance Journal* 34.2 (2003): 126–36.

Association for Corporate Financial Planning, Senior Member.

REFERENCES

References and a portfolio of financial plans are available upon request.

FIGURE R–13. Advanced Résumé (Showing Promotion Within a Single Company)

CAROL ANN WALKER

1436 West Schantz Avenue • Laurel, PA 17322
(717) 399-2712 • caw@yahoo.com

Award-Winning Senior Financial Analyst

Astute senior analyst and corporate financial planner with 11 years of experience and proven success enhancing P&L scenarios by millions of dollars. Demonstrated ability to apply critical thinking and sound strategic/economic analysis to multidimensional business issues. Advanced computer skills include Hyperion, SQL, MS Office, and Crystal Reports.

Financial Analyst of the Year, 2004

Recipient of prestigious national award from the Association for Investment Management and Research (AIMR)

Areas of Expertise

- Financial Analysis & Planning
- Forecasting & Trend Projection
- Trend/Variance Analysis
- Comparative Analysis
- Expense Analysis
- Strategic Planning
- SEC & Financial Reporting
- Risk Assessment

Career Progression

KERFHEIMER CORPORATION–Philadelphia, PA 1998 to Present

Senior Financial Analyst, June 2001 to Present
Financial Analyst, November 1998 to June 2001

Rapidly promoted to lead team of 15 analysts in the management of financial/SEC reporting and analysis for publicly traded, $2.3-billion company. Develop financial/statistical models used to project and maximize corporate financial performance. Support nationwide sales team by providing financial metrics, trends, and forecasts.

Key Accomplishments:

- **Developed long-range funding requirements crucial to firm's subsequent capture of $1 billion** in government and military contracts.
- **Facilitated a 45% decrease in company's long-term debt** during several major building expansions through personally developed computer models for capital acquisition.
- **Jointly led large-scale systems conversion to Hyperion**, including personal upload of database in Essbase. Completed conversion without interrupting business operations.

FIGURE R–14. Advanced Résumé (Combining Functional and Chronological Elements). Prepared by Kim Isaacs, Advanced Career Systems, Inc.

CAROL ANN WALKER

Résumé • Page Two

Career Progression (*continued*)

FIRST BANK, INC.–Bloomington, IN 1993 to 1998

Planning Analyst, September 1993 to November 1998
Compiled and distributed weekly, monthly, quarterly, and annual closings/financial reports, analyzing information for presentation to senior management. Prepared depreciation forecasts, actual-vs.-projected financial statements, key-matrix reports, tax-reporting packages, auditor packages, and balance-sheet reviews.

Key Accomplishments:

- **Devised strategies to acquire over $1 billion at 3% below market rate.**
- **Analyzed financial performance for consistency to plans and forecasts**, investigated trends and variances, and alerted senior management to areas requiring action.
- **Achieved an average 23% return on all personally recommended investments.** Applied critical thinking and sound financial and strategic analysis in all funding options research.

Education

THE WHARTON SCHOOL of the UNIVERSITY OF PENNSYLVANIA–Philadelphia, PA

Ph.D. in Finance Candidate, Expected June 2006

UNIVERSITY OF WISCONSIN–Milwaukee, WI

M.S. in Business Administration, May 1997

INDIANA UNIVERSITY–Bloomington, IN

B.S. in Business Administration, Emphasis in Finance
(*magna cum laude*), May 1993

R

Affiliations

- Association for Investment Management and Research (AIMR), Member, 2000 to Present
- Association for Corporate Financial Planning (ACFP), Senior Member, 1998 to Present

Portfolio of Financial Plans Available on Request
(717) 399-2712 • caw@yahoo.com

FIGURE R–14. Advanced Résumé (Combining Functional and Chronological Elements). Prepared by Kim Isaacs, Advanced Career Systems, Inc. (*continued*)

DAVID B. EDWARDS
6819 Locustview Drive
Topeka, Kansas 66614
(913) 233-1552
dedwards@cpu.fairview.edu

JOB OBJECTIVE
Programmer with writing, editing, and training responsibilities, leading to a career in information design management.

KEYWORDS
Programmer, Operating Systems, Unipro, Newsletter, Graphics, Listserv, Professional Writer, Editor, Trainer, Instructor, Technical Writer, Tutor, Designer, Manager, Information Design.

EDUCATION
** Fairview Community College, Topeka, Kansas
** Associate's Degree, Computer Science, June 2003
** Dean's Honor List Award (six quarters)

RELEVANT COURSE WORK
** Operating Systems Design
** Database Management
** Introduction to Cybernetics
** Technical Writing

EMPLOYMENT EXPERIENCE
** Computer Consultant: September 2003 to Present
Fairview Community College Computer Center: Advised and trained novice users; wrote and maintained Unipro operating system documentation.
** Tutor: January 2002 to June 2003
Fairview Community College: Assisted students in mathematics and computer programming.

SKILLS AND ACTIVITIES
** Unipro Operating System: Thorough knowledge of word-processing, text-editing, and file-formatting programs.
** Writing and Editing Skills: Experience in documenting computer programs for beginning programmers and users.
** Fairview Community Microcomputer Users Group: Cofounder and editor of monthly newsletter ("Compuclub"); listserv manager.

FURTHER INFORMATION
** References, college transcripts, a portfolio of computer programs, and writing samples available upon request.

FIGURE R–15. Electronic Résumé (in ASCII Format)

Use this information to brainstorm any further key details. Then, based on all the details, decide which to include in your résumé and how you can most effectively present your qualifications.

Organizing Your Résumé

A number of different organizational patterns can be used effectively. The following categories are typical—which you choose should depend on your experience and goals, the employer's needs, and any standard practices in your profession.

- Heading (name and contact information)
- Job Objective
- Qualifications Summary
- Education
- Employment Experience
- Related Skills and Abilities
- Honors and Activities
- References

Whether you place education or employment experience first depends on the job you are seeking and on which credentials would strengthen your résumé the most. If you are a recent graduate without much work experience, you would list education first. If you have years of job experience, including jobs directly related to the kind of position you are seeking, you would list employment experience first. In your education and employment sections, use a reverse chronological sequence: List the most recent experience first, the next most recent experience second, and so on.

R

Heading. At the top of your résumé, include your name, address, telephone number (home or cell), and e-mail address. Make sure that your name stands out on the page. If you have both a school address and a permanent home address, place your school address on the left side of the page and your permanent home address on the right side of the page. Place both underneath your name, as shown in Figure R–9. Indicate the dates you can be reached at each address (but do not date the résumé itself).

Job Objective. Some potential employers prefer to see a clear employment objective in résumés. An objective introduces the material in a résumé and helps the reader quickly understand your goal. If you decide to include an objective, use a heading such as "Objective," "Employment Objective," "Career Objective," or "Job Objective." State your immediate goal and, if you know that it will give you an advantage, the direction you hope your career will take. Try to write your objective in no more than three lines, and tailor it to the specific job for which you are applying, as illustrated in the following examples.

- A full-time computer-science position aimed at solving engineering problems and contributing to a management team.
- A position involving meeting the concerns of women, such as family planning, career counseling, or crisis management.
- Full-time management of a high-quality local restaurant.
- A summer research or programming position providing opportunities to use problem-solving skills.

Qualifications Summary. You may wish to include a brief summary of your qualifications to persuade hiring managers to select you for an interview. Sometimes called a *summary statement* or *career summary*, a qualifications summary can include skills, achievements, experience, or personal qualities that make you especially well suited to the position. You may wish to give this section a heading such as "Profile," "Career Highlights," or simply "Qualifications." Or you may use a headline, as in Figure R–14 ("Award-Winning Senior Financial Analyst").

Education. List the school(s) you have attended, the degrees you received and the dates you received them, your major field(s) of study, and any academic honors you have earned. Include your grade point average only if it is 3.0 or higher—or include your average in your major if that is more impressive. List courses only if they are unusually impressive or if your résumé is otherwise sparse (see Figures R–10 and R–11). Mention your high school only if you want to call attention to special high-school achievements, awards, projects, programs, internships, or study abroad.

R

Employment Experience. Organize your employment experience in reverse chronological order, starting with your most recent job and working backward under a single major heading called "Experience," "Employment," "Professional Experience," or the like. You could also organize your experience functionally by clustering similar types of jobs into one or several sections with specific headings such as "Management Experience" or "Major Accomplishments."

One type of arrangement might be more persuasive than the other, depending on the situation. For example, if you are applying for an accounting job but have no specific background in accounting, you would probably do best to list past and present jobs in chronological order, from most to least recent. If you are applying for a supervisory position and have had three supervisory jobs in addition to two nonsupervisory positions, you might choose to create a single section called "Supervisory Experience" and list only your three supervisory jobs. Or you could create two sections—"Supervisory Experience" and "Other Experience"—and include the three supervisory jobs in the first section and your nonsupervisory jobs in the second section.

The functional résumé groups work experience by types of workplace activities or skills rather than by jobs in chronological order. However, many employers are suspicious of functional résumés because they can be used to hide a poor work history, such as excessive job hopping or extended employment gaps. Functional elements can be combined with a chronological arrangement by using a qualifications summary or skills category, as shown in Figure R–14.

In general, follow these conventions when working on the "Experience" section of your résumé.

- Include jobs or internships when they relate directly to the position you are seeking. Although some applicants choose to omit internships and temporary or part-time jobs, including such experiences can make a résumé more persuasive if they have helped you develop specific related skills.
- Include extracurricular experiences, such as taking on a leadership position in a college organization or directing a community-service project, if they demonstrate that you have developed skills valued by potential employers.
- List military service as a job; give the dates served, the duty specialty, and the rank at discharge. Discuss military duties if they relate to the job you are seeking.
- For each job or experience, list both the job and company titles. Throughout each section, consistently begin with either the job or the company title, depending on which will likely be more impressive to potential employers.
- Under each job or experience, provide a concise description of your primary and secondary duties. If a job is not directly relevant, provide only a job title and a brief description of duties that helped you develop skills valued in the position you are seeking. For example, if you were a lifeguard and now seek a management position, focus on supervisory experience or even experience in averting disaster to highlight your management, decision-making, and crisis-control skills.
- Use action verbs (for example, "managed" rather than "as the manager") and state ideas succinctly, as shown in Figure R–12. Even though the résumé is about you, do not use "I" (for example,

instead of "I was promoted to Section Leader," use "Promoted to Section Leader").

- Focus as much as possible on your achievements in your work history (Increased employee retention rate by 16% by developing a training program). Employers want to hire doers and achievers.

Related Skills and Abilities. Employers are interested in hiring applicants with a variety of skills or the ability to learn new ones fairly quickly. Depending on the position, you might list in a skills section items such as fluency in foreign languages, writing and editing abilities, specialized technical knowledge, or computer skills, including knowledge of specific languages, software, and hardware.

Honors and Activities. If you have room on your résumé, list any honors and unique activities near the end. Include items such as student or community activities, professional or club memberships, awards received, and published works. Be selective: Do not duplicate information given in other categories, and include only information that supports your employment objective. Provide a heading for this section that fits its contents, such as "Activities," "Honors," "Professional Affiliations," or "Publications and Memberships."

References. Avoid listing references unless that is standard practice in your profession or your résumé is sparse. You might include a phrase such as "References available upon request" to signal the end of a long résumé, or write "Available upon request" after the heading "References" as a design element to balance a page. In any case, you should have a separate list of references to give to prospective employers after interviews; your list should include the main heading "References for [your name]."

Special Advice for Résumé Preparation

ETHICS NOTE Be truthful. The consequences of giving false information in your résumé are serious. In fact, the truthfulness of your résumé reflects not only on your own ethical stance but also on the integrity with which you would represent the organization. See also ethics in writing. ✦

Salary. Avoid listing the salary you desire in the résumé. On the one hand, you may price yourself out of a job you want if the salary you list is higher than a potential employer is willing to pay. On the other hand, if you list a low salary, you may not get the best possible offer. See salary negotiations.

Returning Job Seekers. If you are returning to the workplace after an absence, most career experts say that it is important to acknowledge the gap in your career. That is particularly true if, for example, you are re-entering the workforce because you have devoted a full-time period to care for children or dependent adults. Do not undervalue such work. Although unpaid, it often provides experience that develops important time-management, problem-solving, organizational, and interpersonal skills. Although gaps in employment can be explained in the application letter, the following examples illustrate how you might reflect such experiences in a résumé. They would be especially appropriate for an applicant seeking employment in a field related to child or health care.

- **Primary Child-Care Provider, 2003 to 2005**
  Provided full-time care to three preschool children at home. Instructed in beginning scholastic skills, time management, basics of nutrition, arts, and swimming. Organized activities, managed household, and served as neighborhood-watch captain.
- **Home Caregiver, 2003 to 2005**
  Provided 60 hours per week in-home care to Alzheimer's patient. Coordinated medical care, developed exercise programs, completed and processed complex medical forms, administered medications, organized budget, and managed home environment.

If you have performed volunteer work during such a period, list that experience. Volunteer work often results in the same experience as does full-time, paid work, a fact that your résumé should reflect, as in the following example.

- **School Association Coordinator, 2003 to 2005**
  Managed special activities of the Briarwood High School Parent-Teacher Association. Planned and coordinated meetings, scheduled events, and supervised fund-drive operations. Raised $70,000 toward refurbishing the school auditorium.

R

Electronic Résumés

In addition to the traditional paper résumé, you can post a Web-based résumé. You may also need to submit a résumé on disk or through e-mail to a potential employer to be included in an organization's database. As Internet and database technologies converge, remain current with the forms and protocols that employers prefer by reviewing popular job-search sites, such as HotJobs at <http://hotjobs.yahoo.com> and Monster.com at <www.monster.com/>.

Web Résumés. If you plan to post your résumé on your own Web site, keep the following points in mind.

- Follow the general advice for Web design, such as viewing your résumé on several browsers to see how it looks.
- Just below your name, you may wish to provide a series of internal page links to such important categories as "experience" and "education."
- Consider building a multipage site for displaying a work portfolio, publications, reference letters, and the like.
- If privacy is an issue, include an e-mail link ("mailto") at the top of the résumé rather than your home address and phone number.

The disadvantage of posting a résumé at your own Web site is that you must attract the attention of employers on your own. As discussed in the entry job search, commercial services can attract recruiters with their large databases.

Scannable and Plain-Text Résumés. A scannable résumé is normally mailed to an employer in paper form, scanned, and downloaded into a company's searchable database. Such a résumé can be well formatted, but it should not contain decorative fonts, underlining, shading, letters that touch each other, and other features that will not scan easily. Scan such a résumé yourself to make sure there are no problems.

Some employers request ASCII (American Standard Code for Information Interchange) or plain-text résumés via e-mail, which can be added directly into the résumé database without scanning. ASCII résumés also allow employers to read the file no matter what type of software they are using. You can copy and paste such a résumé directly into the body of the e-mail message.

DIGITAL TIPS PREPARING AN ASCII RÉSUMÉ

When preparing an ASCII document, proper formatting is critical. For example, you need to insert manual line breaks at 65 characters to prevent long, single lines when the documents are opened in various systems. Further, many word-processing elements such as bullets, underlining, and boldface are incompatible with ASCII, which is limited to letters, numbers, and basic punctuation. For more on this topic, see <*www.bedfordstmartins.com/alredtech*> and select *Digital Tips*, "Preparing an ASCII Résumé."

R

For résumés that will be downloaded into databases, it is good to use nouns rather than verbs to describe experience and skills (*designer* and *management* rather than *designed* and *managed*). You may also include a section in such a résumé titled "Keywords" (or perhaps give a descriptive name, such as "Areas of Expertise"). Keywords, also called *descriptors*, allow potential employers to search the database for qualified candidates. So, be sure to use keywords that are the same as those used in the employer's descriptions of the jobs that best match your interests and qualifications. This section can follow the main heading of your résumé or appear near the end of your résumé. Figure R–15 is an example of an electronic résumé that demonstrates the use of keywords.

E-mail–Attached Résumés. An employer may request or you may prefer to submit a résumé as an e-mail attachment to be printed out by the employer. If so, consider using a relatively plain design and sending the résumé as a rich text format (.rtf) document. Or, if precise design is important, send the résumé as a PDF file that will preserve the fonts, images, graphics, and layout. You can attach this file to an e-mail message that will then serve as your application letter.

WEB LINK **ANNOTATED SAMPLE RÉSUMÉS**

For more examples of résumés with helpful annotations, see <*bedfordstmartins.com/alredtech*> and select *Model Documents Gallery*.

revision

When you revise your draft, read and evaluate it primarily from the point of view of your readers. In fact, revising requires a different frame of mind than writing a draft. To achieve that frame of mind, experienced writers have developed various approaches. For example, some writers print out a draft, double-spaced, and mark up the paper copy. Working on-screen often focuses attention too narrowly on sentence-level problems, causing writers to lose sight of the larger problems of scope and organization. Other tactics include the following:

R

- Allow a "cooling period" between writing the draft and revision in order to evaluate the draft objectively.
- Read your draft aloud—often, hearing the text will enable you to spot problem areas that need improvement.
- Revise in passes by reading through your draft several times, each time searching for and correcting a different set of problems.

When you can no longer spot improvements, you may wish to give the draft to a colleague for review—especially for projects that are crucial for you or your organization as well as for collaborative projects, as described in collaborative writing.

Writer's Checklist: Revising Your Draft

- ☑ *Completeness*. Does the document achieve its primary purpose? Will it fulfill the readers' needs? Your writing should give readers exactly what they need but not overwhelm them.
- ☑ *Appropriate introduction and conclusion*. Check to see that your introduction frames the rest of the document and your conclusion ties the main ideas together. Both should account for revisions to the content of the document.
- ☑ *Accuracy*. Look for any inaccuracies that may have crept into your draft.
- ☑ *Unity and coherence*. Check to see that sentences and ideas are closely tied together (coherence) and contribute directly to the main idea expressed in the topic sentence of each paragraph (unity). Provide transitions where they are missing and strengthen those that are weak.
- ☑ *Consistency*. Make sure that layout and design, visuals, and use of language are consistent. Do not call the same item by one term on one page and a different term on another page.
- ☑ *Conciseness*. Tighten your writing so that it says exactly what you mean. Prune unnecessary words, phrases, sentences, and even paragraphs. See conciseness.
- ☑ *Awkwardness*. Look for awkwardness in sentence construction—especially any garbled sentences.
- ☑ *Ethical writing*. Check for ethics in writing and eliminate biased language.
- ☑ *Active voice*. Use the active voice unless the passive voice is more appropriate.
- ☑ *Word choice*. Delete or replace vague words and unnecessary intensifiers. Check for affectation and unclear pronoun references. See also word choice.
- ☑ *Jargon*. If you have any doubt that all your readers will understand any jargon or special terms you have used, eliminate or define them.
- ☑ *Clichés*. Replace clichés with fresh figures of speech or direct statements.
- ☑ *Grammar*. Check your draft for grammatical errors. Use computer grammar checkers with caution. Because they are not always accurate, treat their recommendations only as *suggestions*.

Writer's Checklist: Revising Your Draft (continued)

☑ *Typographical errors*. Check your final draft for typographical errors both with your spell checker and with thorough proofreading (spell checkers do not catch all errors).

☑ *Other common errors*. Use the search-and-replace command to find and revise wordy phrases, such as *that is*, *there are*, *the fact that*, and *to be*, and unnecessary helping verbs such as *will*.

DIGITAL TIPS **INCORPORATING TRACKED CHANGES**

When colleagues review your document, they can track changes and insert comments within the document itself. Tracking and commenting vary with types and versions of word-processing programs, but in most programs you can view the tracked changes on a single draft or review the multiple drafts of your reviewers' versions. For more on this topic, see <*bedfordstmartins .com/alredtech*> and select *Digital Tips*, "Incorporating Tracked Changes."

rhetorical questions

A rhetorical question is a question to which a specific answer is neither needed nor expected. The question is often intended to make an audience think about the subject from a different perspective; the writer or speaker then answers the question in the article or presentation. The answer to a rhetorical question such as "Is space exploration worth the cost?" may not be a simple yes or no; it might be a detailed explanation of the pros and cons of the value of space exploration.

The rhetorical question can be an effective opening, and it is often used as a title. By its nature, it is somewhat journalistic and indirect, and therefore should be used judiciously in technical writing. For example, although a rhetorical question might be an appropriate opening for a newsletter article, it would not be as effective for a report or an e-mail addressed to a busy manager. When you do use a rhetorical question, be sure it is not trivial, obvious, or forced. The rhetorical question requires that you know your readers.

run-on sentences

A run-on sentence, sometimes called a *fused sentence*, is two or more sentences without punctuation to separate them. The term is also sometimes applied to a pair of independent clauses separated by only a comma, although this variation is usually called a comma splice. See also sentence construction and sentence faults.

Run-on sentences can be corrected, as shown in the following examples, by (1) making two sentences, (2) joining the two clauses with a semicolon (if they are closely related), (3) joining the two clauses with a comma and a coordinating conjunction, or (4) subordinating one clause to the other.

- The client suggested several solutions ~~some~~ *. Some* are impractical.
- The client suggested several solutions *;* some are impractical.
- The client suggested several solutions *, but* some are impractical.
- The client suggested several solutions *, although* some are impractical.

S

salary negotiations

Salary negotiations usually take place either at the end of an interview or after a formal job offer has been made. If possible, delay discussing salary until after you receive a formal written job offer because you will have more negotiating power at that point. See also job search.

Before interviewing for a job, prepare for possible salary negotiations by researching the following:

- The current range of salaries for the work you hope to do at your level (entry? intermediate? advanced?) in your region of the country. Check trade journals and organizations in your field, or ask a reference librarian for help in finding this information. Job listings that include salary can also be helpful.
- Salaries made by last year's graduates from your college or university at your level and in your line of work. Your campus career-development office should have these figures.
- Salaries made by people you know at your level and in your line of work. Attend local organizational meetings in your field or contact officers of local organizations who might have this information or steer you to useful contacts.

If a potential employer requests your salary requirements with a résumé, consider your options carefully. If you provide a salary that is too high, the company might never interview you; if you provide a salary that is too low, you may have no opportunity later in the hiring process to negotiate for a higher salary. However, if you fail to follow the potential employer's directions and omit the requested information, an employer may disqualify you on principle. If you choose to provide salary requirements, always do so in a range (for example, $35,000 to $40,000). See also application letters.

If an interviewer asks your salary requirements toward the end of the job interview, you can try these strategies to delay salary negotiations:

- Say something like "I am sure that this company pays a fair salary for a person with my level of experience and qualifications" or "I am ready to consider your best offer."

- Indicate that you would like to learn more details about the position before discussing salary; point out that your primary goal is to work in a stimulating environment with growth potential, not to earn a specific salary.
- Express generally a strong interest in the position and the organization without referring to the salary.
- Emphasize your unique qualifications (or combination of skills) for the job and what you can do for the company that other candidates cannot.

If the interviewer or company demands to know your salary requirements during a job interview, provide a wide salary range that you know would be reasonable for someone at your level in your line of work in that region of the country. For example, you could say, "I would hope for a salary somewhere between $28,000 and $38,000, but of course this is negotiable."

Once salary negotiations begin, resist the temptation to accept, immediately, the first salary offer you receive. If you have done thorough research, you will know if the first salary offer is at the low, middle, or high end of the salary range for your level of experience in your line of work. If you have little or no experience and receive an offer for a salary at the low end of the range, you will realize that the offer probably is fair and reasonable. Yet, if you receive the same low-end offer but bring considerable experience to the job, you can negotiate for a higher salary that is more reasonable for someone with your background and credentials in your line of work in your region of the country.

Never say that you are unable to accept a salary below a particular figure. To keep negotiations going, simply indicate that you would have trouble accepting the first offer because it was lower than you had expected.

Remember that you are negotiating a package and not just a starting salary. For example, benefits can offer you substantial value. If the starting salary seems low, consider negotiating for some of these possible job perks:

- The chance for an early promotion, thus a higher salary, within a few years
- A particular job title or special job responsibilities that would provide you with impressive chances for career growth
- Tuition credits for continued education
- Payment of relocation costs
- Paid personal leave or paid vacations
- Personal or sick days

- Overtime potential
- Flexible hours
- Health, dental, optical/eye care, disability, and life insurance
- Retirement plans, such as 401(k) and pension plans
- Profit sharing; investment or stock options
- Bonuses or cost-of-living adjustments
- Commuting or parking-cost reimbursement
- Child or elder care
- Discounts on company products and services

You might find it most comfortable to respond to an initial offer in writing and then meet later with the potential employer for further negotiation. If possible, indicate all of your preferences and requirements at one time instead of continually asking for new and different benefits as you negotiate. Throughout this process, focus on what is most important to you (which might differ from what is most important to your friends) and on what you would find acceptable and comfortable.

WEB LINK **SALARY INFORMATION RESOURCES**

For links to Web sites offering useful resources for salary negotiations, see <*bedfordstmartins.com/alredtech*> and select *Links for Handbook Entries*.

scope

S

Scope is the depth and breadth of detail you include in a document as defined by your readers' needs, your purpose, and the context. For example, if you write a trip report about a routine visit to a company facility, your readers may need to know only the basic details and any unusual findings. However, if you prepare a trip report about a visit to a division that has experienced problems and your purpose is to suggest ways to solve those problems, your report will contain many more details, observations, and even recommendations.

You should determine the scope of a document during the preparation stage of the writing process, even though you may refine it later. Defining your scope will expedite your research and can help determine team members' responsibilities in collaborative writing.

Your scope will also be affected by the type of document you are writing as well as the medium you select for your message. For example, government agencies often prescribe the general content and length for proposals, and some organizations set limits for the length of memos and e-mail. See selecting the medium and "Five Steps to Successful Writing."

selecting the medium

The most important considerations in selecting the most appropriate medium for communicating a message are the expectations of your readers and the purpose as well as the context of the communication. If you need to collaborate with someone to solve a problem or if you need to establish rapport, for example, e-mail exchanges may be far less efficient than a phone call or a face-to-face meeting. However, if you need precise wording or a record of a complex message, communicate in writing. Understanding the typical means of communicating on the job will help you select the most appropriate medium.

Letters on Organizational Stationery

Correspondence in the form of printed letters with handwritten signatures are often the most appropriate choice for formal contacts with new business associates or customers. Letters written on an organization's printed letterhead communicate formality, respect, and authority.

Memos

Memos on printed company stationery or attached to an e-mail are appropriate for internal communication among members of the same organization, even when offices are geographically separated. They have many of the same characteristics as letters, such as formality and authority, but they are used for a wider variety of functions—from formal meeting announcements to short reports.

E-mails

E-mail messages can replace letters and memos when writers are sure they are acceptable to their readers. E-mails can also be a less formal medium to exchange information, elicit discussions, collect opinions, and transmit various electronic documents. Because e-mail recipients can print copies of messages and attachments they receive and easily forward them to others, e-mail requires writers to consider and check messages carefully before clicking the send button.

Faxes

A fax is most useful when speed is essential and when the information—a drawing or signed contract, for example—must be viewed in its original form. Faxes are also useful when the recipient either does not have e-mail or prefers faxed documents. Fax machines in offices can be located in shared areas, so call the intended recipient before you send any confidential or sensitive messages.

Telephone Calls

Telephone exchanges are often best used for answering questions or discussing such matters as the conditions of a contract. Because phone calls enable participants to interpret tone of voice, they can make it easier to resolve misunderstandings, although they do not provide the visual cues possible during face-to-face meetings.

For multiple participants, a conference call may be a less expensive alternative to face-to-face meetings requiring travel. Such conference calls work best when the person coordinating the call works from an agenda shared by all the participants and directs the discussion as if chairing a meeting. Telephone calls in which decisions have been reached should be followed with written confirmation.

Voice Mail

Voice mail allows callers to leave brief messages ("Call me about the deadline for the new project" or "I got the package, so you don't need to call the distributor"). If the message is complicated or contains numerous details, use another medium, such as an e-mail message or a letter. If you want to discuss a subject at length, let the recipient know the subject so that he or she can prepare a response before returning your call. When you leave a message, give your name and contact information.

Face-to-Face Meetings

Face-to-face meetings are most appropriate for initial or early contacts with associates and clients with whom you intend to develop an important, long-term relationship. Meetings may also be best for brainstorming, negotiating, solving a technical problem, or handling a controversial issue. For advice on how to record discussions and decisions, see minutes of meetings.

Videoconferencing

Videoconferencing is particularly useful for meetings when travel is impractical. Unlike telephone conference calls, videoconferences have the advantage of allowing participants to see as well as to hear one another. Videoconferences work best with participants who are at ease in front of the camera.

Web Sites

A public Internet or company intranet Web site may be ideal for making available or exchanging information with others. For example, you might make available on your Web site particular documents and files colleagues may need for collaborative writing. Web sites can provide for you and others a home base for resources. See also Web design and writing for the Web.

Instant Messaging

Instant messaging (IM) on a computer or cell phone, although used primarily for personal communication, may be a low-cost alternative to communicating in real time with coworkers, suppliers, and customers—especially in remote locations. Instant messaging often uses online slang and shortened spellings ("u" for "you" and "L8R" for "later"). Instant messaging is obviously limited, and some employers do not consider IM appropriate for business purposes.

semicolons

The semicolon (;) links independent clauses or other sentence elements of equal weight and grammatical rank when they are not joined by a comma and a conjunction. The semicolon indicates a greater pause between clauses than does a comma but not as great a pause as a period.

Independent clauses joined by a semicolon should balance or contrast with each other, and the relationship between the two statements should be so clear that further explanation is not necessary.

- The new Web site was a success*;* every division reported increased online sales.

Do not use a semicolon between a dependent clause and its main clause.

- No one applied for the position; even though it was heavily advertised.

S

With Strong Connectives

In complicated sentences, a semicolon may be used before transitional words or phrases (*that is*, *for example*, *namely*) that introduce examples or further explanation. See also transition.

- The press understands Commissioner Curran's position on the issue*; that is*, local funds should not be used for the highway project.

A semicolon should also be used before conjunctive adverbs (*therefore*, *moreover*, *consequently*, *furthermore*, *indeed*, *in fact*, *however*) that connect independent clauses.

- The test results are not complete*; therefore*, I cannot make a recommendation.
  [The semicolon in the example shows that *therefore* belongs to the second clause.]

For Clarity in Long Sentences

Use a semicolon between two independent clauses connected by a coordinating conjunction (*and*, *but*, *for*, *or*, *nor*, *so*, *yet*) if the clauses are long and contain other punctuation.

- In most cases, these individuals are executives, bankers, or lawyers*; but* they do not, as the press seems to believe, simply push the button of their economic power to affect local politics.

A semicolon may also be used if any items in a series contain commas.

- Among those present were John Howard, president of the Omega Paper Company*;* Carol Delgado, president of Environex Corporation*;* and Larry Stanley, president of Stanley Papers.

Use parentheses or dashes, not semicolons, to enclose a parenthetical element that contains commas.

- All affected job classifications (receptionist, secretary, transcriptionist, and clerk) will be upgraded this month.

Use a colon, not a semicolon, as a mark of anticipation or enumeration.

- Three decontamination methods are under consideration; a zeolite-resin system, an evaporation system, and a filtration system.

The semicolon always appears outside closing quotation marks.

- The attorney said, "You must be accurate"; her client replied, "I will."

sentence construction

DIRECTORY

A sentence is the most fundamental and versatile tool available to writers. Sentences generally flow from a subject to a verb to any objects, complements, or modifiers, but they can be ordered in a variety of ways to achieve emphasis. When shifting word order for emphasis, however, be aware that word order can make a great difference in the meaning of a sentence.

- He was *only* the service technician.
- He was the *only* service technician.

Subjects

The most basic components of sentences are subjects and predicates. The *subject* of a sentence is a noun or pronoun (and its modifiers) about which the predicate of the sentence makes a statement. Although a subject may appear anywhere in a sentence, it most often appears at the beginning. (*The wiring* is defective.) Grammatically, a subject must agree with its verb in number.

- *These departments have* much in common.
- *This department has* several functions.

The subject is the actor in sentences using the active voice.

- *The Webmaster* reported an increase in site visits for May.

A *compound subject* has two or more substantives (nouns or noun equivalents) as the subject of one verb.

- *The doctor* and *the nurse* agreed on the treatment plan.

S

ESL TIPS FOR UNDERSTANDING THE SUBJECT OF A SENTENCE

In English, every sentence, except commands, must have an explicit subject.

- *Paul* worked fast. ~~Established~~ *He established* the parameters for the project.

In commands, the subject *you* is understood and is used only for emphasis.

- (*You*) Show up at the airport at 6:30 tomorrow morning.
- (*You*) Do your homework, young man. [parent to child]

If you move the subject from its normal position (subject-verb-object), English often requires you to replace the subject with an expletive (*there*, *it*). In this construction, the verb agrees with the subject that follows it.

- *There are* two files on the desk. [The subject is *files.*]
- *It is* presumptuous for me to speak for Jim. [The subject is *to speak for Jim.*]

Time, distance, weather, temperature, and environmental expressions use *it* as their subject.

- *It* is ten o'clock.
- *It* is ten miles down the road.
- *It* seldom snows in Florida.
- *It* is very hot in Jorge's office.

Predicates

The *predicate* is the part of a sentence that makes an assertion about the subject and completes the thought of the sentence.

- Bill *has piloted the corporate jet.*

S

The *simple predicate* is the verb and any helping verbs (*has piloted*). The *complete predicate* is the verb and any modifiers, objects, or complements (*has piloted the corporate jet*). A *compound predicate* consists of two or more verbs with the same subject.

- The company *tried* but *did not succeed* in that field.

Such constructions help achieve conciseness in writing. A *predicate nominative* is a noun construction that follows a linking verb and renames the subject.

- She is my *attorney*. [noun]
- His excuse was *that he had been sick*. [noun clause]

Sentence Types

Sentences may be classified according to *structure* (simple, compound, complex, compound-complex); *intention* (declarative, interrogative, imperative, exclamatory); and *stylistic use* (loose, periodic, minor).

Structure. A *simple sentence* consists of one independent clause. At its most basic, a simple sentence contains only a subject and a predicate.

- The power [subject] failed [predicate].

A *compound sentence* consists of two or more independent clauses connected by a comma and a coordinating conjunction, by a semicolon, or by a semicolon and a conjunctive adverb.

- Drilling is the only way to collect samples of the layers of sediment below the ocean floor, *but* it is not the only way to gather information about these strata. [comma and coordinating conjunction]
- The chemical composition of seawater bears little resemblance to that of river water*;* the various elements are present in entirely different proportions. [semicolon]
- It was 500 miles to the site*; therefore,* we made arrangements to fly. [semicolon and conjunctive adverb]

A *complex sentence* contains one independent clause and at least one dependent clause that expresses a subordinate idea.

- The generator will shut off automatically [independent clause] if the temperature rises above a specified point [dependent clause].

A *compound-complex sentence* consists of two or more independent clauses plus at least one dependent clause.

- Productivity is central to controlling inflation [independent clause]; when productivity rises [dependent clause], employers can raise wages without raising prices [independent clause].

Intention. A *declarative sentence* conveys information or makes a factual statement. (The motor powers the conveyor belt.) An *interrogative sentence* asks a direct question. (Does the conveyor belt run constantly?) An *imperative sentence* issues a command. (Restart in MS-DOS mode.) An *exclamatory sentence* is an emphatic expression of feeling, fact, or opinion. It is a declarative sentence that is stated with great feeling. (The files were deleted!)

Stylistic Use. A *loose sentence* makes its major point at the beginning and then adds subordinate phrases and clauses that develop or modify that major point. A loose sentence might seem to end at one or more points before it actually does end, as the periods in brackets illustrate in the following sentence.

- It went up[.], a great ball of fire about a mile in diameter[.], an elemental force freed from its bonds[.] after being chained for billions of years.

A *periodic sentence* delays its main ideas until the end by presenting subordinate ideas or modifiers first.

- During the last century, the attitude of the American citizen toward automation underwent a profound change.

A *minor sentence* is an incomplete sentence that makes sense in its context because the missing element is clearly implied by the preceding sentence.

- In view of these facts, is the service contract really useful? *Or economical?*

Constructing Effective Sentences

The subject-verb-object pattern is effective because it is most familiar to **readers**. In "The writer finished the manual," we know the subject (*writer*) and the object (*manual*) by their positions relative to the verb (*finished*).

An *inverted sentence* places the elements in unexpected order, thus emphasizing the point by attracting the readers' attention.

- A better job I never had. [direct object–subject–verb]
- More optimistic I have never been. [subjective complement–subject–linking verb]
- A poor image we presented. [complement–subject–verb]

S

Use uncomplicated sentences to state complex ideas. If readers have to cope with a complicated sentence in addition to a complex idea, they are likely to become confused. Just as simpler sentences make complex ideas more digestible, a complex sentence construction makes a series of simple ideas more smooth and less choppy.

Avoid loading sentences with a number of thoughts carelessly tacked together. Such sentences are monotonous and hard to read because all the ideas seem to be of equal importance. Rather, distinguish the relative importance of sentence elements with subordination. See also garbled sentences.

LOADED We started the program three years ago, only three members were on the staff, and each member was responsible for a separate state, but it was not an efficient operation.

IMPROVED When we started the program three years ago, only three members were on the staff, each having responsibility for a separate state; however, that arrangement was not efficient.

Express coordinate or equivalent ideas in similar form. The very construction of the sentence helps the reader grasp the similarity of its components, as illustrated in parallel structure.

TIPS FOR UNDERSTANDING THE REQUIREMENTS OF A SENTENCE

- A sentence must start with a capital letter.
- A sentence must end with a period, a question mark, or an exclamation point.
- A sentence must have a subject.
- A sentence must have a verb.
- A sentence must conform to subject-verb-object word order (or inverted word order for questions or emphasis).
- A sentence must express an idea that can stand on its own (called the *main*, or *independent*, *clause*).

sentence faults

S

A number of problems can create sentence faults, including faulty subordination, clauses with no subjects, rambling sentences, omitted verbs, and illogical assertions.

Faulty Subordination

Faulty subordination occurs when a grammatically subordinate element contains the main idea of the sentence or when a subordinate element is so long or detailed that it obscures the main idea. Both of the following sentences are logical, depending on what the writer intends as the main idea and as the subordinate element.

- Although the new filing system saves money, many of the staff are unhappy with it.
  [If the main point is that *many of the staff are unhappy*, this sentence is correct.]
- The new filing system saves money, although many of the staff are unhappy with it.
  [If the main point is that *the new filing system saves money*, this sentence is correct.]

In the following example, the subordinate element overwhelms the main point.

FAULTY	Because the noise level in the assembly area on a typical shift is as loud as a smoke detector's alarm ten feet away, employees often develop hearing problems.
IMPROVED	Employees in the assembly area often develop hearing problems because the noise level on a typical shift is as loud as a smoke detector's alarm ten feet away.

Missing Subjects

Clauses with no subjects occur when writers inappropriately assume a subject that they do not state in the clause. See also sentence fragments.

INCOMPLETE	Your application program can request to end the session after the next command. [Your application program can request *who* or *what* to end the session?]
COMPLETE	Your application program can request *the host program* to end the session after the next command.

S

Rambling Sentences

Rambling sentences contain more information than the reader can comfortably absorb. The obvious remedy for a rambling sentence is to divide it into two or more sentences. When you do that, put the main message of the rambling sentence into the first of the revised sentences.

RAMBLING	The payment to which a subcontractor is entitled should be made promptly in order that in the event of a subsequent contractual dispute we, as general contractors, may not be held in default of our contract by virtue of nonpayment.

DIRECT	Pay subcontractors promptly. Then if a contractual dispute occurs, we cannot be held in default of our contract because of nonpayment.

Missing Verbs

Some sentence faults are created when writers omit a verb.

- I never have *written* and probably never will write the annual report.

Faulty Logic

A sentence is faulty when its predicate makes an illogical assertion about its subject. "Mr. Wilson's *job* is a sales representative" is not logical, but "*Mr. Wilson* is a sales representative" is logical. "Jim's *height* is six feet tall" is not logical, but "*Jim* is six feet tall" is logical. See also logic errors.

sentence fragments

A sentence fragment is an incomplete grammatical unit that is punctuated as a sentence.

FRAGMENT	And quit his job.
SENTENCE	He quit his job.

A sentence fragment lacks either a subject or a verb or is a subordinate clause or phrase. Sentence fragments are often introduced by relative pronouns (*who, whom, which, that*) or subordinating conjunctions (such as *although, because, if, when,* and *while*).

- The new manager instituted several new procedures~~. Although~~ *, although* she didn't clear them with Human Resources.

A sentence must contain a finite verb; verbals (nonfinite) do not function as verbs. The following sentence fragments use verbals (*providing, to work*) that cannot function as finite verbs.

FRAGMENT	*Providing* all employees with disability insurance.
SENTENCE	The company *provides* all employees with disability insurance.
FRAGMENT	*To work* a 40-hour week.
SENTENCE	Most of our employees *must work* a 40-hour week.

Explanatory phrases beginning with *such as*, *for example*, and similar terms often lead writers to create sentence fragments.

, such as
- The staff wants additional benefits~~. For example,~~ the use of company cars.

A hopelessly snarled fragment simply must be rewritten. To rewrite such a fragment, pull the main points out of the fragment, list them in the proper sequence, and then rewrite the sentence as illustrated in garbled sentences. See also sentence construction and sentence faults.

sentence variety

Sentences can vary in length, structure, and complexity. As you revise, make sure your sentences have not become tiresomely alike. See also sentence construction.

Sentence Length

A series of sentences of the same length is monotonous, so varying sentence length makes writing less tedious to the reader. For example, avoid stringing together a number of short independent clauses. Either connect them with subordinating connectives, thereby making some dependent clauses, or make some clauses into separate sentences. See also subordination.

STRING	The river is 60 miles long, and it averages 50 yards in width, and its depth averages 8 feet.
IMPROVED	The river, which is 60 miles long and averages 50 yards in width, has an average depth of 8 feet.
IMPROVED	The river is 60 miles long. It averages 50 yards in width and 8 feet in depth.

S

You can often effectively combine short sentences by converting verbs into adjectives.

failed
- The digital shift indicator ~~failed. It~~ was pulled from the market.

Although too many short sentences make your writing sound choppy and immature, a short sentence can be effective following a long one.

- During the past two decades, many changes have occurred in American life—the extent, durability, and significance of which no one has yet measured. *No one can.*

In general, short sentences are good for emphatic, memorable statements. Long sentences are good for detailed explanations and support. Nothing is inherently wrong with a long sentence, or even with a complicated one, as long as its meaning is clear and direct. Sentence length becomes an element of style when varied for emphasis or contrast; a conspicuously short or long sentence can be used to good effect.

Word Order

When a series of sentences all begin in exactly the same way (usually with an article and a noun), the result is likely to be monotonous. You can make your sentences more interesting by occasionally starting with a modifying word, phrase, or clause.

- *To salvage the project,* she presented alternatives when existing policies failed to produce results. [modifying phrase]

However, overuse of this technique itself can be monotonous, so use it in moderation.

Inverted word order can be an effective way to achieve variety, but be careful not to create an awkward construction.

AWKWARD	Then occurred the event that gained us the contract.
EFFECTIVE	Never have sales been so good.

For variety, you can alter normal sentence order by inserting a phrase or clause.

- Titanium fills the gap, *both in weight and in strength,* between aluminum and steel.

The technique of inserting a phrase or clause is good for achieving emphasis, providing detail, breaking monotony, and regulating pace.

Loose and Periodic Sentences

A loose sentence makes its major point at the beginning and then adds subordinate phrases and clauses that develop or modify the point. A loose sentence could end at one or more points before it actually does end, as the periods in brackets illustrate in the following example.

S

- It went up[.], a great ball of fire about a mile in diameter[.], an elemental force freed from its bonds[.] after being chained for billions of years.

A periodic sentence delays its main idea until the end by presenting modifiers or subordinate ideas first, thus holding the readers' interest until the end.

- During the last century, the attitude of Americans toward technology underwent a profound change.

Experiment with shifts from loose sentences to periodic sentences in your own writing, especially during revision. Avoid the singsong monotony of a long series of loose sentences, particularly a series containing coordinate clauses joined by conjunctions. Subordinating some thoughts to others not only provides emphasis but also makes your sentences more interesting.

sequential method of development

The sequential, or step-by-step, method of development is especially effective for explaining a process or describing a mechanism in operation. (See process explanation.) It is also the logical method for writing instructions, as shown in Figure S–1.

INSTRUCTIONS
How to E-mail a Picture from a Digital Camera

The following step-by-step instructions describe how to transfer images from your camera to your computer, save them, and then e-mail the images to your friends, relatives, or business colleagues.

Transferring Photos from Camera to Computer

Connect the cable from the camera to your computer at the USB, or serial port, to make the camera serve as its own disk drive. If your digital camera includes a docking station, place the camera in the docking station cradle to automatically connect it to the computer. In most cases, an icon of your camera or a folder for it will appear on your screen after you connect it to your computer.

Saving Photos on the Computer

To save your images to your computer, create and name a folder on your hard drive. If you're working on an ongoing project that will have many images, include the dates of the images or other useful identifying information. After you create the folder, open and view the images stored on your camera as described in step 1, select the files you wish to save, highlight them, and drag them into the folder you've created on your hard drive.

FIGURE S–1. Sequential Method of Development

Modifying Photos

Use imaging software, such as Adobe's PhotoShop™, to crop your image, to change the resolution, and to add other effects. Once you're done making changes to your image, use the File/Save As command or, in some imaging programs, the File/Export command, to make a copy of the image. At this point you can give the image a more useful name than that which your camera will have given the file because digital cameras typically assign images a number.

Attaching Photos to E-mail Messages

To attach the file to an e-mail message, open your outgoing e-mail box and click on the Attachment icon or box. When your file directory opens, select the folder where your photo images are stored, highlight the file you wish to send, and click Open. Your photo file is now attached to your e-mail box.

FIGURE S–1. Sequential Method of Development (*continued*)

The main advantage of the sequential method of development is that it is easy to follow because the steps correspond to the process or operation being described. The disadvantages are that it can become monotonous and does not lend itself well to achieving emphasis.

Most methods of development have elements of sequence. The chronological method of development, for example, is also sequential: To describe a trip chronologically, from beginning to end, is also to describe it sequentially.

service

When used as a verb, *service* means "keep up or maintain" as well as "repair." (Our company will *service* your equipment.) If you mean "provide a more general benefit," use *serve*.

- Our company ~~services~~ *serves* the northwest area of the state.

S

set / sit

Sit is an intransitive verb; it does not, therefore, require an object. (I *sit* by a window in the office.) Its past tense is *sat*. (We *sat* around the conference table.) *Set* is usually a transitive verb, meaning "put or place," "establish," or "harden." Its past tense is *set*.

- Please *set* the trophy on the shelf.
- The jeweler *set* the stone carefully.
- Can we *set* a date for the meeting?
- The high temperature *sets* the epoxy quickly.

Set is occasionally an intransitive verb.

- The new adhesive *sets* in five minutes.

shall / will

Although traditionally *shall* was used to express the future tense with *I* and *we*, *will* is generally accepted with all persons. *Shall* is commonly used today only in questions requesting an opinion or a preference (*Shall* we go?) rather than a prediction (*Will* we go?). It is also used in statements expressing determination (I *shall* return!) or in formal regulations (Applicants *shall* provide a proof of certification).

sic (*see* quotations)

slashes

The slash (/)—called a variety of names, including *slant line*, *diagonal*, *virgule*, *bar*, and *solidus*—both separates and shows omission.

The slash can indicate alternatives or combinations.

S

- David's telephone numbers are 549-2278/2235.
- Check the on/off switch before you leave.

The slash often indicates omitted words and letters.

- miles/hour (miles per hour), w/o (without)

In fractions and mathematical expressions, the slash separates the numerator from the denominator (3/4 for three-fourths; x/y for x over y).

Although the slash is used informally with dates (5/11/06), avoid this form in technical writing where the date may not be immediately clear, especially in international correspondence.

The forward slash often separates items in URL (uniform resource locator) addresses for sites on the Internet (<bedfordstmartins.com/alredtech>). The backward slash is used to separate parts of file names (<c:\myfiles\reports\annual06.doc>).

so / so that / such

Avoid *so* as a substitute for *because*. See also as/because/since.

- *Because she* ~~She~~ reads faster, ~~so~~ she finished before I did.

Do not replace the phrase *so that* with *so* or *such that*.

- The report should be written *so that* ~~such that~~ it can be widely understood.

Such, an adjective meaning "of this or that kind," should never be used as a pronoun.

- Our company provides on-site child care, but I do not anticipate using *it.* ~~such.~~

some / somewhat

When *some* functions as an indefinite pronoun for a plural count noun or as an indefinite adjective modifying a plural count noun, use a plural verb.

- *Some* of us *are* prepared to work overtime.
- *Some* people *are* more productive than others.

Some is singular, however, when used with mass nouns.

- *Some* sand *has* trickled through the crack.
- Most of the water evaporated, but *some remains*.

When *some* is used as an adjective or a pronoun meaning "an undetermined quantity" or "certain unspecified persons," it should be replaced by the adverb *somewhat,* which means "to some extent."

- His writing has improved ~~some.~~ *somewhat.*

some time / sometime / sometimes

Some time refers to a duration of time. (We waited for *some time* before making the decision.) *Sometime* refers to an unknown or unspecified time. (We will visit with you *sometime.*) *Sometimes* refers to occasional occurrences at unspecified times. (He *sometimes* visits the branch offices.)

spatial method of development

The spatial method of development describes an object or a process according to the physical arrangement of its features. Depending on the subject, you describe its features from bottom to top, side to side, east to west, outside to inside, and so on. Descriptions of this kind rely mainly on dimension (height, width, length), direction (up, down, north, south), shape (rectangular, square, semicircular), and proportion (one-half, two-thirds). Features are described in relation to one another or to their surroundings, as illustrated in Figure S–2, which might be written by a home inspector or crime-scene investigator. The description in Figure S–2 relies on a bottom-to-top, clockwise (south to west to north to east) sequence, beginning with the front door.

The spatial method of development might be used for descriptions of laboratory equipment, proposals for landscape work, construction-site progress and activity reports, and, in combination with a step-by-step sequence, many types of instructions.

DESCRIPTION

Interior of Two-Story, Six-Room House

Ground Floor

Front hall and stairwell. The front door faces south and opens into a hallway seven feet deep and ten feet wide. At the end of the hallway is a stairwell that begins on the right-hand (east) side of the hallway, rises five steps to a landing, and reverses direction at the left-hand (west) side of the hallway.

Dining room. To the left (west) of the hallway is the dining room, which measures 15 feet along its southern exposure and ten feet along its western exposure.

Kitchen. North of the dining room is the kitchen, which measures ten feet along its western exposure and 15 feet along its northern exposure.

Bathroom. East of the kitchen, along the northern side of the house, is a bathroom that measures ten feet (west to east) by five feet.

Living room. Parallel to the bathroom is a passageway the same size as the bathroom and leading from the kitchen to the living room. The living room (15 feet west to east by 20 feet north to south) occupies the entire eastern end of the floor.

Second Floor

Hallway. On the second floor, at the top of the stairs, is an L-shaped hallway, five feet wide. The base of the L, over the door, is 15 feet long. The vertical arm of the L is 13 feet long.

Southwest bedroom. To the west of the hall is the southwest bedroom, which measures ten feet along its southern exposure and eight feet along its western exposure.

Northwest bedroom. Directly to the north, over the kitchen, is the northwest bedroom, which measures 12 feet along its western exposure and ten feet along its northern exposure.

FIGURE S–2. Spatial Method of Development

specifications

A specification is a detailed and exact statement that prescribes the materials, dimensions, and quality of something to be built, installed, or manufactured.

The two broad categories of specifications—industrial and government—both require precision. A specification must be written clearly and precisely and state *explicitly* what is needed. Because of the stringent requirements of specifications, careful research and preparation are especially important before you begin to write, as is careful revision after you have completed the draft. See also clarity and ambiguity.

Industrial Specifications

Industrial specifications are used, for example, in software development, in which there are no engineering drawings or other means of documentation. An industrial specification is a permanent record that documents the item being developed so that it can be maintained by someone other than the person who designed it and provides detailed technical information about the item to all who need it (engineers, technical writers, technical instructors, etc.).

The industrial specification describes a planned project, a newly completed project, or an old project. The specification for each type of project must contain detailed technical descriptions of all aspects of the project: what was done and how it was done, as well as what is required to use the item, how it is used, what its function is, who would use it, and so on.

Government Specifications

Government agencies are required by law to contract for equipment strictly according to definitions provided in formal specifications. A government specification is a precise definition of exactly what the contractor is to provide. In addition to a technical description of the item to be purchased, the specification normally includes an estimated cost; an estimated delivery date; and standards for the design, manufacture, quality, testing, training of government employees, governing codes, inspection, and delivery of the item.

Government specifications contain details on the scope of the project; documents the contractor is required to furnish with the device; required product characteristics and functional performance of the device; required tests, test equipment, and test procedures; required preparations for delivery; notes; and appendixes. These specifications appear in requests for proposals, which prescribe the content and deadline for government proposals submitted by vendors bidding on a project.

spelling

Because spelling errors in your documents will damage your credibility, careful proofreading is essential. The use of a spell checker is crucial; however, it will not catch all mistakes, especially those in personal and company names. It cannot detect a spelling error if the error results in a valid word; for example, if you mean *to* but inadvertently type *too*, the spell checker will not detect the error. Likewise, spell checkers will not detect errors in the names of people, places, and organizations. If you are unsure about the spelling of a word, do not rely on guesswork or a spell checker—consult a standard dictionary or style guide.

strata / stratum

Strata is the plural form of *stratum*, meaning "a layer of material."

- The land's *strata* are exposed by erosion.
- Each *stratum* is clearly visible in the cliff.

style

A dictionary definition of *style* is "the way in which something is said or done, as distinguished from its substance." Writers' styles are determined by the way writers think and transfer their thoughts to paper—the way they use words, sentences, images, figures of speech, and so on.

A writer's style is the way his or her language functions in particular situations. For example, an e-mail to a friend would be relaxed, even chatty, in tone, whereas a job application letter would be more restrained and formal. Obviously, the style appropriate to one situation would not be appropriate to the other. In both situations, the readers, the purpose, and the context determine the manner or style the writer adopts. Beyond an individual's personal style, various kinds of writing have distinct stylistic traits, such as technical writing style.

Standard English can be divided into two broad categories of style—formal and informal—according to how it functions in certain situations. Understanding the distinction between formal and informal writing styles helps writers use the appropriate style. We must recognize, however, that no clear-cut line divides the two categories and that some writing may call for a combination of the two.

Formal Writing Style

A formal writing style can perhaps best be defined by pointing to certain material that is clearly formal, such as scholarly and scientific articles in professional journals, lectures read at meetings of professional societies, and legal documents. Material written in a formal style is usually the work of a specialist writing to other specialists or writing that embodies laws or regulations. As a result, the vocabulary is specialized and precise. The writer's tone is impersonal and objective because the subject matter looms larger in the writing than does the author's personality. (See point of view.) A formal writing style does not use contractions, slang, or dialect. (See English, varieties of.) Because the material generally examines complex ideas, the sentence construction may be elaborate.

Formal writing need not be dull and lifeless. By using such techniques as the active voice whenever possible, sentence variety, and subordination, a writer can make formal writing lively and interesting, especially if the subject matter is inherently interesting to the reader.

- Although a knowledge of the morphological chemical constitution of cells is necessary to the proper understanding of living things, in the final analysis it is the activities of their cells that distinguish organisms from all other objects in the world. Many of these activities differ greatly among the various types of living things, but some of the basic sorts are shared by all, at least in their essentials. It is these fundamental actions with which we are concerned here. They fall into two major groups—those that are characteristic of the cell in the *steady state*, that is, in the normally functioning cell not engaged in reproducing itself, and those that occur during the process of *cellular reproduction*.

  —Lawrence S. Dillon, *The Principles of Life Sciences*

Whether you should use a formal style in a particular instance depends on your readers and purpose. When writers attempt to force a formal style when it should not be used, their writing is likely to fall victim to affectation, awkwardness, and gobbledygook.

Informal Writing Style

An informal writing style is a relaxed and colloquial way of writing standard English. It is the style found in most personal e-mail and in some business correspondence, nonfiction books of general interest, and mass-circulation magazines. There is less distance between the writer and the reader because the tone is more personal than in a formal writing style. Contractions and elliptical constructions are common. Consider the following passage, written in an informal style, from a nonfiction book.

- Business, like art and science, has been revealed and conceived through the intellect and imagination of people, and it develops or declines because of the intellect and imagination of people.

  In fact, there is no business; there are only people. Business exists only *among* people and *for* people.

  Seems simple enough, and it applies to every aspect of business, but not enough businesspeople seem to get it.

  Reading the economic forecasts and the indicators and the ratios and the rates of this or that, someone from another planet might actually believe that there really are invisible hands at work in the marketplace.

  It's easy to forget what the measurements are measuring. Every number—from productivity rates to salaries—is just a device contrived by people to measure the results of the enterprise of other people. For managers, the most important job is not measurement but motivation. And you can't motivate numbers.

  —James A. Autry, *Love and Profit: The Art of Caring Leadership*

As the example illustrates, the vocabulary of an informal writing style is made up of generally familiar rather than unfamiliar words and expressions, although slang and dialect are usually avoided. An informal style approximates the cadence and structure of spoken English while conforming to the grammatical conventions of written English.

Writers who consciously attempt to create a style usually defeat their purpose. Attempting to impress readers with a flashy writing style can lead to affectation; attempting to impress them with scientific objectivity can produce a style that is dull and lifeless. Technical writing need be neither affected nor dull. It can and should be simple, clear, direct, even interesting—the key is to master basic writing skills and always to keep your readers in mind. What will be both informative and interesting to your audience? When that question is uppermost in your mind as you apply the steps of the writing process, you will achieve an interesting and informative writing style. See "Five Steps to Successful Writing."

Writer's Checklist: Developing an Effective Style

- ☑ Use the active voice—not exclusively but as much as possible without becoming awkward or illogical.
- ☑ Use parallel structure whenever a sentence presents two or more thoughts of equal importance.
- ☑ Use a variety of sentence structures to avoid a monotonous style.
- ☑ Avoid stating positive thoughts in negative terms (write "40 percent responded" instead of "60 percent failed to respond"). See also positive writing and ethics in writing.
- ☑ Concentrate on achieving the proper balance between emphasis and subordination.

subordination

Use subordination to show, by the structure of a sentence, the appropriate relationship between ideas of unequal importance. By putting less important ideas in subordinate clauses or phrases you emphasize your main idea.

- Cigarette smoking is a risk factor for stroke. Smoking is also linked to heart disease and cancer.
  [The two ideas are equally important.]
- Cigarette smoking, *which is a risk factor for stroke*, is linked to heart disease and cancer.
  [The risk factor for stroke is subordinated; the link to heart disease and cancer is emphasized.]
- Cigarette smoking, *which is linked to heart disease and cancer*, is a risk factor for stroke.
  [The link to heart disease and cancer is subordinated; the risk factor for stroke is emphasized.]

S

Effective subordination can be used to achieve conciseness, emphasis, and sentence variety. For example, consider the following sentences.

DEPENDENT CLAUSE	The research report, *which covered eighty pages*, was carefully documented.
PHRASE	The research report, *covering eighty pages*, was carefully documented.
SINGLE MODIFIER	The *eighty-page* research report was carefully documented.

Subordinating conjunctions (*because*, *if*, *while*, *when*, *though*) achieve subordination effectively.

- A buildup of deposits is impossible *because* the pipes are flushed with water every day.

You may use a coordinating conjunction (*and*, *but*, *for*, *nor*, *or*, *so*, *yet*) to concede that an opposite or balancing fact is true; however, a subordinating conjunction (*although*, *since*, *while*) can often make the point more smoothly.

- *Although* their lab is well funded, ours is better equipped.

The relationship between a conditional statement and a statement of consequences is clearer if the condition is expressed as a subordinate clause.

- *Because* the tests were delayed, the surgery was postponed.

Relative pronouns (*who*, *whom*, *which*, *that*) can be used effectively to combine related ideas within sentences.

- The OnlinePro, *which* protects computers from malicious programs and e-mail attachments, makes your system "invisible" to hackers.

Avoid overlapping subordinate constructions that depend on the preceding construction. Overlapping can make the relationship between a relative pronoun and its antecedent less clear.

OVERLAPPING	Shock, *which* often accompanies severe injuries and infections, is a failure of the circulation, *which* is marked by a fall in blood pressure *that* initially affects the skin (*which* explains pallor) and later the vital organs such as the kidneys and brain.
CLEAR	Shock often accompanies severe injuries and infections. Marked by a fall in blood pressure, it is a failure of the circulation, initially to the skin (thus producing pallor) and later to the vital organs like the kidneys and the brain.

S

substantives (*see* **sentence construction**)

suffixes

A suffix is a letter or letters added to the end of a word to change its meaning in some way. Suffixes can change the part of speech of a word.

NO SUFFIX	The proposal was *thorough*. [adjective]
SUFFIX	The *thoroughness* is obvious. [noun]
SUFFIX	The proposal *thoroughly* described the problem. [adverb]

The suffix *-like* is sometimes added to nouns to make them into adjectives. The resulting compound word is hyphenated only if it is unusual or might not immediately be clear (childlike, lifelike, but dictionary-like, robot-like).

surveys (*see* questionnaires)

synonyms

A synonym is a word that means nearly the same thing as another word does (seller, vendor, supplier). The dictionary definitions of synonyms are similar, but the connotations may differ. For example, a *seller* may be the same thing as a *supplier*, but the term *supplier* does not suggest a retail transaction as strongly as *seller* does.

Do not try to impress your readers by finding fancy or obscure synonyms in a thesaurus; the result is likely to be affectation. See also connotation/denotation and antonyms.

syntax

Syntax refers to the way that words, phrases, and clauses are combined to form sentences. In English, the most common structure is the subject-verb-object pattern. For more information about the word order of sentences, see sentence construction, sentence faults, sentence fragments, and sentence variety.

T

tables of contents

A table of contents is typically included in a document longer than ten pages. It previews what the work contains and how it is organized, and it allows readers looking for specific information to locate sections by page number quickly and easily.

When creating a table of contents, use the major headings and subheadings of your document exactly as they appear in the text, as shown in the entry formal reports. (See the table of contents in Figure F–5 on page 203.) Note that the table of contents is typically placed in the front matter so that it follows the title page and abstract, and precedes the list of tables or figures, the foreword, and the preface.

tables

A table can present data, such as statistics, more concisely than text and more accurately than graphs. A table facilitates precise comparisons of data by organizing it into rows and columns. However, overall trends are more easily conveyed in charts, graphs, and other visuals.

Table Elements

Tables typically include the elements shown in Figure T–1 on page 524.

Table Number. Table numbers are usually Arabic and should be assigned sequentially to the tables throughout the document.

Table Title. The title, which is normally placed just above the table, should describe concisely what the table represents.

Box Head. The box head contains the column headings, which should be brief but descriptive. Units of measurement, where necessary, should be either specified as part of the heading or enclosed in parentheses beneath the heading. Standard abbreviations and symbols are acceptable. Avoid vertical lettering whenever possible.

Table 1. Estimated Emissions from Electric Power Generation (tons per gigawatthour)

Fuel	Sulphur Dioxide	Nitrogen Oxides	Parti-culate Matter	Carbon Dioxide	Volatile Organic Compounds
Eastern coal	1.74	2.90	0.10	1,000	0.06
Western coal	0.81	2.20	0.06	1,039	0.09
Gas	0.003	0.57	0.02	640	0.05
Biomass	0.06	1.25	0.11	0*	0.61
Oil	0.51	0.63	0.02	840	0.03
Wind	0	0	0	0	0
Geothermal	0	0	0	0	0
Hydro	0	0	0	0	0
Solar	0	0	0	0	0
Nuclear	0	0	0	0	0

*Net emissions.

SOURCE: Department of Energy

Table number · Table title · Box head · Column headings · Stub · Body · Rule · Footnote · Source line

FIGURE T–1. Elements of a Table

Stub. The stub, the left vertical column of a table, lists the items about which information is given in the body of the table.

Body. The body comprises the data below the column headings and to the right of the stub. Within the body, arrange columns so that the items to be compared appear in adjacent rows and columns. Align the numerical data in columns for ease of comparison, as shown in Figure T–1. Where no information exists for a specific item, substitute a row of dots or a dash to acknowledge the gap.

Rules. Rules are the lines that separate the table into its various parts. Horizontal rules are placed below the title, below the body of the table, and between the column headings and the body of the table. Tables are usually open at the sides. The columns within the table may be separated by vertical rules only if such lines aid clarity.

Footnotes. Footnotes are used for explanations of individual items in the table. Symbols (such as * and †) or lowercase letters (sometimes in parentheses) rather than numbers are ordinarily used to key table footnotes because numbers might be mistaken for numerical data or could be confused with the numbering system for text footnotes. See also documenting sources.

Source Line. The source line identifies where the data originated. When a source line is appropriate, it appears below the table. Many organizations place the source line below the footnotes. See also copyright and plagiarism.

Continued Tables. When a table must be divided so that it can be continued on another page, repeat the column headings and give the table number at the head of each new page with a "continued" label (for example, "Table 3, *continued*").

Informal Tables

To list relatively few items that would be easier for the reader to grasp in tabular form, you can use an informal table, as long as you introduce it properly. Although informal tables do not need titles or table numbers to identify them, they do require column headings that accurately describe the information listed, as shown in Figure T–2.

The sound-intensity levels (decibels) for the three frequency bands (in hertz) were determined to be the following:

Frequency Band (Hz)	*Decibels*
600–1199	68
1200–2399	62
2400–4800	53

FIGURE T–2. Informal Table

technical manuals (*see* manuals)

technical writing style

The goal of technical writing is to enable readers to use a technology or understand a process or concept. Because the subject matter is more important than the writer's voice, technical writing style uses an objective, not a subjective, tone. The writing style is direct and utilitarian, emphasizing exactness and clarity rather than elegance or allusiveness. A technical writer uses figurative language only when a figure of speech would facilitate understanding.

Technical writing is often—but not always—aimed at readers who are not experts in the subject, such as consumers or employees learning to operate unfamiliar equipment. Because such audiences are inexperienced and the procedures described may involve hazardous material or equipment, clarity becomes an ethical as well as a stylistic concern. (See also ethics in writing.) Figure T–3 is an excerpt from a technical manual that instructs eye-care specialists about operating testing equipment. As the figure illustrates, visuals as well as layout and design enhance clarity, providing the reader with both text and visual information.

Technical writing style may use a technical vocabulary appropriate for the reader, as Figure T–3 shows (*monocular electrode*, *saccade*), but it avoids affectation. Good technical writing also avoids overusing the passive voice. See also instructions, organization, and process explanation.

MONOCULAR RECORDING

A monocular electrode configuration might be useful to measure the horizontal eye movement of each eye separately. This configuration is commonly used during the voluntary saccade tests.

1. Place electrodes at the inner canthi of both eyes (**A and B**, Figure 51).
2. Place electrodes at the outer canthi of both eyes (**C and D**).
3. Place the last electrode anywhere on the forehead (**E**).
4. Plug the electrodes into the jacks described in Figure 52 on the following page.

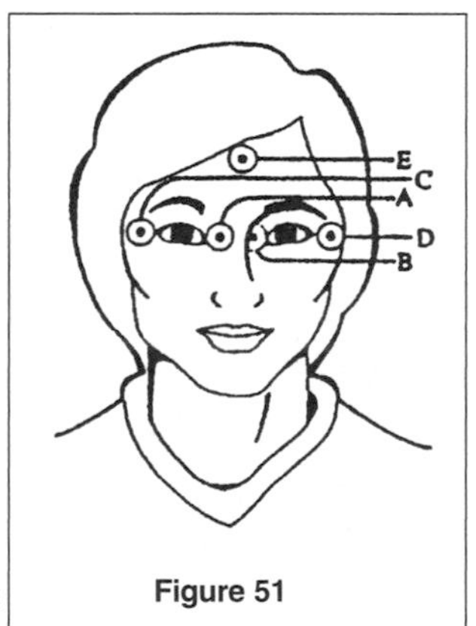

Figure 51

With this configuration, the Channel A display moves upward when the patient's left eye moves nasally and moves downward when the patient's left eye moves temporally. The Channel B display moves upward when the patient's right eye moves temporally and moves downward when the patient's right eye moves nasally.

NOTE: *During all testing, the Nystar Plus identifies and eliminates eye blinks without vertical electrode recordings.*

FIGURE T–3. Technical Writing Style

telegraphic style

Telegraphic style condenses writing by omitting articles, pronouns, conjunctions, and transitions. Although conciseness is important, especially in instructions, writers sometimes try to achieve conciseness by omitting necessary words. Telegraphic style forces readers to supply the missing words mentally, thus creating the potential for misunderstandings. Compare the following two passages and notice how much easier the revised version reads (the added words are italicized).

TELEGRAPHIC Take following action when treating serious burns. Remove loose clothing on or near burn. Cover injury with clean dressing and wash area around burn. Secure dressing with tape. Separate fingers/toes with gauze/cloth to prevent sticking. Do not apply medication unless doctor prescribes.

CLEAR Take *the* following action when treating *a* serious burn. Remove *any* loose clothing on or near *the* burn. Cover *the* injury with *a* clean dressing and wash *the* area around *the* burn. *Then* secure *the* dressing with tape. Separate *the* fingers *or* toes with gauze *or* cloth to prevent *them from* sticking *together.* Do not apply medication unless *a* doctor prescribes *it.*

Telegraphic style can also produce ambiguity, as the following example, written for a crane operator, demonstrates.

AMBIGUOUS Grasp knob and adjust lever before raising boom. [Does this sentence mean that the reader should *adjust the lever* or also *grasp an "adjust lever"*?]

CLEAR Grasp *the* knob and adjust *the* lever before raising *the* boom.

Although you may save yourself work by writing telegraphically, your readers will have to work that much harder to decipher your meaning.

T

tenant / tenet

A *tenant* is one who holds or temporarily occupies a property owned by another person. (The *tenant* was upset by the rent increase.) A *tenet* is an opinion or principle held by a person, an organization, or a system. (Competition is a central *tenet* of capitalism.)

tense

DIRECTORY

Tense is the grammatical term for verb forms that indicate time distinctions. The six tenses in English are past, past perfect, present, present perfect, future, and future perfect. Each tense also has a corresponding progressive form.

TENSE	BASIC FORM	PROGRESSIVE FORM
Past	I began	I was beginning
Past perfect	I had begun	I had been beginning
Present	I begin	I am beginning
Present perfect	I have begun	I have been beginning
Future	I will begin	I will be beginning
Future perfect	I will have begun	I will have been beginning

Perfect tenses allow you to express a prior action or condition that continues in a present, past, or future time.

PRESENT PERFECT — I *have begun* to compile the survey results and will continue for the rest of the month.

PAST PERFECT — I *had begun* to read the manual when the lights went out.

FUTURE PERFECT — I *will have begun* this project by the time funds are allocated.

Progressive tenses allow you to describe some ongoing action or condition in the present, past, or future.

PRESENT PROGRESSIVE — I *am beginning* to be concerned that we will not meet the deadline.

PAST PROGRESSIVE — I *was beginning* to think we would not finish by the deadline.

FUTURE PROGRESSIVE — I *will be requesting* a leave of absence when this project is finished.

Past Tense

The simple past tense indicates that an action took place in its entirety in the past. The past tense is usually formed by adding *-d* or *-ed* to the root form of the verb. (We *closed* the office early yesterday.)

Past Perfect Tense

The past perfect tense (also called *pluperfect*) indicates that one past event preceded another. It is formed by combining the helping verb *had* with the past-participle form of the main verb. (He *had finished* by the time I arrived.)

Present Tense

The simple present tense represents action occurring in the present, without any indication of time duration. (I *ride* the train.)

A general truth is always expressed in the present tense. (He believes the saying "Time *heals* all wounds.") The present tense can be used to present actions or conditions that have no time restrictions. (Water *boils* at 212 degrees Fahrenheit.) Similarly, the present tense can be used to indicate habitual action. (I *pass* the coffee shop every day.) The present tense is also used for the "historical present," as in newspaper headlines (FDA *Approves* New Cancer Drug) or as in references to an author's opinion or a work's contents—even though it was written in the past.

- In her 1979 article, Carolyn Miller *argues* that technical writing possesses significant humanistic value.

Present Perfect Tense

The present perfect tense describes something from the recent past that has a bearing on the present—a period of time before the present but after the simple past. The present perfect tense is formed by combining a form of the helping verb *have* with the past-participle form of the main verb. (We *have finished* the draft and can now revise it.)

Future Tense

The simple future tense indicates a time that will occur after the present. It uses the helping verb *will* (or *shall*) plus the main verb. (I *will finish* the job tomorrow.) Do not use the future tense needlessly; doing so merely adds complexity.

- This system ~~will be~~ *is* explained on page 3.
- When you press this button, the hoist ~~will move~~ *moves* the plate into position.

Future Perfect Tense

The future perfect tense indicates action that will have been completed at the time of or before another future action. It combines *will have* and the past participle of the main verb. (She *will have driven* 1,400 miles by the time she returns.)

TIPS FOR USING THE PROGRESSIVE FORM

The progressive form of the verb is composed of two features: a form of the helping verb *be* and the *-ing* form of the base verb.

PRESENT PROGRESSIVE	I *am updating* the Web site.
PAST PROGRESSIVE	I *was updating* the Web site last week.
FUTURE PROGRESSIVE	I *will be updating* the Web site regularly.

The present progressive is used in three ways:

1. To refer to an action that is in progress at the moment of speaking or writing:
   - The secretary *is taking* the meeting minutes.
2. To highlight that a state or an action is not permanent:
   - The office temp *is helping* us for a few weeks.
3. To express future plans:
   - The summer intern *is leaving* to return to school this Friday.

The past progressive is used to refer to a continuing action or condition in the past, usually with specified limits.

- I *was failing* calculus until I got a tutor.

The future progressive is used to refer to a continuous action or condition in the future.

- We *will be monitoring* his condition all night.

Verbs that express mental activity (*believe*, *know*, *see*, and so on) are generally not used in the progressive.

- I ~~am believing~~ *believe* the defendant's testimony.

Shift in Tense

Be consistent in your use of tense. The only legitimate shift in tense records a real change in time. Illogical shifts in tense will only confuse your readers.

- Before he installed the circuit, the technician ~~cleans~~ [*cleaned*] the contacts.

test reports

The test report differs from the more formal laboratory report in both size and scope. Considerably smaller, less formal, and more routine than the laboratory report, the test report can be presented as a memo or a letter, depending on whether its recipient is inside or outside the organization. Either way, the report should have a subject line at the beginning to identify the test being discussed. See also correspondence.

The opening of a test report should state the test's purpose, unless it is obvious from the subject line. The body of the report presents the data and, if relevant, describes procedures used to conduct the test. State the results of the test and, if necessary, interpret them. Conclude the report, if appropriate, with any recommendations made as a result of the test. Figure T–4 on page 532 shows a report of tests required by a government agency to monitor asbestos in the air at a highway construction site.

that / which / who

The word *that* is often overused and can foster wordiness.

- [*When*] ~~I think that when~~ this project is finished [*, I think*] ~~that~~ you should publish the results.

However, include *that* in a sentence if it avoids ambiguity or it improves the pace. See also conciseness.

- Some designers fail to appreciate [*that*] the workers who operate equipment constitute an important safety system.

Use *which*, not *that*, with nonrestrictive clauses (clauses that do not change the meaning of the basic sentence). See also restrictive and nonrestrictive elements.

Biospherics, Inc.

4928 Wyaconda Road
Rockville, MD 20852

Phone: 301-598-9011
Fax: 301-598-9570

September 11, 2006

Safety Committee
The Angle Company, Inc.
1869 Slauson Boulevard
Waynesville, VA 23927

Subject: Monitoring Airborne Asbestos at the Route 66 Site

On August 29, Biospherics, Inc., performed asbestos-in-air monitoring at your Route 66 construction site, near Front Royal, Virginia. Six people and three construction areas were monitored.

All monitoring and analyses were performed in accordance with "Occupational Exposure to Asbestos," U.S. Department of Health and Human Services, Public Health Service, National Institute for Occupational Safety and Health, 2004. Each worker or area was fitted with a battery-powered personal sampler pump operating at a flow rate of approximately two liters per minute. The airborne asbestos was collected on a 37 mm Millipore-type AA filter mounted in an open-face filter holder. Samples were collected over an 8-hour period.

In all cases, the workers and areas monitored were exposed to levels of asbestos fibers well below the standard set by OSHA. The highest exposure found was that of a driller exposed to 0.21 fibers per cubic centimeter. The driller's sample was analyzed by scanning electron microscopy followed by energy-dispersive X-ray techniques that identify the chemical nature of each fiber, to identify the fibers as asbestos or other fiber types. Results from these analyses show that the fibers present were tremolite asbestos. No nonasbestos fibers were found.

If you need more details, please let me know.

Yours truly,

Gary Geirelach

Gary Geirelach
Chemist
gg2@Bios.org

T

FIGURE T–4. Test Report

NONRESTRICTIVE	After John left the law firm, *which* is the largest in the region, he started a private practice.
RESTRICTIVE	Companies *that* diversify usually succeed.

That and *which* should refer to animals and things; *who* should refer to people.

- Companies *that* fund basic research must not expect immediate results.
- The jet stream, *which* flows west to east, usually travels in excess of 67 miles per hour.
- Diane Stoltzfus, *who* retires tomorrow, worked 23 years for the company.

there / their / they're

There is an expletive (a word that fills the position of another word, phrase, or clause) or an adverb.

EXPLETIVE	*There* were more than 1,500 people at the conference.
ADVERB	More than 1,500 people were *there.*

Their is the possessive case form of *they.* (Managers should check *their* e-mail regularly.) *They're* is a contraction of *they are.* (Clients tell us *they're* pleased with our services.)

thesaurus

A thesaurus is a book or an electronic file of words and their synonyms and antonyms, arranged or retrievable by categories. Thoughtfully used, a thesaurus can help you with word choice during the revision phase of the writing process. However, the variety of words it offers may tempt you to choose an inappropriate word for the context or to use an obscure synonym just because it is available. Use a thesaurus only to clarify or refine your meaning, not to impress your readers. (See also affectation.) Never use a word unless you are sure of its meanings; its connotations might be unknown to you and could mislead your readers.

titles

Titles are important because many readers decide whether to read documents—such as reports, e-mails, and newsletter articles—based on their titles. This entry treats both creating titles and referring to them in your writing. For advice on labeling figures and tables, see visuals.

Reports and Long Documents

Titles for reports, proposals, articles, and similar documents should state the document's topic, reflect its tone, and indicate its scope and purpose, as in the following title of a scientific article.

- "Effects of 60-Hertz Electric Fields on Embryo Chick Development, Growth, and Behavior"

Such titles should be concise but not so short that they are not specific. For example, the title "Electric Fields and Living Organisms" announces the topic and might be appropriate for a book, but it does not answer important questions that readers of an article would expect, such as "What is the relationship between electric fields and living organisms?" and "What aspects of the organisms are related to electric fields?"

Avoid titles with such redundancies as "Notes on," "Studies on," or "A Report on." However, works like annual reports or feasibility reports should be identified as such in the title because this information specifies the purpose and scope of the report. For titles of periodic or progress reports, indicate the dates in a subtitle ("Quarterly Report on Hospital Admission Rates: January–March 2006"). Avoid using abbreviations, chemical formulas, and the like in your title unless the work is addressed exclusively to specialists in the field. For multivolume publications, repeat the title on each volume and include the subtitle and number of each volume.

Titles do not use the sentence form, except for titles of articles in newsletters and magazines that ask a rhetorical question.

- "Is Online Learning Right for You?"

T

Memos and E-mails

Subject lines of memos and e-mails function as titles and should concisely describe the topic of the message. Such titles must be specific and accurate to aid in filing and later retrieval.

VAGUE Subject: Tuition Reimbursement

SPECIFIC Subject: Tuition Reimbursement for Time-Management Seminar

Recipients often use subject-line titles to prioritize and file their incoming messages. Although the title in the subject line announces your topic, you should still begin the memo or e-mail with an opening that provides context for the message.

Formatting Titles

Capitalization. Capitalize the initial letters of the first and last words of a title as well as all major words in the title. Do not capitalize articles (*a*, *an*, *the*), coordinating conjunctions (*and*, *but*), or short prepositions (*at*, *in*, *on*, *of*) unless they begin or end the title (*The Lives of a Cell*). Capitalize prepositions in titles if they contain more than four letters (*Between*, *Because*, *Until*, *After*).

Italics. Use italics or underlining when referring to titles of separately published works, such as books, periodicals, newspapers, pamphlets, brochures, legal cases, movies, and television programs.

- *Turning Workplace Conflict into Collaboration* by Joyce Richards was reviewed in the *New York Times*. [book, newspaper]

Abbreviations of such titles are italicized if their spelled-out forms would be italicized.

- The *NYT* is one of the nation's oldest newspapers.

Italicize the titles of compact discs, videotapes, plays, long poems, paintings, sculptures, and long musical works.

CD-ROM	*Computer Security Tutorial on CD-ROM*
PLAY	Arthur Miller's *Death of a Salesman*
LONG POEM	T. S. Eliot's *The Wasteland*
PAINTING	M. C. Escher's *Drawing Hands*
SCULPTURE	Auguste Rodin's *The Thinker*
MUSICAL WORK	Gershwin's *Porgy and Bess*

Quotation Marks. Use quotation marks when referring to parts of publications, such as chapters of books and articles or sections within periodicals.

- His report, "Effects of Government Regulations on Motorcycle Safety," cited the article "No-Fault Insurance and Motorcycles" published in *American Motorcyclist* magazine.

Titles of reports, essays, short poems, short musical works (including songs), short stories, and single episodes of radio and television programs are also enclosed in quotation marks.

REPORT	"Analysis of Ethics Cases at CGE Corporation"
ESSAY	Ralph Waldo Emerson's "The American Scholar"
SHORT POEM	Robert Frost's "The Road Not Taken"
SONG	Bob Dylan's "Like a Rolling Stone"
SHORT STORY	Nathaniel Hawthorne's "The Birthmark"
TV PROGRAM	"In Depth: Jacques Barzun" on C-Span [episode of *In Depth*]

Special Cases. Some titles, by convention, are not set off by quotation marks, underlining, or italics. Such titles follow standard practice for capitalization.

- Technical Communication [college course title], Amazon.com, Old Testament, Magna Carta, the Constitution, Lincoln's Gettysburg Address, the Lands' End Catalog

to / too / two

To, *too*, and *two* are confused only because they sound alike. *To* is used as a preposition or to mark an infinitive. See verbs.

- Send the report *to* the district manager. [preposition]
- I do not wish *to* attend. [mark of the infinitive]

Too is an adverb meaning "excessively" or "also."

- The price was *too* high. [excessively]
- I, *too*, thought it was high. [also]

Two is a number (*two* buildings, *two* concepts).

tone

Tone is the attitude a writer expresses toward the subject and his or her readers. In workplace writing, tone may range widely—depending on the purpose, situation, context, audience, and even the medium of a communication. For example, in an e-mail message to be read only by an associate who is also a friend, your tone might be casual.

- Your proposal to Smith and Kline is super. We'll just need to hammer out the schedule. If we get the contract, I owe you lunch!

In a memo to your manager or superior, however, your tone might be more formal and respectful.

- I think your proposal to Smith and Kline is excellent. I have marked a couple of places where I'm concerned that we are committing ourselves to a schedule that we might not be able to keep. If I can help in any other way, please let me know.

In a message that serves as a report to numerous readers, the tone would be professional, without the more personal style that you would use with an individual reader.

- The Smith and Kline proposal appears complete and thorough, based on our department's evaluation. Several small revisions, however, would ensure that the company is not committing itself to an unrealistic schedule. These revisions are marked on the copy of the report attached to this message.

The word choice, the introduction, and even the title contribute to the overall tone of your document. For instance, a title such as "Ecological Consequences of Diminishing Water Resources in California" clearly sets a different tone from "What Happens When We've Drained California Dry?" The first title would be appropriate for a report; the second title would be appropriate for a popular magazine or newsletter article. See also correspondence.

trade journal articles

DIRECTORY

T

Trade journal articles are written for professional periodicals that aim to further the knowledge in a field among specialists in a field. (See readers.) Such periodicals, commonly known as trade journals (or professional or scholarly journals), are often the official publications of professional societies. *Technical Communication*, for example, is an official voice of the Society for Technical Communication. Other professional publications include *IEEE Transactions on Communications*, *Chemical Engineering*, *Nucleonics Week*, and hundreds of similar titles. Professional staff people, such as engineers, scientists, educators, and legal professionals, regularly contribute articles to trade journals.

Writing a trade journal article in your field can make your work more widely known, provide publicity for your employer, give you a sense of satisfaction, and even improve your chances for professional advancement.

Planning the Article

When you are thinking about writing an article for a trade journal, consider the following questions:

- Is your work or your knowledge of the subject original? If not, what is there about your approach that justifies publication?
- Will the significance of the article justify the time and effort needed to write it?
- What parts of your work, project, or study are most appropriate to include in the article?

To help you answer those questions, learn as much as possible about the periodical or periodicals to which you plan to submit an article and consult your colleagues for advice. Once you have decided on several journals, consider the following factors about each one:

- The professional interests and size of its readership
- The professional reputation of the journal
- The appropriateness of your article to the journal's goals, as stated on its masthead page or in a mission statement on its Web site
- The frequency with which its articles are cited in other journals

Next, read back issues of the journal or journals to find out information such as the amount and kind of details that the articles include, their length, and the typical writing style. (See also context.)

If your subject involves a particular project, begin work on your article when the project is in progress. That allows you to write the draft in manageable increments and record the details of the project while they are fresh in your mind. It also makes the writing integral to the project and may even reveal any weaknesses in the design or details, such as the need for more data.

As you plan your article, decide whether to invite one or more coauthors to join you. Doing so can add strength and substance to an article. However, as with all collaborative writing, you should establish a schedule, assign tasks, and designate a primary author to ensure that the finished article reads smoothly.

Gathering the Data

As you gather information, take notes from all the sources of primary and secondary research available to you. Begin your research with a careful review of the literature to establish what has been published about your topic. A review of the relevant information in your field can be insurance against writing an article that has already been published. (Some articles, in fact, begin with a literature review.) As you compile that information, record your references in full; include all the information you need to document the source. See also documenting sources.

Organizing the Draft

Some trade journals use a prescribed organization for the major sections. The following organization is common in scientific journals: introduction, materials and methods, results, discussion, and sources cited. Look closely at several issues of the journals that you target to determine how those or other sections are developed. If the major organization is not prescribed, choose and arrange the various sections of the draft in a way that shows your results to best advantage.

The best guarantee of a logically organized article is a good outline. (See outlining.) If you have coauthors, work from a common, well-developed outline to coordinate the various writers' work and make sure the parts fit together logically.

Preparing Sections of the Article

As you prepare an article, pay particular attention to a number of key sections and elements: the abstract, the introduction, the conclusion, visuals and tables, headings, and references.

Abstract. Although it will appear at the beginning of the article, write the abstract only after you have finished writing the body of the manuscript. Follow any instructions provided by the journal on writing abstracts and review those previously published in that journal. Prepare your abstract carefully—it will be the basis on which other researchers decide whether to read your article in full. Abstracts are often published independently in abstract journals and are a source of terms (called *keywords*) used to index, by subject, the original article for computerized information-retrieval systems.

Introduction. The introduction should discuss these aspects of the article:

- The purpose of the article
- A definition of the problem examined
- The scope of the article
- The rationale for your approach to the problem or project and the reasons you rejected alternative approaches
- Previous work in the field, including other approaches described in previously published articles

Above all, your introduction should emphasize what is new and different about your approach, especially if you are not dealing with a new concept. It should also demonstrate the overall significance of your project or approach by explaining how it fills a need, solves a current problem, or offers a useful application.

Conclusion. The conclusion section pulls together your results and interprets them in relation to the purpose of your study and the methods used to conduct it. Your conclusion must grow out of the evidence for the findings in the body of the article.

Visuals and Tables. Use visuals and tables wherever they are appropriate, but design each for a specific purpose: to describe a function, to show an external appearance, to show internal construction, to display statistical data, or to indicate trends. Used effectively and appropriately, visuals can clarify information and reinforce the point you are making in the article.

Headings. Headings are important for an article because they break the text into manageable portions. They also allow journal readers to understand the development of your topic and pinpoint sections of particular interest to them. Check the use of headings when you review back issues of the journal.

T

References. The references section (often titled "Works Cited") of the article lists the sources you used in the article. The specific format for listing sources varies from field to field. Usually the journal to which you submit an article will specify the form the editors require for citing sources. For a detailed discussion of using and citing sources, see documenting sources and quotations.

Preparing the Manuscript

Some trade journals recommend a particular style guide, like *The Chicago Manual of Style*, or offer a style sheet with detailed guidelines on style and format. Such style sheets often include specific instructions about how to format the manuscript, how many copies to submit, and how to handle abbreviations, symbols, mathematical equations, and the like. The following guidelines are typical.

- Double-space the manuscript, leaving one-inch margins all around, and number each page.
- Provide specific, accurate, and self-explanatory captions for all figures and tables. Add *call-outs* (labels) to those illustrations that need them.
- Provide clear and accurately worded labels for drawings and other illustrations.
- Place any mathematical equations on separate lines in the text and number them consecutively.
- Include only high-quality photographs or scanned images. If you submit an original photograph, check with the editor about any special requirements. Identify each photograph by writing lightly in pencil on the back. If necessary, indicate which edge of the photograph is the top.

Obtaining Publication Clearance

After you have finalized your article, submit a copy to your employer for review before sending it to the journal. A review will ensure that you have not inadvertently revealed any proprietary information. Likewise, secure permission ahead of time to print information for which someone else holds the copyright. See also plagiarism.

transition

Transition is the means of achieving a smooth flow of ideas from sentence to sentence, paragraph to paragraph, and subject to subject. Transition is a two-way indicator of what has been said and what will be said; it provides readers with guideposts for linking ideas and clarifying the relationship between them.

Transition can be obvious.

- *Having considered* the technical problems outlined in this proposal, *we move next* to the question of adequate staffing.

Transition can be subtle.

- *Even if* the technical problems could be solved, there *still remains* the problem of adequate staffing.

Either way, you now have your readers' attention fastened on the problem of adequate staffing, exactly what you set out to do.

Methods of Transition

Transition can be achieved in many ways: (1) using transitional words and phrases, (2) repeating keywords or key ideas, (3) using pronouns with clear antecedents, (4) using enumeration (1, 2, 3, or first, second, third), (5) summarizing a previous paragraph, (6) asking a question, and (7) using a transitional paragraph.

Certain words and phrases are inherently transitional. Consider the following terms and their functions:

FUNCTION	TERMS
Result	*therefore, as a result, consequently, thus, hence*
Example	*for example, for instance, specifically, as an illustration*
Comparison	*similarly, likewise, in comparison*
Contrast	*but, yet, still, however, nevertheless, on the other hand*
Addition	*moreover, furthermore, also, too, besides, in addition*
Time	*now, later, meanwhile, since then, after that, before that time*
Sequence	*first, second, third, initially, then, next, finally*

Within a paragraph, such transitional expressions clarify and smooth the movement from idea to idea. Conversely, the lack of transitional devices can make for disjointed reading. See also telegraphic style.

Transition Between Sentences

You can achieve effective transition between sentences by repeating keywords or key ideas from preceding sentences and by using pronouns that refer to antecedents in previous sentences. Consider the following short paragraph, which uses both of those means.

- Representative of many American university towns is Middletown. *This midwestern town*, formerly *a sleepy farming community*, is today the home of a large and vibrant *academic community*. Attracting students from all over the Midwest, *this university town* has grown very rapidly in the last ten years.

Enumeration is another device for achieving transition.

- The recommendation rests on *two conditions. First*, the department staff must be expanded to handle the increased workload. *Second*, sufficient time must be provided for training the new staff.

Transition Between Paragraphs

The means discussed so far for achieving transition between sentences can also be effective for achieving transition between paragraphs. For paragraphs, however, longer transitional elements are often required. One technique is to use an opening sentence that summarizes the preceding paragraph and then moves on to a new paragraph.

- One property of material considered for manufacturing processes is hardness. Hardness is the internal resistance of the material to the forcing apart or closing together of its molecules. Another property is ductility, the characteristic of material that permits it to be drawn into a wire. Material also may possess malleability, the property that makes it capable of being rolled or hammered into thin sheets of various shapes. Engineers must consider these properties before selecting manufacturing materials for use in production.

  The requirements of hardness, ductility, and malleability account for the high cost of such materials. . . .

Another technique is to ask a question at the end of one paragraph and answer it at the beginning of the next.

- New technology has always been feared because it has at times displaced some jobs. However, it invariably creates many more jobs than it eliminates. Almost always, the jobs eliminated by technological advances have been menial, unskilled jobs, and workers who have been displaced have been forced to increase their skills, which resulted in better and higher-paying jobs for them. *In view of these facts, is new technology really bad?*

  Certainly technology has given us unparalleled access to information and created many new roles for employees. . . .

A purely transitional paragraph may be inserted to aid readability.

- The problem of poor management was a key factor that has caused the weak performance of the company.

  Two other setbacks to the company's fortunes also marked the company's decline: the loss of many skilled workers through the early retirement program and the intensification of the rate of employee turnover.

  The early retirement program caused the failure . . .

If you provide logical organization and have prepared an outline, your transitional needs will easily be satisfied and your writing will have unity and coherence. During revision, look for places where transition is missing and add it. Look for places where it is weak and strengthen it.

trip reports

A trip report provides a permanent record of a business trip and its accomplishments. It provides managers with essential information about the results of the trip and can enable many employees to benefit from the information. See also reports.

A trip report is normally written as a memo or an e-mail and addressed to an immediate superior, as shown in Figure T–5. The subject line identifies the destination and dates of the trip. The body of the report explains why you made the trip, whom you visited, and what you accomplished. The report should devote a brief section to each major event and may include a heading for each section. You need not give

From: James D. Kerson <jdkerson@psys.com>
To: Roberto Camacho <rcamacho@psys.com>
Sent: Fri, 14 Jan 2005 12:16:30 EST
Subject: Trip to Smith Electric Co., Huntington, West Virginia, January 2005

Attachments: Expense Report.xls (25 KB)

I visited the Smith Electric Company in Huntington, West Virginia, to determine the cause of a recurring failure in a Model 247 printer and to fix it.

Problem
The printer stopped printing periodically for no apparent reason. Repeated efforts to bring it back online eventually succeeded, but the problem recurred at irregular intervals. Neither customer personnel operating the printer nor the local maintenance specialist was able to solve the problem.

Action
On January 3, I met with Ms. Ruth Bernardi, the Office Manager, who explained the problem. My troubleshooting did not reveal the cause of the problem then or on January 4.

Only when I tested the logic cable did I find that it contained a broken wire. I replaced the logic cable and then ran all the normal printer test patterns to make sure no other problems existed. All patterns were positive, so I turned the printer over to the customer.

Conclusion
There are over 12,000 of these printers in the field, and to my knowledge this is the first occurrence of a bad cable. Therefore, I do not believe the logic cable problem found at Smith Electric Company warrants further investigation.

==================================
James D. Kerson, Maintenance Specialist
Printer Systems, Inc.
1366 Federal St., Allentown, PA 18101
(610) 747-9955 Fax: (610) 747-9956
jdkerson@psys.com
www.psys.com
==================================

FIGURE T–5. Trip Report Sent as E-mail (with Attachment)

equal space to each event—instead, elaborate on the more important events. Follow the body of the report with the appropriate conclusions and recommendations. Finally, if required, attach a record of expenses to the trip report.

trite language (*see* clichés)

trouble reports

The trouble report is used to analyze such events as accidents, equipment failures, or health emergencies. For example, the report shown in Figure T–6 describes an accident involving personal injury. The report assesses the causes of the problem and suggests changes necessary to prevent its recurrence. Because it is usually an internal document, the trouble report normally follows a memo format. See also reports.

Consolidated Energy, Inc.

To: Marvin Lundquist, Vice President
Administrative Services
From: Kalo Katarlan, Safety Officer KK
Field Service Operations
Date: August 19, 2005
Subject: Field Service Employee Accident on August 5, 2005

The following is an initial report of an accident on Friday, August 5, 2005, involving John Markley that resulted in two days of lost time.

Accident Summary
John Markley stopped by a rewiring job on German Road. Chico Ruiz was working there, stringing new wire, and John was checking with Chico about the materials he wanted for framing a pole. Some tree trimming had been done in the area, and John offered to help remove some of the debris by loading it into the pickup truck he was driving. While John was loading branches into the bed of the truck, a piece broke off in his right hand and struck his right eye.

Accident Details
1. John's right eye was struck by a piece of tree branch. John had just undergone laser surgery on his right eye on Wednesday, August 3, to reattach his retina.
2. John immediately covered his right eye with his hand, and Chico Ruiz gave him a paper towel with ice to cover his eye and help ease the pain.

FIGURE T–6. Trouble Report

T

7. On Thursday, August 11, John returned to his eye surgeon. Although bruised, his eye was not damaged, and the surgically reattached retina was still in place.

Recommendations
To prevent a recurrence of such an accident, the Safety Department will require the following actions in the future:

- When working around and moving debris such as tree limbs or branches, all service crew employees must wear safety eyewear with side shields.
- All service crew employees must always consider the possibility of shock for an injured employee. If crew members cannot leave the job site to care for the injured employee, someone on the crew must call for assistance from the Service Center. The Service Center phone number is printed in each service crew member's handbook.

FIGURE T–6. Trouble Report (*continued*)

In the subject line of the memo, state the precise problem you are reporting. Then, in the body of the report, provide a detailed, precise description of the problem. What happened? Where and when did the problem occur? Was anybody hurt? Was there any property damage? Was there a work stoppage?

ETHICS NOTE Because insurance claims, workers' compensation awards, and even lawsuits may hinge on the information contained in a trouble report, be sure to include precise times, dates, locations, treatment of injuries, names of any witnesses, and any other crucial information. (Notice the careful use of language and factual detail in Figure T–6.) Be thorough and accurate in your analysis of the problem and support any judgments or conclusions with facts. Be objective: Always use a neutral tone and avoid assigning blame. If you speculate about the cause of the problem, make it clear to your reader that you are speculating. See also ethics in writing. ✦

In your conclusion, state what has been or will be done to correct the conditions that led to the problem. That may include training in safety practices, improved equipment, protective clothing, and so on.

T

try to

The phrase *try and* is colloquial for *try to.* For technical writing, use *try to.*

- Please try ~~and~~ *to* finish the report on time.

U

unity

Unity is singleness of purpose and treatment; a unified paragraph or document has a central idea and does not digress into unrelated topics.

The logical sequence provided through outlining is essential to achieving unity. An outline enables you to lay out the most direct route from introduction to conclusion, and it enables you to build each paragraph around a topic sentence that expresses a single idea.

Effective transition helps build unity, as well as coherence, because transitional terms clarify the relationship of each part to what precedes it.

up

Adding the word *up* to verbs often creates a redundant phrase. See also conciseness.

- You must open ~~up~~ the exhaust valve.

usability testing

Usability refers to whether readers can use a document to easily fulfill their goals or accomplish tasks. For example, the usability of a technical manual might refer to the ability of readers to perform a procedure accurately and smoothly with the aid of the instructions. The usability of a brochure might refer to the ability of readers to understand the contents quickly and decide whether to buy a product or use a service.

Ideally, usability is built into documents or products from the beginning of their life cycle during the early planning, development, and design phases. To ensure that a document is usable, it is important to focus on users throughout the evolution of a document. *Usability engineering*, or *user-centered design* (UCD), refers to the process of designing usable products and systems, with the user's needs in mind. Everything involved with a product or service, including its goals, context, and environment, are approached from the user's viewpoint.

Usability testing helps produce a document that reduces the learning curve, allows more functionality with less effort, and increases productivity. At the same time, by focusing development on users, companies also reap benefits in reduced costs and increased customer satisfaction in the following ways:

- The documents are functional and more likely to be used.
- Product and document changes can be made before they become expensive.
- Efficient document development and training are facilitated.
- The need for updates and maintenance releases is minimized.
- More documents (and products) are sold.
- The organization's reputation is enhanced.

Usability testing is a common way to involve users in the development process—for example, by testing periodic drafts on users and then revising the document in response to the test results. The process typically begins with the establishment of specific, quantitative, and measurable goals for documents. Next, the documents are designed to fulfill those goals. Finally, tests must be conducted with representative users to detect problems and determine whether the established goals have been achieved.

Usability testing involves teams of skilled usability specialists, interface designers, and technical writers. Usability testing has three main goals:

- To create a document that is easy to use and allows users to accomplish the tasks outlined in that document.
- To detect potential problems for users as well as to guide designers in resolving such issues. That process minimizes the risk of releasing ineffective documents and poorly designed products.
- To enable companies to avoid repeating mistakes when developing future documents.

Test participants typically are members of the actual target audience for the document. If the participants encounter problems with the document, it is likely that actual users will experience similar problems. For example, if page 15 of a tax form is unclear to test participants, it is likely to be confusing to most taxpayers.

Usability tests can rely on one or more of the following methods.

- *User testing.* Testers observe and record the actions of test participants who perform real tasks using the document.

- *Protocols.* Test participants make comments aloud as they read documents and perform document tasks, to reveal their thought processes, attitudes, and reasons for decision-making.
- *Comprehension tests.* Users complete tests to determine whether they understand and can recall document features, such as visuals or tables.
- *Surveys and interviews.* Testers interview users both before and after they read documents to determine their comprehension and attitudes and the clarity of the documents.

Analyzing the results of those test methods can help document designers detect problems and determine whether the document's purposes have been met. If test participants have trouble navigating a document or quickly locating specific items, for example, the organization as well as layout and design are revised. See also forms design, interviewing for information, and questionnaires.

usage

Usage describes the choices we make among the various words and expressions available in our language. The lines between standard English and nonstandard English and between formal and informal English are determined by those choices. Your guideline in any situation requiring such choices should be appropriateness: Is the word or expression you use appropriate to your readers and your subject? When it is, you are practicing good usage.

This book contains many entries that focus on various usages. For a complete list of the usage entries, which appear in italics throughout the book, see "Commonly Misused Words and Phrases." An up-to-date dictionary is also an invaluable aid in your selection of the right word.

WEB LINK ONLINE USAGE AND STYLE GUIDES

Bartleby.com provides classic reference books online, including the *American Heritage Book of English Usage*. For this and additional useful links, see *<bedfordstmartins.com/alredtech>* and select *Links for Handbook Entries.*

U

utilize

Do not use *utilize* as a long variant of *use*, which is the general word for "employ for some purpose." *Use* will almost always be clearer and less pretentious. See affectation.

V

vague words

A vague word is one that is imprecise in the context in which it is used. Some words encompass such a broad range of meanings that there is no focus for their definition. Words such as *real, nice, important, good, bad, contact, thing*, and *fine* are often called *omnibus words* because they can have so many meanings and interpretations. In speech, our vocal inflections help make the meanings of such words clear. Because you cannot rely on vocal inflections when you are writing, avoid using vague words. Be concrete and specific. See also abstract/concrete words and word choice.

VAGUE	It was a *good* meeting. [Why was it good?]
SPECIFIC	The meeting resolved three questions: pay scales, fringe benefits, and workloads.

verbals

Verbals are derived from verbs but function as nouns, adjectives, and adverbs. The three types of verbals are gerunds, infinitives, and participles.

Gerunds

A gerund is a verbal ending in *-ing* that is used as a noun. A gerund can be used as a subject, a direct object, the object of a preposition, a subjective complement, or an appositive.

- *Estimating* is an important managerial skill. [subject]
- I find *estimating* difficult. [direct object]
- We were unprepared for their *coming.* [object of preposition]
- Seeing is *believing.* [subjective complement]
- My primary departmental function, *programming*, occupies about two-thirds of my time on the job. [appositive]

Only the possessive form of a noun or pronoun should precede a gerund.

- *John's* working has not affected his grades.
- *His* working has not affected his grades.

Infinitives

An infinitive is the bare, or uninflected, form of a verb (for example, *go*, *run*, *fall*, *talk*, *dress*, *shout*) without the restrictions imposed by person and number. Along with the gerund and the participle, it is one of the nonfinite verb forms. The infinitive is generally preceded by the word *to*, which, although not an inherent part of the infinitive, is considered to be the sign of an infinitive. An infinitive is a verbal and can function as a noun, an adjective, or an adverb.

- *To expand* is not the only objective. [noun]
- These are the instructions *to follow.* [adjective]
- The company struggled *to survive.* [adverb]

The infinitive can reflect two tenses: the present and (with a helping verb) the present perfect.

- to go [present tense]
- to have gone [present perfect tense]

The most common mistake made with infinitives is using the present perfect tense when the simple present tense is sufficient.

- I should not have tried to *go* ~~have gone~~ so early.

Infinitives formed with the root form of transitive verbs can express both active and (with a helping verb) passive voice.

- to hit [present tense, active voice]
- to have hit [present perfect tense, active voice]
- to be hit [present tense, passive voice]
- to have been hit [present perfect tense, passive voice]

A split infinitive is one in which an adverb is placed between the sign of the infinitive, *to*, and the infinitive itself. Because they make up a grammatical unit, the infinitive and its sign are better left intact than separated by an intervening adverb.

- To ~~initially~~ build the table in the file, you could input transaction records containing the data necessary to construct the record and table.

However, it may occasionally be better to split an infinitive than to allow a sentence to become awkward, ambiguous, or incoherent.

AMBIGUOUS	She agreed immediately *to deliver* the toxic materials. [This sentence could be interpreted to mean that she agreed immediately.]
CLEAR	She agreed *to* immediately *deliver* the toxic materials. [This sentence is no longer ambiguous.]

Participles

A participle is a verb form that functions as an adjective. Present participles end in *-ing.*

- *Declining* sales forced us to close one branch office.

Past participles end in *-ed*, *-t*, *-en*, *-n*, or *-d.*

- What are the *estimated* costs?
- Repair the *bent* lever.
- Return the *broken* part.
- What are the metal's *known* properties?
- The story, *told* many times before, was still interesting.

The perfect participle is formed with the present participle of the helping verb *have* plus the past participle of the main verb.

- *Having gotten* [perfect participle] a large bonus, the *smiling* [present participle], *contented* [past participle] sales representative worked harder than ever.

A participle cannot be used as the verb of a sentence. Inexperienced writers sometimes make that mistake, and the result is a <u>sentence fragment</u>.

- The committee chairperson was responsible~~. His~~ *, his* vote being the decisive one.

- The committee chairperson was responsible. His vote ~~being~~ *was* the decisive one.

For information on participial and infinitive phrases, see <u>phrases</u>.

verbs

DIRECTORY

A verb is a word or group of words that describes an action (The copier *jammed* at the beginning of the job), states how something or someone is affected by an action (He *was disappointed* that the proposal was rejected), or affirms a state of existence (She *is* a district manager now).

Types of Verbs

Verbs are either transitive or intransitive. A *transitive verb* requires a direct object to complete its meaning.

- They *laid* the foundation on October 24.
  [*Foundation* is the direct object of the transitive verb *laid.*]
- Rosalie Anderson *wrote* the treasurer a letter.
  [*Letter* is the direct object of the transitive verb *wrote.*]

An *intransitive verb* does not require an object to complete its meaning. It makes a full assertion about the subject without assistance (although it may have modifiers).

- The engine *ran.*
- The engine *ran* smoothly and quietly.

A *linking verb* is an intransitive verb that links a complement to the subject.

- The carpet *is* stained.
  [*Is* is a linking verb; *stained* is a subjective complement.]

Some intransitive verbs, such as *be, become, seem,* and *appear,* are almost always linking verbs. A number of others, such as *look, sound, taste, smell,* and *feel,* can function as either linking verbs or simple intransitive verbs. If you are unsure about whether one of those verbs is a linking verb, try substituting *seem;* if the sentence still makes sense, the verb is probably a linking verb.

- Their antennae *feel* delicate.
  [*Seem* can be substituted for *feel*—thus *feel* is a linking verb.]
- Their antennae *feel* delicately for their prey.
  [*Seem* cannot be substituted for *feel*; in this case, *feel* is a simple intransitive verb.]

Forms of Verbs

Verbs are described as being either finite or nonfinite.

Finite Verbs. A finite verb is the main verb of a clause or sentence. It makes an assertion about its subject and often serves as the only verb in its clause or sentence. (The telephone *rang*, and the receptionist *answered* it.) See also sentence construction.

A helping verb (sometimes called an *auxiliary verb*) is used in a verb phrase to help indicate mood, tense, and voice. (The phone *had* rung.) Phrases that function as helping verbs are often made up of combinations with the sign of the infinitive, *to* (for example, *am going to*, *is about to*, *has to*, and *ought to*). The helping verb always precedes the main verb, although other words may intervene. (Machines *will* never completely *replace* people.)

Nonfinite Verbs. Nonfinite verbs are verbals—verb forms that function as nouns, adjectives, or adverbs.

A *gerund* is a noun that is derived from the *-ing* form of a verb. (*Seeing* is *believing*.) An *infinitive*, which uses the root form of a verb (usually preceded by *to*), can function as a noun, an adverb, or an adjective.

- He hates *to complain*. [noun, direct object of *hates*]
- The valve closes *to stop* the flow. [adverb, modifies *closes*]
- This is the proposal *to consider*. [adjective, modifies *proposal*]

A *participle* is a verb form that can function as an adjective.

- The *rejected* proposal was ours.
  [*Rejected* is a verb form that is used as an adjective modifying *proposal*.]

Properties of Verbs

Verbs must (1) agree in person with personal pronouns functioning as subjects, (2) agree in tense and number with their subjects, and (3) be in the appropriate voice.

 TIPS FOR AVOIDING SHIFTS IN VOICE, MOOD, OR TENSE

To achieve clarity in your writing, you must maintain consistency and avoid shifts. A shift is an abrupt change in voice, mood, or tense. Pay special attention when you edit your writing to check for the following types of shifts.

VOICE

- The captain permits his crew to go ashore, but ~~they are not~~ *he does not permit* ~~permitted~~ *them* to go downtown.

  [The entire sentence is now in the active voice.]

MOOD

- Reboot your computer,/ and ~~you should~~ empty the cache.

  [The entire sentence is now in the imperative mood.]

TENSE

- I was working quickly, and suddenly a box ~~falls~~ *fell* off the conveyor belt and ~~breaks~~ *broke* my foot.

  [The entire sentence is now in the past tense.]

Person is the term for the form of a personal pronoun that indicates whether the pronoun refers to the speaker, the person spoken to, or the person (or thing) spoken about. Verbs change their forms to agree in person with their subjects.

- I *see* [first person] a yellow tint, but she *sees* [third person] a yellow-green hue.

Tense refers to verb forms that indicate time distinctions. The six tenses are past, past perfect, present, present perfect, future, and future perfect.

Number refers to the two forms of a verb that indicate whether the subject of a verb is singular (The copier *was* repaired) or plural (The copiers *were* repaired).

Most verbs show the singular of the present tense by adding *-s* or *-es* (he *stands*, she *works*, it *goes*), and they show the plural without *-s* or *-es* (they *stand*, we *work*, they *go*). The verb *to be*, however, normally changes form to indicate the singular (I *am* ready) or plural (We *are* ready).

Voice refers to the two forms of a verb that indicate whether the subject of the verb acts or receives the action. The verb is in the *active voice* if the subject of the verb acts (The bacteria *grow*); the verb is in the passive voice if it receives the action (The bacteria *are grown* in a petri dish).

Conjugation of Verbs

The conjugation of a verb arranges all forms of the verb so that the differences caused by the changing of the tense, number, person, and voice are readily apparent. Figure V–1 shows the conjugation of the verb *drive.*

TENSE	NUMBER	PERSON	ACTIVE VOICE	PASSIVE VOICE
Present	Singular	1st	I drive	I am driven
		2nd	You drive	You are driven
		3rd	He/she drives	He/she is driven
	Plural	1st	We drive	We are driven
		2nd	You drive	You are driven
		3rd	They drive	They are driven
Present progressive	Singular	1st	I am driving	I am being driven
		2nd	You are driving	You are being driven
		3rd	He/she is driving	He/she is being driven
	Plural	1st	We are driving	We are being driven
		2nd	You are driving	You are being driven
		3rd	They are driving	They are being driven
Present perfect	Singular	1st	I have driven	I have been driven
		2nd	You have driven	You have been driven
		3rd	He/she has driven	He/she has been driven
	Plural	1st	We have driven	We have been driven
		2nd	You have driven	You have been driven
		3rd	They have driven	They have been driven
Present perfect progressive	Singular	1st	I have been driving	I have been being driven
		2nd	You have been driving	You have been being driven
		3rd	He/she has been driving	He/she has been being driven
	Plural	1st	We have been driving	We have been being driven
		2nd	You have been driving	You have been being driven
		3rd	They have been driving	They have been being driven

FIGURE V–1. Verb-Conjugation Chart (***continued on next page***)

TENSE	NUMBER	PERSON	ACTIVE VOICE	PASSIVE VOICE
Past	Singular	1st	I drove	I was driven
		2nd	You drove	You were driven
		3rd	He/she drove	He/she was driven
	Plural	1st	We drove	We were driven
		2nd	You drove	You were driven
		3rd	They drove	They were driven
Past progressive	Singular	1st	I was driving	I was being driven
		2nd	You were driving	You were being driven
		3rd	He/she was driving	He/she was being driven
	Plural	1st	We were driving	We were being driven
		2nd	You were driving	You were being driven
		3rd	They were driving	They were being driven
Past perfect	Singular	1st	I had driven	I had been driven
		2nd	You had driven	You had been driven
		3rd	He/she had driven	He/she had been driven
	Plural	1st	We had driven	We had been driven
		2nd	You had driven	You had been driven
		3rd	They had driven	They had been driven
Past perfect progressive	Singular	1st	I had been driving	I had been being driven
		2nd	You had been driving	You had been being driven
		3rd	He/she had been driving	He/she had been being driven
	Plural	1st	We had been driving	We had been being driven
		2nd	You had been driving	You had been being driven
		3rd	They had been driving	They had been being driven
Future	Singular	1st	I will drive	I will be driven
		2nd	You will drive	You will be driven
		3rd	He/she will drive	He/she will be driven
	Plural	1st	We will drive	We will be driven
		2nd	You will drive	You will be driven
		3rd	They will drive	They will be driven
Future progressive	Singular	1st	I will be driving	I will be being driven
		2nd	You will be driving	You will be being driven
		3rd	He/she will be driving	He/she will be being driven
	Plural	1st	We will be driving	We will be being driven
		2nd	You will be driving	You will be being driven
		3rd	They will be driving	They will be being driven

FIGURE V–1. Verb-Conjugation Chart (*continued*)

TENSE	NUMBER	PERSON	ACTIVE VOICE	PASSIVE VOICE
Future perfect	Singular	1st	I will have driven	I will have been driven
		2nd	You will have driven	You will have been driven
		3rd	He/she will have driven	He/she will have been driven
	Plural	1st	We will have driven	We will have been driven
		2nd	You will have driven	You will have been driven
		3rd	They will have driven	They will have been driven
Future perfect progressive	Singular	1st	I will have been driving	I will have been being driven
		2nd	You will have been driving	You will have been being driven
		3rd	He/she will have been driving	He/she will have been being driven
	Plural	1st	We will have been driving	We will have been being driven
		2nd	You will have been driving	You will have been being driven
		3rd	They will have been driving	They will have been being driven

FIGURE V–1. Verb-Conjugation Chart (*continued*)

very

The temptation to overuse intensifiers like *very* is great. In many sentences, the word can simply be deleted.

- The board was ~~very~~ angry about the newspaper report.

When you do use intensifiers, clarify their meaning. (The patient's resting heart rate was *very* fast; it was measured at 180 beats per minute.)

via

Via is Latin for "by way of." The term should be used only in routing instructions.

- The package was shipped *via* FedEx.
- Her project was funded ~~via~~ *as a result of* the recent legislation.

visuals

Visuals can express ideas or convey information in ways that words alone cannot by making abstract concepts and relationships concrete. Visuals can show how things look (drawings, photographs, maps), represent numbers and quantities (graphs, tables), depict relationships (flowcharts, schematic diagrams), and show hierarchical relationships (organizational charts). They also highlight important information and emphasize key concepts succinctly and clearly.

Many of the qualities of good writing—simplicity, clarity, conciseness, directness—are equally important in the creation and use of visuals. Presented with clarity and consistency, visuals can help readers focus on key portions of your document, presentation, or Web site. Be aware, though, that even the best visual will not be effective without context—and most often context is provided by the text that introduces the visual and clarifies its purpose.

The following entries in this book are related to specific visuals and their use in printed and online documents, as well as in presentations (see that entry for presentation graphics).

Selecting Visuals

Consider your audience and your purpose carefully in selecting visuals. You would need different illustrations for an automobile owner's manual or an auto dealer's Web site, for example, than you would for a mechanic's diagnostic guide. Figure V–2 on page 560 can help you select the most appropriate visuals, based on their purposes and special features. Jot down visual options when you are considering your scope and organization.

ETHICS NOTE Be aware that visuals have the potential of misleading readers when data are selectively omitted or distorted. For example, Figure G–5 on page 235 shows a graph that gives the appearance of a dramatic decrease in accidents because the scale is unevenly compressed with some of the years selectively omitted. Visuals that mislead readers at the least call the credibility of you and your organization into question—and they are unethical. The use of misleading visuals can even subject you and your organization to lawsuits. ✦

CHOOSING APPROPRIATE VISUALS

TO SHOW OBJECTS AND SPATIAL RELATIONSHIPS

DRAWINGS CAN . . .

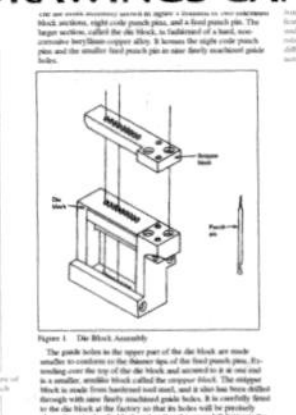

- Depict real objects difficult to photograph
- Depict imaginary objects
- Highlight only parts viewers need to see
- Show internal parts of equipment in cutaway views
- Show how equipment parts fit together in exploded views

PHOTOGRAPHS CAN . . .

- Show actual physical images of subjects
- Record an event in process
- Record the development of phenomena over time
- Record the as-found condition of a situation for an investigation

TO DISPLAY GEOGRAPHIC INFORMATION

MAPS CAN . . .

- Show specific geographic features of an area
- Show distance, routes, or locations of sites
- Show the geographic distribution of information (e.g., populations by region)

TO SHOW NUMERICAL AND OTHER RELATIONSHIPS

TABLES CAN . . .

Divisions	*Employees*
Research	*1,052*
Marketing	*2,782*
Automotive	*13,251*
Consumer Products	*2,227*

- Organize information systematically in rows and columns
- Present large numerical quantities concisely
- Facilitate item-to-item comparisons
- Clarify trends and other graphical information with precise data

BAR & COLUMN GRAPHS CAN . . .

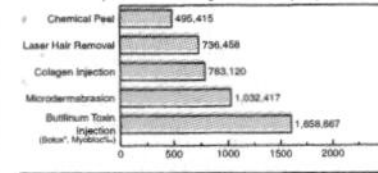

- Depict data in vertical bars and horizontal columns for comparison
- Show quantities that make up a whole
- Visually represent data shown in tables

LINE GRAPHS CAN . . .

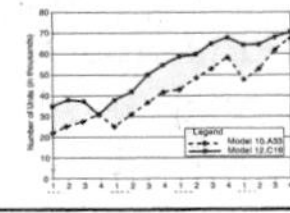

- Show trends over time in amounts, sizes, rates, and other measurements
- Give an at-a-glance impression of trends, forecasts, and extrapolations of data

FIGURE V–2. Chart for Choosing Appropriate Visuals

- Compare more than one kind of data over the same time period
- Visually represent data shown in tables

PICTURE GRAPHS CAN . . .

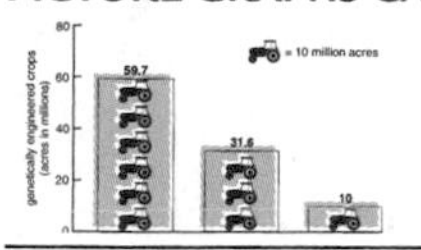

- Use recognizable images to represent specific quantities
- Help nonexpert readers grasp the information
- Visually represent data shown in tables

PIE GRAPHS CAN . . .

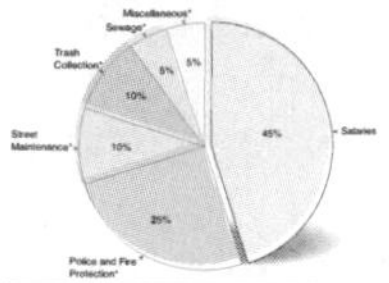

- Show quantities that make up a whole
- Give an immediate visual impression of the parts and their significance
- Visually represent data shown in tables or lists

TO SHOW STEPS IN A PROCESS OR RELATIONSHIPS IN A SYSTEM

FLOWCHARTS CAN . . .

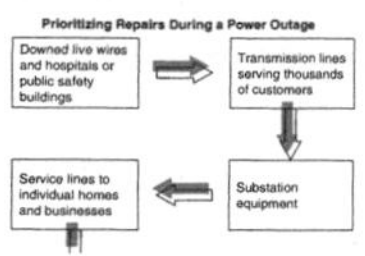

- Show how the parts or steps in a process or system interact
- Show the stages of an actual or a hypothetical process in the correct direction, including recursive steps

SCHEMATIC DIAGRAMS CAN...

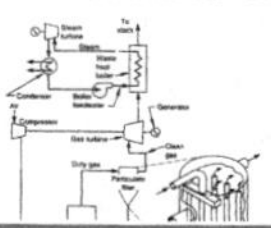

- Show how the components in electronic, chemical, electrical, and mechanical systems interact and are interrelated
- Use symbolic representations rather than realistic depictions of system components

TO SHOW RELATIONSHIPS IN A HIERARCHY

ORGANIZATION CHARTS CAN . . .

- Give an overview of an organization's departmental components
- Show how the components relate to one another
- Depict lines of authority within an organization

TO SUPPLEMENT OR REPLACE WORDS

SYMBOLS OR ICONS CAN . . .

- Convey ideas without words
- Save space and add visual appeal
- Transcend individual languages to communicate ideas effectively for international readers

FIGURE V–2. Chart for Choosing Appropriate Visuals (*continued*)

Integrating Visuals with Text

Consider the best locations for visuals before you begin writing a draft. Your goal should be to use visuals where they will best advance your purpose, aid your readers, and integrate smoothly within your text. One way to use visuals wisely is to make their placement a part of your outlining process. At appropriate points in your outline, either make a rough sketch of the visual, if you can, or write "illustration of . . . ," noting the source of the visual and enclosing each suggestion in a text box.

When you write the draft, place visuals as close as possible to the text where they are discussed—in fact, no visual should precede its first text mention. Clarify for readers why each visual is included in the text. The amount of description you should provide will vary, depending on your readers' backgrounds. For example, nonexperts may require lengthier explanations than do experts.

Give each visual a concise title that clearly describes its content. Assign figure and table numbers, particularly if your document contains more than one illustration or table. Then refer to visuals in the text of your document by their figure or table numbers. Refer to graphical illustrations (such as drawings, maps, and photographs) as "figures" and to tables as "tables."

ETHICS NOTE Obtain written permission to use copyrighted visuals—including images and multimedia material from Web sites—and acknowledge borrowed material in a source line below the caption for a figure and in a footnote at the bottom of a table. Use a site's "Contact Us" page to request approval. Acknowledge your use of any public (uncopyrighted) information, such as demographic or economic data from government publications and Web sites, with a source line. See also copyright, documenting sources, and plagiarism. ✦

Writer's Checklist: Creating and Integrating Visuals

CREATING VISUALS

- ☑ Keep visuals simple: Include only information needed for discussion in the text and eliminate unneeded labels, arrows, boxes, and lines.
- ☑ Position the lettering of any explanatory text or labels horizontally; allow adequate white space within and around the visual.
- ☑ Specify the units of measurement used, make sure relative sizes are clear, and indicate distance with a scale, when appropriate.
- ☑ Use consistent terminology; for example, do not refer to a "proportion" in the text and a "percentage" in the visual.
- ☑ Define abbreviations the first time they appear in the text and in figures and tables. If any symbols are not self-explanatory, include a key.

Writer's Checklist: Creating and Integrating Visuals (continued)

☑ Give each visual a concise title that clearly describes its content, and assign figure and table numbers if your document contains more than one illustration or table.

INTEGRATING VISUALS

☑ Clarify for readers why each visual is included in the text and provide an appropriate description.

☑ Place visuals as close as possible to the text where they are discussed, but after their first text mention.

☑ Refer to visuals in the text of your document as "figures" or "tables" and by their figure or table numbers.

☑ Consider placing lengthy, detailed visuals in an appendix and refer to it in the text.

☑ In documents with more than five illustrations or tables, include a section following the table of contents titled "List of Figures" or "List of Tables" that identifies each by number, title, and page number.

☑ Check the editorial guidelines or recommended style manual when preparing visuals for a trade journal article.

voice

In grammar, *voice* indicates the relation of the subject to the action of the verb. When the verb is in the *active voice*, the subject acts; when it is in the *passive voice*, the subject is acted upon.

ACTIVE David Cohen *wrote* the newsletter article. [The subject, *David Cohen*, performs the action; the verb, *wrote*, describes the action.]

PASSIVE The newsletter article *was written* by David Cohen. [The subject, *the newsletter article*, is acted upon; the verb, *was written*, describes the action.]

The two sentences say the same thing, but each has a different emphasis: The first emphasizes *David Cohen*; the second emphasizes *the newsletter article*. In technical writing, it is often important to emphasize who or what performs an action. Further, the passive-voice version is indirect because it places the performer of the action behind the verb instead of in front of it. Because the active voice is generally more direct, more concise, and easier for readers to understand, use the active voice unless the passive voice is more appropriate, as described on

V

page 565. Whether you use the active voice or the passive voice, be careful not to shift voices in a sentence.

- David Cohen corrected the inaccuracy as soon as ~~it was identified by~~ the editor *identified it*.

Using the Active Voice

Improving Clarity. The active voice improves clarity and avoids confusion, especially in instructions and policies and procedures.

PASSIVE	Sections B and C *should be checked* for errors. [Are they already checked?]
ACTIVE	*Check* sections B and C for errors. [The performer of the action, *you*, is understood: (You) *Check* the sections.]

Active voice can also help avoid dangling modifiers.

PASSIVE	Hurrying to complete the work, the cables *were connected* improperly. [*Who* was hurrying? The implication is the cables were hurrying!]
ACTIVE	Hurrying to complete the work, the technician *connected* the cables improperly. [Here, *hurrying to complete the work* properly modifies the performer of the action: *the technician.*]

Highlighting Subjects. One difficulty with passive sentences is that they can bury the performer of the action in expletives and prepositional phrases.

PASSIVE	It *was reported by* the testing facility that the new model is defective.
ACTIVE	The testing facility *reported* that the new model is defective.

Sometimes writers using the passive voice fail to name the performer—information that might be missed.

PASSIVE	The problem *was discovered* yesterday.
ACTIVE	The Maintenance Department *discovered* the problem yesterday.

Achieving Conciseness. The active voice helps achieve concise-ness because it eliminates the need for an additional helping verb as well as an extra preposition to identify the performer of the action.

PASSIVE Arbitrary changes in policy *are resented by* employees.

ACTIVE Employees *resent* arbitrary changes in policy.

The active-voice version takes one verb (*resent*); the passive-voice version takes two verbs (*are resented*) and an extra preposition (*by*).

Using the Passive Voice

The passive voice is sometimes effective or even necessary. Indeed, for reasons of tact and diplomacy, you might need to use the passive voice to avoid accusing others.

ACTIVE Your staff *did not meet* the quota last month.

PASSIVE The quota *was not met* last month.

ETHICS NOTE Be careful, however, not to use the passive voice to evade responsibility or obscure an issue or information that readers should know.

- Several mistakes *were made.* [*Who* made the mistakes?]
- It *has been decided.* [*Who* has decided?]

See also ethics in writing. ✦

When the performer of the action is either unknown or unimportant, of course, use the passive voice. (The copper mine *was discovered* in 1929.) When the performer of the action is less important than the receiver of that action, the passive voice is sometimes more appropriate. (Ann Bryant *was presented* with an award by the president.)

When you are explaining an operation in which the reader is not actively involved or when you are explaining a process or a procedure, the passive voice may be more appropriate. In the following example, anyone—it really does not matter who—could be the performer of the action.

- Area strip mining *is used* in regions of flat–to–gently rolling terrain, like that found in the Midwest. Depending on applicable reclamation laws, the topsoil *may be removed* from the area *to be mined, stored,* and later *reapplied* as surface material during reclamation of the mined land. After the removal of the topsoil, a trench *is cut* through the overburden to expose the upper surface of the coal to be mined. The overburden from the first cut *is placed* on the

> unmined land adjacent to the cut. After the first cut *has been completed*, the coal *is removed.*

Do not, however, simply assume that any such explanation should be in the passive voice; in fact, as in the following example, the active voice is often more effective.

- In the operation of an internal combustion engine, an explosion in the combustion chamber *forces* the pistons down in the cylinders. The movement of the pistons in the cylinders *turns* the crankshaft.

Ask yourself, "Would it be of any advantage to the reader to know the performer of the action?" If the answer is yes, use the active voice, as in the previous example.

TIPS FOR CHOOSING VOICE

Different languages place different values on active-voice and passive-voice constructions. In some languages, the passive is used frequently; in others, hardly at all. As a nonnative speaker of English, you may have a tendency to follow the pattern of your native language. But remember, even though technical writing may sometimes require the passive voice, active verbs are highly valued in English.

wait for / wait on

Wait on should be restricted in writing to the activities of hospitality and service employees. (We need extra staff to *wait on* customers.) Otherwise, use *wait for.* (Be sure to *wait for* Ms. Sturgess's approval.) See also idioms.

Web design

DIRECTORY

You can apply many of the principles covered throughout this book to designing Web sites. As you prepare, carefully consider your purpose and readers' needs when designing Web pages and assembling those pages to create Web sites. Most organizational sites have well-defined goals, such as reference, marketing, education, or advocacy. Before you begin building your Web pages and sites, create a clear statement of purpose that identifies your target audience.

EXTERNAL SITE The purpose of this site is to enable our customers to locate product information, place online orders, and contact our customer-service department.

INTERNAL SITE The purpose of this site is to provide SNR Security Corporation employees, suppliers, and partner organizations with a single, consistent, and up-to-date resource for materials about SNR Security.

W

As the examples suggest, the two general kinds of sites are external and internal. *External sites* target an audience from the entire Internet. *Internal sites* are designed for audiences either on an *intranet* (a network within an organization) that is not accessible to audiences outside that organization or an *extranet* (a password-protected site) available exclusively to trusted partners and suppliers to provide them with common data and documents.

Navigation and Pages

Your main goal when designing a Web site is to establish a predictable environment in which users can comfortably navigate and easily find the information they need. This information should be logically accessible to the user in the fewest possible steps (or clicks). See also methods of development.

Navigation Plan. To design an efficient navigation plan, draft a navigation chart or map of your site early in the process. Figure W–1 shows an initial navigation plan for a Web site for a U.S. Department of Energy National Laboratory.

In Figure W–1, each higher-level page (for example, Basic Science) is linked to related subject areas (Biology, Chemistry, Sensors, Materials, Mathematics, Physics). Under Scientific Facilities, related content areas are shown, with a specific site for the laboratory's Auto Shredder Facility, which in turn is linked to a site with information about recovering plastics from scrapped autos. That page has an e-mail link to the principal investigator for the Recovering Plastics project. The pattern is repeated, with greater or lesser detail, throughout the site. The site map is preliminary, so it does not yet include areas such as a Privacy Statement, What's New, About Us, or a site search window.

Page Links. When viewers access Web pages randomly, they often have no context for where they are in the site. One way to orient users is to provide a set of links in graphics or words called a toolbar or button bar, as shown here.

[Home | Search | Order | What's New | About Us]

A toolbar at the top, bottom, or sides of each page serves as a table of contents and shows visitors the top-level structure of your site. Toolbars on each page help you avoid sending users to a dead-end page (one with no link); in fact, every page should, if nothing else, link back to your homepage. You may also incorporate the name of your company or organization and its logo in a banner at the top of each page that links back to your homepage.

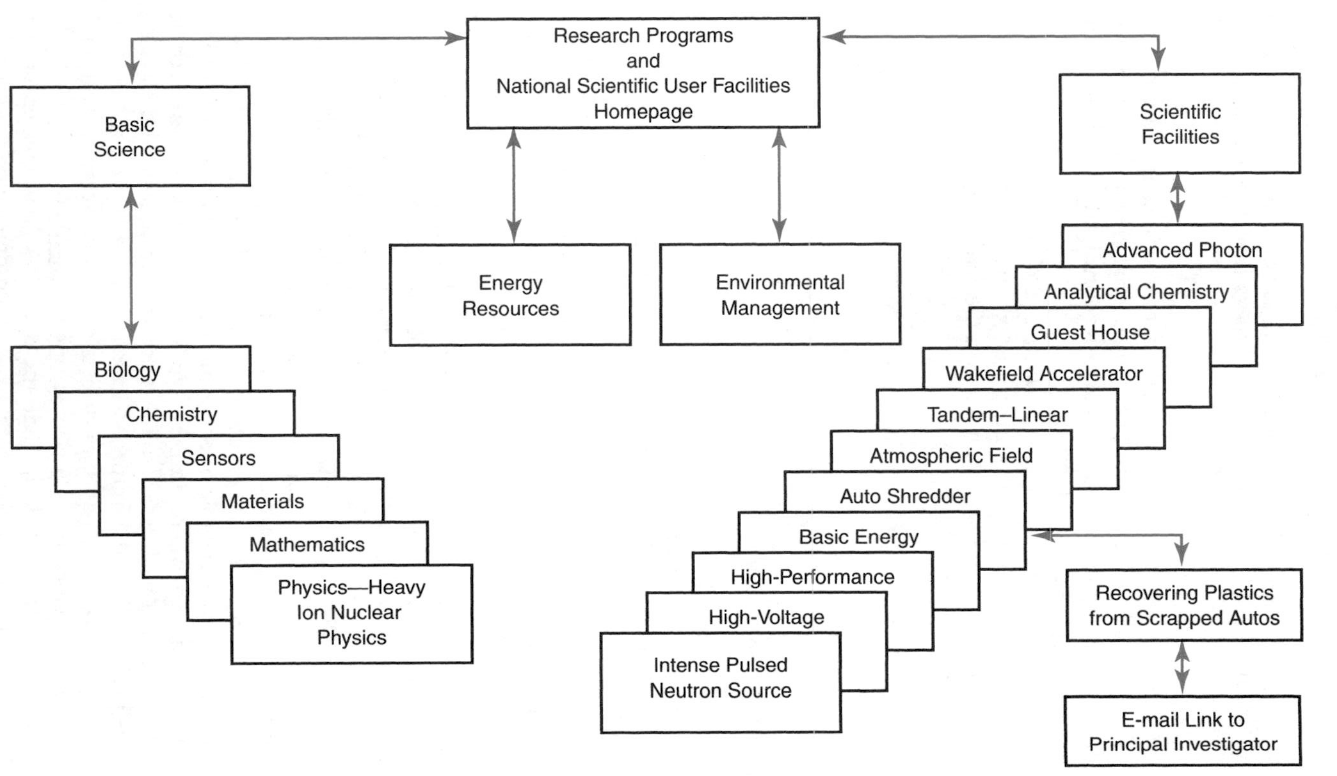

FIGURE W–1. Site Map

Check links to external sites routinely because these sites can change or disappear without your knowledge. Some sites place an icon or EXIT next to outside links to inform users that they are leaving the host site.

Link Design. Create links to other pages and sites that are related to your visitors' organizational or professional needs and interests. Links to customer service and online order forms are especially useful for cultivating potential customers.

When you identify links, do not write out "Click here for more information." Instead, write the sentence as you normally would, and anchor the link on the most relevant word in the sentence, as illustrated in the following example.

- For information about employment opportunities, visit Human Resources.

Avoid writing paragraphs that are dense with links. Instead, list links alphabetically in groups of about four to seven to make them easier for the viewer to see at a glance.

▶ Basic Science
▶ Energy Resources
▶ Environmental Management
▶ Scientific Facilities

In addition to identifying links with words, you can use icons and graphics, as shown in the preceding list; they not only are easy to use but also provide the visual cues Web users expect.

Page Layout. Most users do not like scrolling down long pages, so pages should contain no more than one or two screens of information. Keep homepages, in particular, to one screen. Many computer monitors and settings do not display more than half of a typical Web page at any one time, allowing only the top four or five inches to be visible. For that reason, place titles, banners, and important graphic elements and information in the upper half of the page (or top 300 pixels).

When a greater amount of information must be covered on a single page, include a concise table of contents at the top of the page linked to specific sections elsewhere on the page. Viewers can select the section they want without having to scroll through the entire page. You can also provide periodic links that return visitors to the top of the page.

Use short narrative passages or lists so that viewers can scan and access the information quickly. See writing for the Web.

Your Web pages should include the date they were created or last updated (usually in a footer) so that users can determine whether the information is current. You might also place a New icon next to each new or updated item to alert users to these changes. Be sure to remove the icon after a suitable interval, such as 30 days.

Finally, test your design by viewing it on several different Web browsers as well as on different platforms (including UNIX, Linux, Mac, and Windows). Different software and different platforms can interact unpredictably, and you may be surprised at how different your page can look on systems unlike your own.

Homepages. A homepage (or *start page*) is an important focus for all pages in your Web site. Many organizational homepages include an image map that functions like a toolbar, linking to other pages in the site. The image map introduces visitors to the overall site design, identifies the purpose of the site, and provides an overview of major content areas. Include links to your site's Privacy Statement and About Us page, which describes your organization. You should also consider including a Contact Us link that provides visitors with your e-mail address and other contact information, as appropriate.

Because a complex homepage graphic can take a minute or longer to download, consider using relatively small graphics files on your homepage, gradually increasing the size of the graphics for pages deeper into your site. Users who go beyond a page or two into a site are more committed and therefore more willing to tolerate longer delays, especially if you offer them warnings that particular pages contain graphics that may take time to download.*

Graphics and Typography

Your audience and purpose should determine the graphic style and theme of your Web site. Although graphic elements provide visual relief from dense text, do not overdo graphics—use graphics that fit the particular audience, purpose, and context. Complex graphics and motion can clutter or slow access to your site. Avoid bold and multiple colors; instead, use lighter colors, especially for backgrounds. Keep in mind that large or high-resolution graphics, like color photographs, can cause long delays as they download to the user's system. For high-resolution graphics, you might use thumbnails (images reduced to 10 to 15 percent of the original file size) that link to the original-size image. Consider as well providing a graphics-free option for quicker access.

Aim for consistency in your graphics and typography to establish a sense of unity and to provide consistent visual cues that help visitors find information. Consider creating a cascading style sheet to implement the

*The current industry "best practices" recommendation is that a page load in ten seconds or less on a 56K modem.

specifications for fonts, spacing between paragraphs, heading styles, and other elements. *Cascading style sheets* are codes, linked from or included in your HTML (hypertext markup language) that work like templates. You can define the style (color, font, size, emphasis, etc.) for a particular set of HTML tags once, and thereafter every time you use those tags in your site they will look the same. Cascading style sheets are especially helpful for large sites with numerous pages, as you can quickly and easily change your headings by simply changing the style sheet.

DIGITAL TIPS **TESTING YOUR WEB SITE**

You might take advantage of Web sites that will test a limited sample of your site's features for free. (They also offer greatly expanded testing for a fee.) The services can test your site's browser and hardware platform compatibility, page-load speed, and links. NetMechanic.com, for example, will test pages at your site using 16 different browser and computer combinations. It will also show page-load speeds in seconds on 14.4K, 28.8K, and 56K modems, as well as on high-speed lines. You can then print the test results and work to eliminate any problems identified. For more on this topic, see <*bedfordstmartins.com/alredtech*> and select *Digital Tips*, "Testing Your Web Site."

Access for People with Disabilities

Design elements like colorful graphics, animation, and streaming video and audio can be barriers to people with impaired vision or hearing or those who are color-blind. Use the following strategies to meet the needs of such audiences.

- Avoid frames, complex tables, animation, JavaScript, and other design elements incompatible with text-only browsers and adaptive technologies, such as voice or large-print software.
- Provide HTML versions of pages and documents whenever possible because this format is most compatible with the current generation of screen readers.
- Attach text equivalents for graphic or audio elements.
- Design for the color-blind reader by making meaning independent of color. For example, rather than asking users to "Click on the green button for more information," label the button (Click Here) or embed a link in a sentence: "See our catalog for more information."

You may want to consider offering different options for site visitors, such as full-graphics, light-graphics, and text-only versions.

Finally, test your pages using "Bobby," a site that will provide an on-screen analysis of a Web page to ensure its accessibility: <http://bobby.watchfire.com/>.

Writer's Checklist: Designing Web Pages

Many of the guidelines on typography in layout and design are applicable to Web pages; however, not all apply, so keep the following in mind:

- ☑ Limit the number of typeface styles as well as colors.
- ☑ Use sans serif fonts for text passages; they are generally legible on computer screens.
- ☑ Use uppercase letters or boldface type sparingly.
- ☑ Use typeface colors that contrast (but do not clash) with background colors.
- ☑ Use underlined text only for links.
- ☑ Separate text from graphics with generous blank space (the equivalent of white space).
- ☑ Limit your text lines to 50 to 70 characters (or 10 to 12 words) for readability.
- ☑ Use **heading** and subheading styles sparingly and consistently.
- ☑ Block-indent text sections that you expect viewers to read in detail.
- ☑ Test your design by viewing it on alternate browsers and platforms.
- ☑ Check that all your links work, particularly after site changes.
- ☑ Test the load time of your pages using a slow modem.

WEB LINK **WEB DESIGN RESOURCES**

For Mike Markel's helpful overview of the process and principles of designing Web sites, see *<bedfordstmartins.com/alredtech>* and select *Tutorials*, "Designing for the Web." For links to more tutorials as well as to sites that provide advice for improving accessibility for people with disabilities, select *Links for Handbook Entries.*

when / where / that

When and if (or *if and when*) is a colloquial expression that should not be used in writing.

W

- When ~~and if~~ funding is approved, you will get the position.

- *If* ~~When and if~~ funding is approved, you will get the position.

In phrases using the *where . . . at* construction, *at* is unnecessary and should be omitted.

- Where is his office ~~at~~?

Do not substitute *where* for *that* to anticipate an idea or fact to follow.

- I read in the newsletter *that* ~~where~~ research funding will increase.

whether

Whether communicates the notion of a choice. When *whether or not* is used to indicate a choice between alternatives, omit *or not*; it is redundant.

- The client asked whether ~~or not~~ the proposal was finished.

The phrase *as to whether* is clumsy and redundant. Either omit it altogether or use only *whether*.

- *We have decided to* ~~As to whether we will~~ commit to a long-term *contract.* ~~contract, we have decided to do so.~~

while

While, meaning "during an interval of time," is sometimes substituted for connectives like *and*, *but*, *although*, and *whereas*. Used as a connective in that way, *while* often causes ambiguity.

- Ian Evans is sales manager, *and* ~~while~~ Joan Thomas is director of research.

Do not use *while* to mean *although* or *whereas*.

- *Although* ~~While~~ Ryan Patterson wants the job of engineering manager, he has not yet applied for it.

Restrict *while* to its meaning of "during the time that."

- I'll have to catch up on my reading *while* I am on vacation.

white papers

A white paper is a document that announces a new product or a newly developed technical or business solution to a common problem. White papers function as (1) educational tools that provide credibility for an organization and (2) marketing tools to establish a company's market position as an industry leader. White papers are often posted on a company's Web site, distributed to potential customers at trade shows and conferences, and published at Web sites that feature white papers from various sources.

Using Persuasion

White papers are promotional documents, so they must be interesting to read, reflect strong empathy with the reader, and use effective persuasion. White papers require that writers shift from a technology-centered point of view to the "you" viewpoint, as suggested in the following example from Darren K. Barefoot's article "Ten Tips on Writing White Papers."*

TECHNOLOGY CENTERED	Buy a DVD player because it has a 128-bit oversampling and advanced virtual surround sound.
READER CENTERED	Buy a DVD player because you want better sound and a crisper picture.

If your white paper is targeted to nonspecialists, leave the details to the technical manuals, sales brochures, and press releases. Focus on your readers' concerns and explain clearly how your company or product can solve their specific problems.

Preparing to Write

Begin by checking with management to determine their goals for the white paper and how it fits into their overall strategy. Then analyze your audience to determine whether your readers are technical experts or nonexperts such as executive management. Plan your language and approach appropriately. For example, experts understand and appreciate technical terminology, whereas executives tend to focus on the bottom line and the benefits of solving a problem.

Decide on the most appropriate method of development and then create an outline that offers your product or solution as the most effective way to solve the readers' problem. An outline helps provide coherence

*Darren K. Barefoot, "Ten Tips on Writing White Papers," *Intercom* February 2002, 12–13.

and transition to your draft so that readers can follow the development of a new technology or product as the white paper moves smoothly from the introduction to the conclusion. Define any new terms, avoid affectation, and follow the principles of effective technical writing style.

As you consider your scope, make certain that your white paper deals with only one product or solution; for multiple topics, use multiple white papers. The length of your white paper depends on the complexity of your product, but try to keep it as brief as possible. If your document runs more than ten pages, include a table of contents.

Planning Layout and Design

A good design can make the most complex information look accessible and enable your readers to retrieve information easily. A white paper should make good use of visuals to help readers grasp and retain your message. Use such visuals carefully, however. A complex visual that requires paragraphs of explanation is more likely to hinder your reader's understanding than enhance it. Make the best use as well of headings, lists, call-outs, text boxes, and other devices that are described in layout and design.

Developing Major Sections

The introduction to a white paper should attract your readers' interest by explaining the problem that needs to be solved. Your readers might be persuaded to accept your message more readily if they believe your company is developing innovative, cutting-edge products, as in the following example.

- The rings of Saturn have puzzled astronomers ever since Galileo discovered them in 1610 using the first telescope. Recently, even more rings have been discovered. . . .

  Our company's Scientific Instruments Division designs and manufactures research-quality, computer-controlled telescopes that promise to solve the puzzles of Saturn's rings by enabling scientists to use multicolor differential photometry to determine the rings' origins and composition.

Use the body of a white paper to provide supporting evidence that your product or solution is the best way to solve the readers' problem. Be careful not to be too biased toward your own product, however, especially early in your white paper. You might start by explaining what led to the development of the problem and how its solution impacts all customers in the industry. Describe the kind of technology that can solve the problem, and develop a high standard for the needed solution. Then explain precisely how your way of solving the customer's problem is the best way to meet that standard. Provide persuasive supporting evidence from outside sources. See research and documenting sources.

The conclusion of a white paper should summarize the problem and how your solution best solves it. As a marketing tool, your white paper should conclude by telling readers how to contact your company for more information.

Revising and Getting Clearances

Ask management, subject-matter experts, and perhaps your legal department to read the finished draft to make certain it meets management's objectives, is technically accurate, and stays within legal requirements. Be sure to cite any outside sources used and get permission from any source quoted. See copyright, documenting sources, and plagiarism.

WEB LINK **WHITE PAPERS**

For links to online collections of white papers, see <*bedfordstmartins.com/alredtech*> and select *Links for Handbook Entries.*

who / whom

Writers are often unsure whether to use *who* or *whom. Who* is the subjective case form, whereas *whom* is the objective case form. When in doubt about which form to use, substitute a personal pronoun to see which one fits. If *he*, *she*, or *they* fits, use *who.*

- *Who* is the service manager?
  [You would say, "*She* is the service manager."]

If *him*, *her*, or *them* fits, use *whom.*

- It depends on *whom*?
  [You would say, "It depends on *them.*"]

who's / whose / of which

Who's is the contraction of *who is.* (*Who's* scheduled today?) *Whose* is the possessive case of *who.* (Consider *whose* budget should be cut.)

Normally, *whose* is used with persons, and *of which* is used with inanimate objects.

- The employee *whose* car had been towed away was angry.
- Completing a Ph.D. in engineering is an achievement *of which* to be proud.

If *of which* causes a sentence to sound awkward, *whose* may be used with inanimate objects. (Compare "The business *whose* profits steadily declined" with "The business the profits *of which* steadily declined.")

-wise

Although the suffix *-wise* often seems to provide a tempting shortcut in writing, it leads more often to inept than to economical expression. It is better to rephrase the sentence.

- Our department ~~rates~~ *has a* high ~~efficiencywise.~~ *efficiency rating.*

The *-wise* suffix is appropriate, however, in instructions that indicate certain space or directional requirements (lengthwise, clockwise).

word choice

Mark Twain once said, "The difference between the right word and almost the right word is the difference between 'lightning' and 'lightning bug.'" The most important goal in choosing the right word in technical writing is the preciseness implied by Twain's comment. Vague words and abstract words defeat preciseness because they do not convey the writer's meaning directly and clearly.

VAGUE It was a *productive* meeting.

PRECISE The meeting resulted in the approval of the health-care benefits package.

In the first sentence, *productive* sounds specific but conveys little; the revised sentence says specifically what made the meeting "productive." Although abstract words may at times be appropriate to your topic, using them unnecessarily will make your writing difficult to understand.

Being aware of the connotations and denotations of words will help you anticipate readers' reactions to the words you choose. Understanding antonyms (fresh/stale) and synonyms (notorious/infamous) will increase your ability to choose the proper word. Make other usage decisions carefully, especially in technical contexts, such as average/median/mean and biannual/biennial.

Although many of the entries throughout this book will help you improve your word choices and avoid impreciseness, the following entries should be particularly helpful:

affectation 24
biased language 49
buzzwords 58
clichés 72
conciseness 90
euphemisms 181
idioms 252
jargon 288
logic errors 309
vague words 550

A key to choosing the correct and precise word is to keep current in your reading and to be aware of new words in your profession and in the language. In your quest for the right word, remember that there is no substitute for an up-to-date dictionary. See also English as a second language.

WEB LINK **WISE WORD CHOICES**

For online exercises on word choice, see <*bedfordstmartins.com/alredtech*> and select *Exercise Central*.

writing a draft

You are well prepared to write a rough draft when you have established your purpose and readers' needs, considered the context, defined your scope, completed adequate research, and prepared an outline (whether rough or developed). (See also outlining.) Writing a draft is simply transcribing and expanding the notes from your outline into paragraphs, without worrying about grammar, refinements of language, or spelling. Refinement will come with revision and proofreading. See also "Five Steps to Successful Writing."

Writing and revising are different activities. Do not let worrying about a good opening slow you down. Instead, concentrate on getting your ideas on paper—now is not the time to polish or revise. Do not wait for inspiration—treat writing a draft as you would any other on-the-job task.

Writer's Checklist: Writing a Rough Draft

☑ Set up your writing area with whatever supplies you need (notepads, pens and pencils, reference material, etc.) to keep going once you get started. Then hang out the "Do Not Disturb" sign.

Writer's Checklist: Writing a Rough Draft (continued)

- ☑ Avoid the temptation of writing first drafts on the computer without any planning by using a good outline as a springboard to start and keep going.
- ☑ Keep in mind your readers' needs, expectations, and knowledge of the subject. Doing so will help you write directly to your readers and suggest which ideas need further development.
- ☑ When you are trying to write quickly and you come to something difficult to explain, try to relate the new concept to something with which the readers are already familiar, as discussed in figures of speech.
- ☑ Start with the section that seems easiest. Your readers will neither know nor care that the middle section of the document was the first section you wrote.
- ☑ Give yourself a set time (ten or fifteen minutes, for example) in which you write continuously, regardless of how good or bad your writing seems to be. Don't stop when you are rolling along easily—if you stop and come back, you may lose momentum.
- ☑ Routinely save to your hard drive and create a backup copy of your documents on separate disks or on the company network.
- ☑ Give yourself a small reward—a short walk, a soft drink, a brief chat with a friend, an easy task—after you have finished a section.
- ☑ Reread what you have written when you return to your writing. Seeing what you have already written can return you to a productive frame of mind.

writing for the Web

In the workplace, those who write content for Web sites are not always the same people who design the sites. This entry provides guidelines only for those writing for an online audience; for more information on site design, see Web design.

ETHICS NOTE Work with your Webmaster or site administrator to comply with Web policy guidelines as well as technical issues of access and design standards. On campus, consult your instructor or campus computer support staff about standards for posting content. ✦

Crafting Content for the Web

W

Because most readers scan Web sites for specific information, the way you write and organize your information will greatly affect your readers' understanding. State your important points first, before any detailed supporting information. (This method is called *inverted pyramid organization.*) Keep your writing style simple and straightforward, and avoid such directional cues as "as shown in the example below" that make sense on the printed page but not on a Web screen. Use the following techniques to make your content more accessible to readers.

Headings. Divide your text into short passages, each focusing on one facet of your topic. Use informative headings to help the reader decide at a glance whether to read a passage. Headings also simplify text by highlighting structure and organization, and they signal transitions from one topic to the next.

Lists. Use bullets and numbered lists to break up dense paragraphs, reduce text length, and highlight important content.

Keywords. To help search engines and your audience find your site, use keywords in the first 50 or so words of your text.

WITHOUT KEYWORDS	We are proud to introduce a new commemorative coin honoring our bank's founder and president. The item will be available on this Web site after December 3, 2007, which is the 100th anniversary of our first deposit.
WITH KEYWORDS	The new *Reynolds* commemorative coin features a portrait of *George G. Reynolds*, the founder and president of *Reynolds Bank*. The coin can be purchased after December 3, 2007, in honor of the 100th anniversary of the sale of the first *Reynolds* deposit.

Graphics. Graphics provide visual relief from text and make your site attractive and appealing. Use only visuals that are appropriate for your audience and purpose, however, and use graphics that load quickly.

Hyperlinks. Use hyperlinks to help readers navigate the information in your site. If a passage of text is longer than two or three screens, create a table of contents of hyperlinks for it at the top of the Web page and link each item to the relevant content further down the page. Avoid hyperlinks within text paragraphs because they can distract readers, make scanning the text difficult, and tempt readers to leave before reaching the end of your content.

Fonts. Font sizes and styles affect screen legibility. Because computer screens display fonts at lower resolutions than does printed text, sans serif fonts work better for online text passages. Consult your

Webmaster about your site's font preferences. Do not use capital letters or boldface type for blocks of text because they slow the reader. For content that contains special characters (such as for mathematical or chemical content), consult your Webmaster about the best way to submit the files for HTML (hypertext markup language) coding or post them as PDF (portable document format) files.

Line Length. Line length also affects readability because short line lengths reduce the amount of eye movement necessary to scan text. Optimal length is approximately half the width of the screen. To achieve this length, draft text that is between 50 and 70 characters (or between 10 and 12 words) to a line.

Writing for a Global Audience

When you write for public-access sites, eliminate expressions and references that make sense only to someone very familiar with American English. Express dates, clock times, and measurements consistent with international practices. For visuals, choose symbols and icons, colors, representations of human beings, and captions that can be easily understood, as described in global communication and global graphics. See also biased language and English as a second language.

Linking to External Sites

Links to outside sites can expand your content. However, review such sites carefully before linking to them. Is the site's author or sponsoring organization reputable? Is its content accurate, current, and unbiased? Does the site date-stamp its content with notices such as "This page was last updated on January 1, 2007"? (For more advice on evaluating Web sites, see research.) Link directly to the page or specific area of an outside site that is relevant to your users, and be sure that you provide a clear context for why you are sending your readers there.

Posting an Existing Document

If you post files for existing paper documents to a Web site, try to retain the document's original sequence and page layout. If you shorten or revise the original document for posting to the Web, add a notice informing readers that it differs from the printed original.

Regardless of format or version used, be sure to:

- Obtain permission from the copyright holder for the use of copyrighted text, tables, or images. See also plagiarism.

- Ask the Webmaster or site administrator about the preferred file format to submit for coding and posting. On campus, consult your instructor or campus computer support staff.
- Request that the Webmaster optimize any slow-loading graphics files for quick access.
- Review the document internally before it is posted to the public site: Is it the correct version? Is any information missing? Do all links work and go to the right places? See proofreading.
- Ask the Webmaster to create a single-file version of the document (a version formatted as a single, long Web page) for readers who will print it to read offline.

ETHICS NOTE Document sources of information or of help received—text, images, streaming video, and other multimedia material. Seek prior approval before using any copyrighted information. Documenting your sources is not only required, it also bolsters the credibility of your site. To do so, either provide links to your source or use a citation as described in documenting sources. ✦

DIGITAL TIPS USING PDF FILES

Converting documents such as reports, articles, and brochures to PDF files allows you to retain the identical look of the printed documents. The PDF pages will display on-screen exactly as they appear on the printed page. Readers can read the document online, download and save it, or print it in whole or in part. For more on this topic, see <*bedfordstmartins.com/alredtech*> and select *Digital Tips*, "Using PDF Files."

Protecting the Privacy of Users

Put a link on your page to the site's privacy statement, particularly if you solicit comments about your content or have an e-mail link for unsolicited comments for site users. A Privacy Statement informs site visitors about how the site sponsor handles solicited and unsolicited information from individuals, its policy on the use of cookies,* and legal action it takes against hackers. Inform users if you intend to use their information for marketing or to share their information with third parties, and give visitors the option of refusing you permission to use or share their information.

*"Cookies" are small files that are downloaded to your computer when you browse certain Web pages. Cookies hold information, such as your user name and password, so you do not need to re-enter it each time you visit the site.

"you" viewpoint

The "you" viewpoint places your readers' interest and perspective foremost. It is based on the principle that your readers are naturally more concerned about their own needs than they are about those of the writer or organization.

The "you" viewpoint often, but not always, means using the words *you* and *your* rather than *we*, *our*, *I*, and *mine*. Consider the following sentence that focuses on the needs of the writer and organization (we) rather than on those of the reader.

- *We must receive* your signed invoice before *we can process* your payment.

Even though the sentence uses *your* twice, the words in italics suggest that the point of view centers on the writer's need to receive the invoice in order to process the payment. Consider the following revision, written with the "you" viewpoint.

- *So you can receive* your payment promptly, please send your signed invoice.

Because the benefit to the reader is stressed, the writer is more likely to motivate the reader to act. See also persuasion.

In some instances, as suggested earlier, you may need to avoid using the pronouns *you* and *your* to achieve a positive tone and maintain goodwill. Notice how the first of the following examples—with *your*—seems to accuse the reader, while the second—without *your*—achieves the goals of the "you" viewpoint with positive writing.

ACCUSATORY	*Your* budget makes no allowance for setup costs.
POSITIVE	The budget should include an allowance for setup costs to meet all the concerns of our client.

As this example illustrates, the "you" viewpoint means more than using particular pronouns or adopting a particular writing style. By placing the readers' interests at the center, you can achieve your purpose not only in correspondence, e-mail, and memos but also in proposals, brochures, many reports, and presentations.

your / you're

Your is a possessive pronoun (*your* wallet); *you're* is the contraction of *you are* (*You're* late for the meeting). If you tend to confuse *your* with *you're*, use the search function of your word processor to review both terms during proofreading.

Acknowledgments (continued)

Figure D–13: "Cutaway Drawing (Hard Drive)" from P. D. Moulton and Timothy S. Stanley, *Hard Disk Quick Reference*. Copyright © 1989 by Que Publishing. Reprinted with the permission of Pearson Education, Inc., Upper Saddle River, NJ.

Figure E–2: Source: U.S. Nuclear Regulatory Commission.

Figure F–2: Information from PEPCO, "How We Restore Power" from *Lines* 33, no. 7 (July 2004). Reprinted with permission.

Figure F–3: Infographic: FDA/Renée Gordon.

Figure F–5: Reprinted with the permission of Susan Litzinger, a student at Pennsylvania State University, Altoona.

Figure G–3: Source: U.S. Food and Drug Administration.

Figure G–5: Copyright © 2003 by ISO. Reprinted with permission.

Figure G–11: Information from National Agricultural Statistics Service/USDA.

Figure I–6: "Char-Broil Use and Care Manual." Reprinted with the permission of W. C. Bradley Company, 1997, Columbus, GA.

Figure L–4: From Gerald J. Alred et al., *The Professional Writer*. Copyright © 1991 by Bedford/St. Martin's. Reprinted with permission of Bedford/St. Martin's.

Figure L–6: Excerpt from Pauline Slade, "Relationships Between Cleft Severity and Attractiveness of Newborns with Unrepaired Clefts" from *Cleft Palate–Craniofacial Journal* 32, no. 4 (July 1995): 318-322. Reprinted with permission.

Figure M–1: Source: U.S. Nuclear Regulatory Commission.

Figure N–1: Source: National Transportation Safety Board.

Figures N–2, N–3, and P–2: Copyright © 1998, 2000 by Ken Cook Co. Reprinted with the permission of Ken Cook Company.

Figure O–3: Source: U.S. Food and Drug Administration.

Figure P–5: "The Treatment Process" from Steve Coffel and Karyn Feden, *Indoor Pollution*. Copyright © 1991 by Steve Coffel and Karyn Feden. Reprinted with the permission of Ballantine Books, a division of Random House, Inc.

Figure R–3: "Library Homepage." Reprinted with the permission of Columbia University Libraries.

Figure R–4: "InfoTrac Search Page." Reprinted with permission.

Figure R–5: "Advanced Google Search." Reprinted with the permission of Google, Inc.

Figure R–6: "Google's Main Subject Directory." Reprinted with the permission of Google, Inc.

Figure R–10: Prepared by Kim Isaacs, Advanced Career Systems, Inc.

Figure R–14: Prepared by Kim Isaacs, Advanced Career Systems, Inc.
"Four Criteria for Evaluating Sources" from Leigh Ryan, *The Bedford Guide for Writing Tutors,* Third Edition. Copyright © 2002 by Bedford/St. Martin's. Reprinted with the permission of Bedford/St. Martin's.

Index

COMMONLY MISUSED WORDS AND PHRASES

Many entries in this book focus on the standard usage of various words, phrases, and constructions. Use this complete list as a quick reference for finding usage entries, which appear in italics throughout the book.

MODEL DOCUMENTS AND FIGURES BY TOPIC

Use the following list as a quick reference for finding samples of technical writing and visuals by topic. See also the Contents by Topic on pages xxiii–xxvi. For additional model documents and resources, see the companion Web site at <*bedfordstmartins.com/alredtech*>.

TECHNICAL WRITING DOCUMENTS AND ELEMENTS

DESIGN AND VISUALS

CORRESPONDENCE

JOB SEARCH AND APPLICATION

(continued)

PRESENTATIONS AND MEETINGS

RESEARCH AND DOCUMENTATION

ORGANIZATION, WRITING, AND REVISION

GRAMMAR, STYLE, AND LANGUAGE